By-Products: Characterisation and Use as Food

By-Products: Characterisation and Use as Food

Editors

Simona Grasso
Konstantinos Papoutsis
Claudia Ruiz-Capillas
Ana Herrero Herranz

MDPI • Basel • Beijing • Wuhan • Barcelona • Belgrade • Manchester • Tokyo • Cluj • Tianjin

Editors
Simona Grasso
School of Agriculture and Food Science,
University College Dublin, Dublin, Ireland
simona.grasso@ucdconnect.ie

Konstantinos Papoutsis
School of Agriculture and Food Science,
University College Dublin, Dublin, Ireland
kostas.papoutsis@ucd.ie

Claudia Ruiz-Capillas
Research Scientist
Instituto de Ciencia y Tecnología de
Alimentos y Nutrición (ICTAN)
Consejo Superior de Investigaciones
Científicas (CSIC)
Department of Products
José Antonio Novais, 10
28040-Madrid. Spain
claudia@ictan.csic.es

Ana Herrero Herranz
Spanish National Research Council, CSIC,
Department of Products, Madrid, Spain
ana.herrero@ictan.csic.es

Editorial Office
MDPI
St. Alban-Anlage 66
4052 Basel, Switzerland

This is a reprint of articles from the Special Issue published online in the open access journal *Foods* (ISSN 2304-8158) (available at: https://www.mdpi.com/journal/foods/special_issues/food_byproducts).

For citation purposes, cite each article independently as indicated on the article page online and as indicated below:

LastName, A.A.; LastName, B.B.; LastName, C.C. Article Title. *Journal Name* **Year**, *Volume Number*, Page Range.

ISBN 978-3-0365-3705-4 (Hbk)
ISBN 978-3-0365-3706-1 (PDF)

Cover image courtesy of Simona Grasso

© 2022 by the authors. Articles in this book are Open Access and distributed under the Creative Commons Attribution (CC BY) license, which allows users to download, copy and build upon published articles, as long as the author and publisher are properly credited, which ensures maximum dissemination and a wider impact of our publications.

The book as a whole is distributed by MDPI under the terms and conditions of the Creative Commons license CC BY-NC-ND.

Contents

Preface to "By-Products: Characterisation and Use as Food" ix

Marta Barral-Martinez, Maria Fraga-Corral, Pascual Garcia-Perez, Jesus Simal-Gandara and Miguel A. Prieto
Almond By-Products: Valorization for Sustainability and Competitiveness of the Industry
Reprinted from: *Foods* **2021**, *10*, 1793, doi:10.3390/foods10081793 1

Zein Najjar, Maitha Alkaabi, Khulood Alketbi, Constantinos Stathopoulos and Meththa Ranasinghe
Physical Chemical and Textural Characteristics and Sensory Evaluation of Cookies Formulated with Date Seed Powder
Reprinted from: *Foods* **2022**, *11*, 305, doi:10.3390/foods11030305 23

Simona Grasso, Tatiana Pintado, Jara Pérez-Jiménez, Claudia Ruiz-Capillas and Ana Maria Herrero
Characterisation of Muffins with Upcycled Sunflower Flour
Reprinted from: *Foods* **2021**, *10*, 426, doi:10.3390/foods10020426 37

Juliana Mandha, Habtu Shumoy, Athanasia O. Matemu and Katleen Raes
Valorization of Mango By-Products to Enhance the Nutritional Content of Maize Complementary Porridges
Reprinted from: *Foods* **2021**, *10*, 1635, doi:10.3390/foods10071635 45

Marina Jovanović, Snežana Zlatanović, Darko Micić, Dragan Bacić, Dragana Mitić-Ćulafić, Mihal Đuriš and Stanislava Gorjanović
Functionality and Palatability of Yogurt Produced Using Beetroot Pomace Flour Granulated with Lactic Acid Bacteria
Reprinted from: *Foods* **2021**, *10*, 1696, doi:10.3390/foods10081696 59

Raimondo Gaglio, Pietro Barbaccia, Marcella Barbera, Ignazio Restivo, Alessandro Attanzio, Giuseppe Maniaci, Antonino Di Grigoli, Nicola Francesca, Luisa Tesoriere, Adriana Bonanno, Giancarlo Moschetti and Luca Settanni
The Use of Winery by-Products to Enhance the Functional Aspects of the Fresh Ovine "Primosale" Cheese
Reprinted from: *Foods* **2021**, *10*, 461, doi:10.3390/foods10020461 75

Salvador Hernández-Macias, Núria Ferrer-Bustins, Oriol Comas-Basté, Anna Jofré, Mariluz Latorre-Moratalla, Sara Bover-Cid and María del Carmen Vidal-Carou
Revalorization of Cava Lees to Improve the Safety of Fermented Sausages
Reprinted from: *Foods* **2021**, *10*, 1916, doi:10.3390/foods10081916 93

Dolors Parés, Mònica Toldrà, Estel Camps, Juan Geli, Elena Saguer and Carmen Carretero
RSM Optimization for the Recovery of Technofunctional Protein Extracts from Porcine Hearts
Reprinted from: *Foods* **2020**, *9*, 1733, doi:10.3390/foods9121733 107

Lucia Ferron, Chiara Milanese, Raffaella Colombo and Adele Papetti
Development of an Accelerated Stability Model to Estimate Purple Corn Cob Extract Powder (Moradyn) Shelf-Life
Reprinted from: *Foods* **2021**, *10*, 1617, doi:10.3390/foods10071617 123

Zhiqiang Wang, Zhaoyang Wu, Guanglei Zuo, Soon Sung Lim and Hongyuan Yan
Defatted Seeds of *Oenothera biennis* as a Potential Functional Food Ingredient for Diabetes
Reprinted from: *Foods* **2021**, *10*, 538, doi:10.3390/foods10030538 141

María del Mar Contreras, Irene Gómez-Cruz, Inmaculada Romero and Eulogio Castro
Olive Pomace-Derived Biomasses Fractionation through a Two-Step Extraction Based on the Use of Ultrasounds: Chemical Characteristics
Reprinted from: *Foods* **2021**, *10*, 11, doi:10.3390/foods10010111 159

Gabriella Di Lena, Jose Sanchez del Pulgar, Ginevra Lombardi Boccia, Irene Casini and Stefano Ferrari Nicoli
Corn Bioethanol Side Streams: A Potential Sustainable Source of Fat-Soluble Bioactive Molecules for High-Value Applications
Reprinted from: *Foods* **2020**, *9*, 1788, doi:10.3390/foods9121788 183

Sneh Punia, Kawaljit Singh Sandhu, Simona Grasso, Sukhvinder Singh Purewal, Maninder Kaur, Anil Kumar Siroha, Krishan Kumar, Vikas Kumar and Manoj Kumar
Aspergillus oryzae Fermented Rice Bran: A Byproduct with Enhanced Bioactive Compounds and Antioxidant Potential
Reprinted from: *Foods* **2021**, *10*, 70, doi:10.3390/foods10010070 199

Maria Irakli, Athina Lazaridou and Costas G. Biliaderis
Comparative Evaluation of the Nutritional, Antinutritional, Functional, and Bioactivity Attributes of Rice Bran Stabilized by Different Heat Treatments
Reprinted from: *Foods* **2021**, *10*, 57, doi:10.3390/foods10010057 211

Sabrina Spartano and Simona Grasso
Consumers' Perspectives on Eggs from Insect-Fed Hens: A UK Focus Group Study
Reprinted from: *Foods* **2021**, *10*, 420, doi:10.3390/foods10020420 229

Preface to "By-Products: Characterisation and Use as Food"

Each year, the agricultural industry produces large amounts of animal and plant by-products, which in many cases are not valorised, thus generating a major environmental problem. For both sustainable reasons and for economic motives, the interest in adding value to these by-products has recently increased. Many by-products can be transformed into edible ingredients, becoming natural sources of nutritionally enhanced compounds with positive health effects, such as polyphenols, proteins, fibres, lipids, etc. These agro-industrial by-products and co-products can be used as ingredients in food, upcycling them to produce different eco-friendly products such as baked goods, cereals, snacks, meat, fish, dairy products, etc. This Special Issue collected recent advances in the agro-industry regarding by-products and co-products, specifically the use of by-products as food and their use as food-grade ingredients in the formulation of different food products. Specifically, this Special Issue focused on the extraction, processing, and characterisation of by-products and associated technological, nutritional, and sensory quality. It included articles on new approaches and the use of technologies such as fermentation to improve the delivery of extracted bioactive compounds into foods. The Editors welcomed contributions on consumer attitudes toward the use of upcycled ingredients and circular economy initiatives in foods.

This Special Issue collected a total of 15 articles: 1 review on almond by-products, 1 short communication on muffins with upcycled sunflower flour, and 13 research articles (https://www.mdpi.com/journal/foods/special_issues/food_byproducts).

These included five articles on the application of by-products into new foods: cookies with date seed powder, sausages with cava lees, yogurt with beetroot pomace flour, maize complementary porridges with mango by-products, and the use of winery by-products in fresh ovine "Primosale" cheese.

A total of seven research articles were exploratory in nature, focusing more on by-product characterisation than their specific applications. The topics covered were varied and included the shelf life of purple corn cob extract, the use of defatted seeds from *Oenothera biennis* as a potential functional food ingredient for diabetes, the fermentation of rice bran with *Aspergillus oryzae*, the stabilisation of rice bran with different heat treatments, the valorisation of corn bioethanol side streams, the fractionation of olive pomace-derived biomasses, and finally the protein extraction from porcine hearts.

There was only one article specifically looking at consumers: a focus group study on eggs from insect-fed hens.

This Special Issue will be valuable for all those involved in by-product valorisation and circular economy initiatives from farm to fork (primary producers, processing companies, researchers, scientists, consumers, etc.). Many of the studies featured in this Special Issue will encourage future research in this area of growing interest.

Lastly, we acknowledge the efforts of all of the authors who contributed their hard work, expertise, and novel approaches to the publications in this Special Issue.

Simona Grasso, Konstantinos Papoutsis, Claudia Ruiz-Capillas, and Ana Herrero Herranz
Editors

Review

Almond By-Products: Valorization for Sustainability and Competitiveness of the Industry

Marta Barral-Martinez [1], Maria Fraga-Corral [1,2], Pascual Garcia-Perez [1], Jesus Simal-Gandara [1,*] and Miguel A. Prieto [1,2,*]

[1] Nutrition and Bromatology Group, Analytical and Food Chemistry Department, Faculty of Food Science and Technology, Ourense Campus, University of Vigo, E-32004 Ourense, Spain; marta.barral@uvigo.es (M.B.-M.); mfraga@uvigo.es (M.F.-C.); pasgarcia@uvigo.es (P.G.-P.)

[2] Centro de Investigação de Montanha (CIMO), Instituto Politécnico de Bragança, Campus de Santa Apolonia, 5300-253 Bragança, Portugal

* Correspondence: jsimal@uvigo.es (J.S.-G.); mprieto@uvigo.es (M.A.P.)

Citation: Barral-Martinez, M.; Fraga-Corral, M.; Garcia-Perez, P.; Simal-Gandara, J.; Prieto, M.A. Almond By-Products: Valorization for Sustainability and Competitiveness of the Industry. *Foods* **2021**, *10*, 1793. https://doi.org/10.3390/foods10081793

Academic Editors: Simona Grasso, Konstantinos Papoutsis, Claudia Ruiz-Capillas and Ana Herrero Herranz

Received: 30 June 2021
Accepted: 30 July 2021
Published: 2 August 2021

Publisher's Note: MDPI stays neutral with regard to jurisdictional claims in published maps and institutional affiliations.

Copyright: © 2021 by the authors. Licensee MDPI, Basel, Switzerland. This article is an open access article distributed under the terms and conditions of the Creative Commons Attribution (CC BY) license (https://creativecommons.org/licenses/by/4.0/).

Abstract: The search for waste minimization and the valorization of by-products are key practices for good management and improved sustainability in the food industry. The production of almonds generates a large amount of waste, most of which is not used. Until now, almonds have been used for their high nutritional value as food, especially almond meat. The other remaining parts (skin, shell, hulls, etc.) are still little explored, even though they have been used as fuel by burning or as livestock feed. However, interest in these by-products has been increasing as they possess beneficial properties (caused mainly by polyphenols and unsaturated fatty acids) and can be used as new ingredients for the food, cosmetic, and pharmaceutical industries. Therefore, it is important to explore almond's valorization of by-products for the development of new added-value products that would contribute to the reduction of environmental impact and an improvement in the sustainability and competitiveness of the almond industry.

Keywords: *Prunus dulcis*; almond skins; almond hulls; almond shells; almond blanch water; bioactive compounds; bioactivities; agri-waste management

1. Introduction

Global food supply is an ongoing concern and, above all, a major scientific challenge as population, food products, and especially waste generated continue to grow. To ensure a global food supply that complies with food safety standards and minimizes the production of residues, it is necessary to implement sustainable food systems. This paradigm leads to the search for alternative food sources without depleting the agricultural sector. One of the proposed strategies is the use of by-products derived from industrial processes with interesting applications as a source of food ingredients [1].

Nuts constitute a nutritious food with high lipid content that provide a wide range of bioactive compounds beneficial to health. They are well recognized in gastronomy for their distinguished flavor, and their consumption is recommended by nutritionists and doctors for providing antioxidants, bioactive molecules, and nutrients [2,3]. In fact, moderate doses of these foods have been shown to reduce blood levels of total cholesterol and low-density lipoprotein (LDL), both of which are linked to the development of cardiovascular diseases [4]. Several studies have demonstrated these different health properties when consumed, but there is still a need for a deeper research into the full spectra of the benefits they possess. Among these nuts, it is worth mentioning the almond [2].

Almond is a stone fruit known under the scientific name *Prunus dulcis* (Mill.) D.A. Webb (herein referred to as *P. dulcis*). However, unlike other stone fruits, such as apricots or plums, where the pulp (mesocarp) is eaten and the seed (endocarp) discarded, in the case of almonds, the opposite occurs. The almond is the edible part, along the full ripening cycle of the fruit, while the mesocarp and endocarp can be eaten only at the beginning of its ripening. The main structural parts of the almond include an outer greenish cover named the hull; an intermediate shell; a brownish skin; and finally the edible seed, referred to as kernel, meat, or nut (Figure 1) [3,5]. Almond hulls have a green appearance and variable ripening cycles, which may vary due to environmental conditions such as intense heat, ultraviolet radiation, or pest infestation [3]. It is the heaviest part of the kernel, representing around 52% of the total fresh weight [6]. The common almond shell can appear in different shapes and sizes, with modifications in its appearance (wrinkles and pores mainly, among others), and its hardness is also very variable [7]. Almond skins have a brownish appearance, and it represents a light portion of the fruit with a 4% of the total weight of the almond. It is a protective layer that prevents from the oxidation and microbial contamination of the kernel [3].

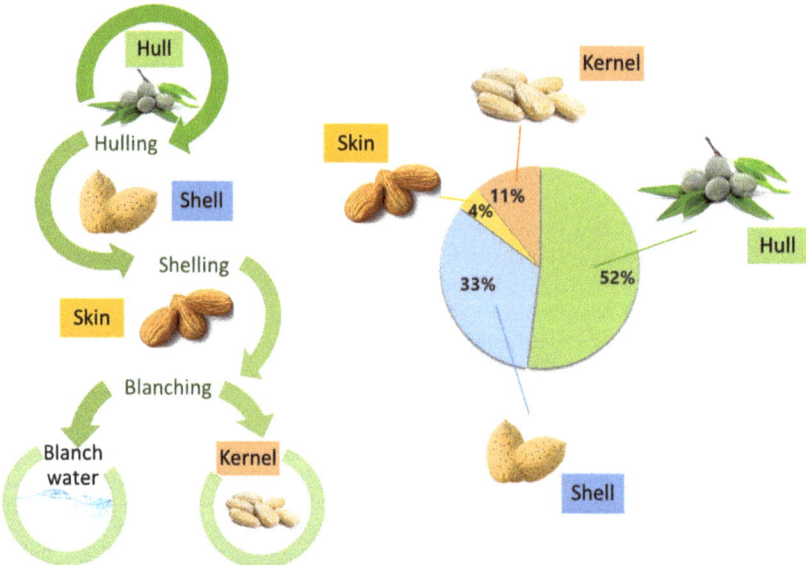

Figure 1. Generation of almond by-products (left) and proportion of constituents of almond fruits (right).

As explained for other nuts, the consumption of whole almonds has been described to possess certain health benefits as it reduces postprandial glycaemia, insulinemia, and others [8]. Extracts obtained from this species have been traditionally applied for treating some diseases, neurological disorders, or respiratory and urinary tract affections. In addition, pharmacological studies indicated that *P. dulcis* has various biological activities such as prebiotic, antimicrobial, antioxidant, anti-inflammatory, anticancer, laxative, hepatoprotective, cardiometabolic protective, neotropic, anxiolytic, sedative, hypnotic, and neuroprotective effects [9].

Almond is the most produced nut worldwide in recent years. *P. dulcis* belongs to the Rosaceae family, and it is characterized for being cultivated in arid areas with dry conditions, thus motivating the establishment of sustainable irrigation practices together with good awareness to ensure high production and, at the same time, carry out an efficient development [10].

The largest almond producer is the United States, and its cultivation is generated in California in particular, with approximately 283,000 ha of cultivation [3]. According to the International Nuts and Dried Fruits Statistical Yearbook, this world production leader will have a 79% of the global almond harvest in 2020/2021, followed by Australia, in second place, which shares the 7% of the market. Spain is in third place, behind Australia and the USA, and is the country with the largest area under almond cultivation, accounting for more than 700,000 ha [11]. There is a worldwide trend towards an increase in the production of this product, which is accompanied of a parallel increase in the amount of by-products that can result in being environmentally and economically harmful if they are not properly managed [10].

The industrial process of treating almonds for consumption involves the elimination of the external parts of almond fruit in a sequential process, obtaining the almond meat or almond kernel followed by different processes: roasting, in which almond skin is detached but not eliminated, and blanching, in which the peeled almonds are generated, thus discarding the peel and generating another by-product known as blanch water [4,5]. In these processes, different forms of equipment are used to separate the meat apart from the outer part of the kernel, including shelling and peeling machines. This type of machinery is usually found in facilities close to the harvesting areas, causing less environmental impact with the transport of the product. In the final processing, almonds are packaged and stored for their distribution.

Around 10 to 25% of the total biomass that reaches the process facilities comes from the remains of the field, such as garden waste, soil, and pebbles. To these residues, one must add the by-products resulting from the dehulling process, which are separated into several piles in different areas of the facility and sorted (shell, hull, and sticks). Both hulls and sticks are predominant among all these fractions, although shells are also an important output [12]. Indeed, the processing of almonds to obtain a kilogram of nuts usually involves the production of 0.6 kg and 2.5 kg of shell and hull in terms of dry weight (dw), respectively, thus representing an annual production of shells and hulls of 0.5 and 2.2 million dry metric tons, respectively [12].

In the past, shells were used as animal feed, while the remaining almond by-products were burned for fuel production and energy use. However, almond by-products, including leaves and flowers, have been described to be of great interest as source of bioactive phytochemicals [10].

Almond hulls possess higher flavonoid content compared to other fruits. This by-product that may present variable concentration of biomolecules depending on the ripening conditions, has been demonstrated to be a rich source of triterpenoids, such as betulinic, ursolic, and oleanoic acids, as well as flavonol glycosides and phenolic acids [3]. Among its applications, almond hulls have been used as an additional source of ingredients to formulate feed diets in dairy industries, which can pay more than~$110 per ton of hulls [12].

The worldwide annual production of almond shells can account from 0.8 to 1.7 million tons. This by-product has high cellulose and lignin contents [13]. Shells can be reused in the industry itself as fuel because of its high calorific value. In detail, almond shells have been utilized during the process of scalding grains for heating water [7], and, similarly, have been exploited by other industries to generate energy as part of boiler fuel. Almond husks can also have structural functions—for instance, in the dairy industry, shells form part of the animal bedding material [12] and they can also be part of the composition of chipboard, used to polish certain metals, or used as a natural dye in wool. Nevertheless, recent reports have indicated their use as organic inclusions in ceramic bodies, as additives, or as a substrate in soilless crops [7].

Almond skins present between 70 and 100% of the total phenols present in the whole almond fruit. Many studies have demonstrated that they contain beneficial phytochemical properties, as they are a great source of phenolic compounds, such as quercetin glycosides, kaempferol, naringenin, catechin, protocatechuic acid, and vanillic acid [3,4]. In addition, polyphenols from almond skins have been shown to have synergistic effects when combined with vitamins C and E, since they protect against LDL oxidation and improve the antioxidant defense. Besides polyphenols, almond skins present different quantities of triterpenoids, especially betulinic acid, oleanoic acid, and ursolic acid, which have been described to possess anti-inflammatory, anticancer, and antiviral activity against human immunodeficiency virus (HIV).

Hence, almond by-products currently represent an issue of waste management, being potentially converted into revalorized products or ingredients in order to ensure their reduction, which would also minimize their environmental impact. In fact, as mentioned above, most of these by-products have been characterized and different bioactive properties have been identified, such as phenolic compounds that play an important role in the prevention of degenerative and cardiovascular diseases [3]. Therefore, a deeper study of the phytochemical and physicochemical composition of this fruit and by-products may allow the recovery of natural and functional ingredients with a potential application in pharmaceutical, cosmetic, or food industries for the development of innovative and sustainable added-value products.

2. Bioactive Composition of Almond by-Products

2.1. Phenolic Compounds

Kernels of *P. dulcis* represent a rich dietary source of polyphenols that provide an important intake of condensed and hydrolysable tannins, flavonoids, and phenolic acids [14] (Figure 2). The total phenolic content of almond nuts has been found to be variable depending on the cultivar selected (Table 1) and the ripening degree, whereas the harvest year has a slight impact [15–17]. Several scientific works support this high content of phenolic compounds in almonds; however, the amounts are highly variable depending on the study checked. It is noticeable that the number of total phenolic compounds detected, the extraction protocol applied, the quantification units employed (determined as percentage or weight, either referred to dry or fresh biomass), the standards used for expressing final concentrations, and the detection method chosen for analyzing the phenolic content of almond kernels influence final results. For instance, in a study where 18 specific polyphenols were evaluated, these following molecules were underlined as the major ones: catechin, epicatechin, naringenin-7-O-glucoside, kaempferol-3-O-rutinoside, dihydroxykaempferol, isorhamnetin-3-O-rutinoside, isorhamnetin-3-O-glucoside, and naringenin [16]. In another assay, up to 28 polyphenols (9 phenolic acids and 19 flavonoids) were determined and their relative presence quantified. This work also points to the flavonoid catechin as a major compound (ranging from 3 to 81%, on average 46%) followed by chlorogenic acid (1–21%, 4.1% on average), naringenin (0.2–16%, 4.3% on average), rutin (0–11%, 2.1% on average), apigenin (0.1–10%, 2.9% on average), and astragalin (0–9%, 2.4% on average) [18]. While these two studies ([17,19]) employed LC–MS data for the final quantification of the targeted polyphenols analyzed, other authors [15,17] quantified the polyphenolic content using spectrophotometric techniques based on absorbance measurements, and thus they did not provide specific information on the chemical profile of the samples.

Regarding tannins, almonds have been described to contain both hydrolysable and condensed tannins. The extractable condensed tannins mostly consist of bound units of (+)-catechin and (−)-epicatechin as dimmers or trimers but also creating greater oligomers. Those can be constituted of units of (epi)afzelechin, (epi)catechin, and (epi)gallocatechin bound by A- and B-linkages that may easily reach a polymerization degree of 13, although when the polymerization is higher than six, only the B-bound is present [19]. Tannins quantified in three Californian varieties (Nonpareil, Carmel, and Butte), where the number of condensed tannins were expressed as proanthocyanidin B2 equivalents, strongly varied among varieties in a range from 322 to 1073 µg/g almond. The quantification of hydrolysable tannins provided narrower ranges for ellagic acid (487–632 µg/g) and gallic acid (141–406 µg/g) [19].

Another interesting group of polyphenolic compounds are stilbenes, although their presence in almonds is less abundant than other polyphenols. Polydatin was detected in almond kernels, almond skins, and the blanch water (0.7, 1.8, and 72 ng/g), while piceatannol and oxyresveratrol were found in almond blanch water (17 ng/g) [14].

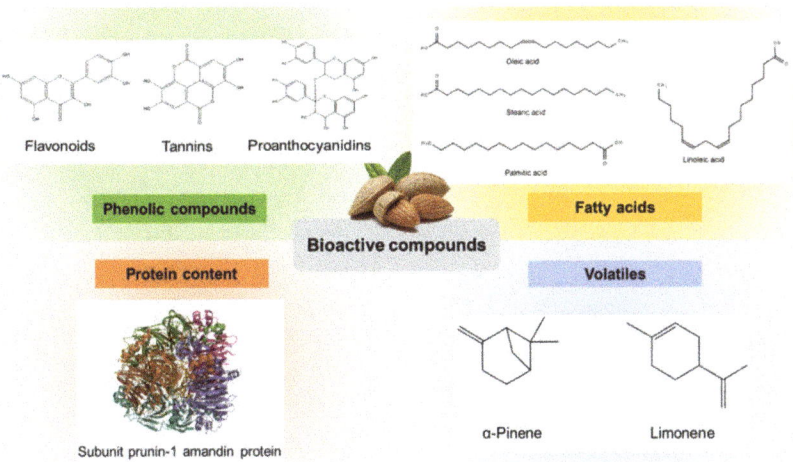

Figure 2. Major bioactive compounds isolated from almond and almond by-products.

Table 1. Total content of polyphenols of different almond varieties from diverse geographical origins (expressed as mg of gallic acid/g dw).

Geographical Origin	Almond Variety	Phenolic Content (mg/g)	References
Australian	Johnston Prolific	0.8–0.9	[15]
Californian	Butte	6.1	[16]
	Carmel	6.1	
	Fritz	3.6	
	Mission	6.1	
	Monterrey	4.3	
	Nonpareil	5.3	
	Texas	1.5	
	Thompson	0.8–1.5	[15]
Italian	Duro	164	[17]
	Filippo Ceo	1.3–2.7	
	Genco	2.2–11.0	[15]
	Tuono	1.3–4.9	
Spanish	Desmajo Largueta	1.7–1.8	[15]
	Marcona	0.3–1.0	
Italian and Spanish	Ferragnès	2.5–4.2	[15]
	Francolì	2.7–7.3	
Portuguese	Pegarinhos	34.2	[17]

Moreover, almond by-products such as skins have been chemically characterized and found to contain high levels of phenolic acids and flavonoids that can account for more than 30 molecular structures [14]. In fact, the variety of phenolic compounds found in almond skins included hydroxybenzoic and hydroxycinnamic acids; protocatechuic aldehydes; glycosides of flavonols and flavanones; and aglycones of flavonols, dihydroflavonols, and flavanones (Table 2). In a work based on the analysis of blanched almond skins, the authors revealed that the most abundant compounds detected belong to the flavanol class (20–38 μg of (+)-catechin per g dw of almond skin and 7–26 μg/g of (−)-epicatechin) and to the flavonol glycosides family (5–41 μg/g of kaempferol-3-O-rutinoside and 5–58 μg/g of isorhamnetin-3-O-rutinoside). These contents of phenolic compounds in blanched skins were observed to get increased after the application of dehydration treatments, such as roasting or drying. In the same work, authors also explored the presence of A- and B-type procyanidins (homopolymers of (epi)catechins), propelargonidins (heteropolymers with one unit of (epi)afzelechin, and multiple (epi)catechins) and prodelphinidin (heteropolymers with one unit of (epi)gallocatechin combined with several (epi)catechins). Procyanidins and propelargonidins were analyzed up to heptamers and A- and B-type prodelphinidins up to hexamers. Other interesting molecules found in plasma and urine as metabolites after the ingestion of almond skins were glucuronide and a few derivatives, including conjugates of naringenin and isorhamnetin, sulfate derivatives of (epi)catechin, sulfate conjugates of isorhamnetin, and conjugates of hydroxyphenylvalerolactones [4].

Table 2. Phenolic content in almond products and by-products. Quantitative data extracted from [4,14,20] [1].

Compounds	Blanched Skin (µg/g)	Whole Almonds (µg/g)	Blanch Water (µg/g)
Hydroxybenzoic acids and aldehydes	**26.3–57.3**	**2.4–7.0**	
p-Hydroxybenzoic acid	3.1–7.2	0.03–0.06	
Vanillic acid	11.1–19.2	1.1–2.5	
Protocatechuic acid	6.7–17.2	1.3–4.4	
Protocatechuic aldehyde	5.4–13.7	-	
Hydroxycinnamic acids	**1.8–4.9**		
trans p-Coumaric acid	0–1.1	-	
3-O-Caffeoylquinic acid	1.8–3.8	-	
Flavan-3-ols	**61.5–134**	**12.7–51.3**	
(+)-Catechin	20.1–38.3	9.5–38.6	
(−)-Epicatechin	7.2–26.5	3.2–12.7	
Procyanidin B1-3, B5, B7, C1	25.0–46.1	-	
Unknown dimers/trimers A [(epi)catechin → A → (epi)catechin]	9.2–23.0		
Flavonol glycosides	**15.6–130**	**113–231**	
Kaempferol-3-O-rutinoside	5.3–40.7	7.1–14.3	
Kaempferol-3-O-glucoside	0–14.2	0.08–0.2	
Kaempferol-3-O-galactoside	-	0.05–0.2	
Isorhamnetin-3-O-rutinoside	5.3–58.0	99–191	
Isorhamnetin-3-O-glucoside	5.0–15.3	-	
Isorhamnetin-3-O-galactoside	-	3.0–7.1	
Quercetin-3-O-glucoside	0–1.7	0.5–1.6	
Quercetin-3-O-galactoside	-	3.1–12.6	
Quercetin-3-O-rutinoside	-	0.6–3.7	
Flavanone glycosides	**4.8–28.6**	**0.9–1.9**	
Naringenin-7-O-glucoside	2.3–25.9	0.9–1.9	
Eriodictyol-7-O-glucoside	2.5–2.7	-	
Flavonol aglycones	**9.6–33.0**	**1.1–4.9**	
Kaempferol	2.7–12.1	0–0.04	
Quercetin	1.0–4.9	0.2–0.3	
Isorhamnetin	5.9–16.0	0.9–4.6	
Dihydroflavonol aglycones	**0–10.3**	**0.4–3.0**	
Dihydroquercetin	0–10.3	-	
Dihydroxykaempferol	-	0.4–3.0	
Flavanone aglycones	**6.7–19.9**	**1.2–6.9**	
Naringenin	3.7–12.1	0.4–1.2	
Eriodictyol	3.0–7.8	0.8–5.7	
Stilbenes	**0.002**	**0.008**	**0.07–0.11**
Polydatin	1.5–2.2 ng/g	7.2–8.5 ng/g	63–84 ng/g
Piceatannol + oxyresveratrol	ND	ND	9.1–25.5 ng/g
trans-Resveratrol	ND	<LOD	
Pterostilbene	ND	<LOD	
Total	**111–418**	**132–306**	

[1] Bold letters indicate the sum of total concentration referred to each subfamily of phenolic compounds. LOD: limit of detection; ND: not determined.

2.2. Fatty Acids

The total content of fatty acids in almonds is dependent on the cultivar, growing year, and its conditions, as it has been revealed for other biomolecules [21]. Nevertheless, different works point to similar ranges than those provided by European Food Safety Authority (EFSA), showing that the lipid content of almonds from different cultivars and locations is quite constant and in the range of 36 and 63% [19,22,23]. EFSA suggests that the fat content of almonds is around 40–50%, half of which consist of unsaturated fatty acids (UFAs), from which polyunsaturated fatty acids (PUFAs) represent 22%, mainly comprised of linoleic acid; 70% are monounsaturated fatty acids (MUFAs), with oleic acid as the major representative; and the remaining 8% is composed by saturated fatty acids (SFAs) [24]. These data are consistent with later published works that underlined that the sum of oleic (C18:1), linoleic (C18:2), palmitic (C16:0), and stearic (C18:0) acids account for over 99% of the total fatty acid content in almonds (Figure 2). The average content of UFAs, as the most abundant lipids in almonds, possess quite constant percentual values: for oleic acid (C18:1) from 58 to 77%, for linoleic acid (C18:2) 16–30%, for palmitoleic acid (C16:1) 0.2–0.6%, for vaccenic acid (C18:1) 0.7–2.2%, and for heptadecenoic acid (C17:1) 0.11%. These values demonstrate that almonds have a high amount of UFAs, whereas the presence of saturated acids is less relevant, although they also contribute to their total lipidic content. Palmitic acid (C16:0) accounts for 4.7–7.0%, stearic acid (C18:0) 1.4–2.6%, arachidic acid (C20:0) 0.06–0.12%, and myristic acid (C14:0) 0.01–0.22%. Other fatty acids present at trace levels include the SFAs pentadecanoic (C15:0), margaric (C17:0), heneicosylic (C21:0), behenic (C22:0), tricosanoic acid (C23:0), and lignoceric (C24:0) acids and the UFAs paullinic (C20:1), α-linolenic (C18:3), dihomo-γ-linolenic (C20:3), and docosadienoic (C22:2 cis 13, 16) acids [18,23].

Moreover, almonds are regarded as an important source of triacylglycerols (TAGs), since their content has been estimated to reach high percentages, about 98%. The main TAGs identified in almonds correspond to OOO (O: oleic acid) and OLO (L: linoleic acid), accounting for more than 60%. Other TAGs present in the lipidic fraction of almonds include POO (P: palmitic acid), OLL, PLO, SOO (S: stearic acid), LnOO (Ln: linolenic acid), LLL, LLP, PLP, and POP [25–27].

The proportion of oil in almonds is highly dependent on the ripening state of the almond kernel. The lipidic fraction becomes incremented due to a higher oil synthesis, but also because of the dehydration of the kernel, and after 14–18 weeks of maturity, almonds develop oil bodies. The maximum accumulation rate obtained for oleic acid, and thus in general for the total fatty acid content, since this is one of the major fatty acids, was between the week 14 and 17 post-anthesis (after the flower opening). Along this period of days, the maximum accumulation rate was 18 µg/day/g of fresh weight (fw), and then it decreased [26,27]. Therefore, the almond kernel development is a key phenomenon for maximizing the accumulation of lipids, both in terms of fatty acids and TAGs.

2.3. Volatiles

The most common volatile terpenes described in almonds are α-pinene and limonene, which have been detected in raw almonds at low levels (around 17 ng/g) [14]. Recently, different terpenes were analyzed in almond leaves, flowers, and young fruits. In leaves, the most abundant terpenes were eugenol, which reached 675 ng/g fw, and the monoterpene geraniol, for which higher recorded contents were between 31 and 54 ng/g fw. In young almond fruits, researchers quantified similar amounts for linalool (around 30 ng/g fw). The lowest amounts of terpene compounds were found in almond flowers, where other compounds like trans-linalool oxide, carvacrol, or β-cyclocitral were punctually detected but present at much lower levels (1–4 ng/g fw) [28].

Apart from terpenes, other kinds of volatiles that mainly include carbonyls, pyrazines, and alcohols had been identified in almond kernels, leaves, or flowers. Nevertheless, the most abundant amounts were described in kernels, especially after longer roasting treatments. Benzaldehyde was one of the most abundant compounds in almond leaves, where it may reach levels of 2.4 µg/g, similar to its concentration in raw almonds, around 3 µg/g. However, when raw almonds were submitted to a roasting process, benzaldehyde strongly reduced its concentration to 0.3 µg/g. Other volatiles affected by roasting were 2-methyl-1-propanol, 3-methyl butanol, 2-phenylethyl alcohol, α-pinene, and methylsulfanylmethane, whose concentration was diminished. Instead, roasting increased the amounts of other volatiles. For instance, hexanal, which was an abundant volatile in raw almonds (0.4 µg/g) and leaves (0.2 µg/g), increased its content nearly three times after the roasting of raw almonds. A similar pattern was found for 1,2-propanediol, with initial levels of 0.2 µg/g in raw almonds that got triplicated after short roasting times. The most relevant increments of volatiles after the roasting of raw almonds were found for 2- and 3-methylbutanal that increased their presence more than 100 and 400 times, respectively [28,29]. The effect of roasting on the volatile profile of almonds was also recorded on several Portuguese varieties, showing a marked increase of hexanal and benzaldehyde after roasting, whereas benzaldehyde together with 3-methyl-1-butanol were mostly reported on raw almond fruits [30].

2.4. Protein Content

Almonds have been described as a good source of high-quality proteins with relative amounts between 20 and 25% [24,31,32]. Regarding the protein content of almond and its by-products, the kernel, together with almond cake, possess quite abundant amounts of protein with percentages of 8.4–35% and 37%, respectively [33,34]. Other by-products have lower protein levels, such as almond skins with 10–13% [35], and almond hulls, which have the lowest values at 5.4–6.7% [36].

Almond proteins possess a chemical profile characterized to contain most of the essential amino acids, even though almonds represent a limited source of sulfur amino acids (methionine and cysteine), lysine, and threonine for children below 5 years. Instead, in adults, almonds constitute a more complete protein source since the sulfur amino acids alone are considered limiting [32].

This limiting amino acid profile is mainly provided for the major protein in almonds, amandin. Amandin is a legumin-type protein formed by two subunits, prunin-1 (Figure 2) and prunin-2, with a total molecular weight of 427 kDa, approximately. Amandin may account for nearly the 70% of total almond proteins. Regarding its amino acid profile, it is consistent with that from almonds, since its essential limiting amino acids are methionine, lysine, and threonine [31]. The majority of proteins present in almonds are hydrosoluble, and their digestibility, at least in the varieties Carmel, Mission, and Nonpareil, was shown to be higher than 82% [32]. Thus, almond and its by-products may result in an efficient source of vegetal proteins.

3. Biological Activities from Almond By-Products

Almond has been largely reported for its associated bioactivities, mainly focusing on the phytochemical characterization of the edible kernel. However, due to the interest regarding the valorization of almond by-products obtained along the productive workflow (mostly skin, shell, hull, and blanch water), an increasing number of studies focused on the bioactivities attributed to almond residues have appeared. Thus, the revalorization of almond by-products constitutes a promising approach during almond waste management, thus enabling the design of added-value products. Table 3 shows a general overview of the bioactivities attributed to almond and its by-products.

Table 3. Bioactivities of almond and its by-products and their mechanisms of action.

Mechanisms of Action	References
Antioxidant activity	
Free-radical scavenging activity: DPPH, ORAC, ABTS	[37,38]
Reducing power: FRAP	[37,39]
Antioxidant enzymes induction: SOD, CAT, GPx, APX	[40–42]
Cell antioxidant response modulation: Nrf2, ARE expression	[40]
Inhibition of lipid oxidation TBARS	[3]
Depletion of oxidative stress markers: ROS, GSH, DNA, and protein degradation	[42,43]
Anticancer activity	
Effectiveness against MCF-7, MDA-MB-468, HepG2, HCT-116, Saos-2, Colo-320, Colo-741, Caco-2, and B-16 cancer cell lines	[44–48]
Oxidative stress alleviation	[49]
Cell cycle arrest	[50]
Impairment of mitochondrial function and induction of caspase-mediated apoptosis	[46]
Inhibition of tumor migration, metastasis, and cell malignancy	[47]
Anti-inflammatory activity	
Inhibition of immune cell infiltration	[51]
Reduction of pro-inflammatory CKs: IL-1β, IL-6, TNF-α, CINC-1, MCP-1	[51,52]
Depletion of inflammatory mediators: PGE2, NFκB, NO, ICAM-1, selectins	[53–55]
Inhibition of pro-inflammatory enzyme activity: iNOS, COX-2, MPO, PARP	[51,52]
Antimicrobial activity	
Bacteriostatic effect against both pathogenic Gram-positive and Gram-negative bacteria	[39]
Antifungal activity against *C. albicans*	[39]
Antiviral activity HSV-1 and HSV-2: inhibition of viral penetration, suppression of early viral proteins and viral DNA accumulation, enhancement of antiviral immune cell response	[39]
Prebiotic activity	
Enhancement of bifidobacterial and lactobacilli populations via butyrate production	[56,57]
Promotion of β-galactosidase activity and inhibition of β-glucuronidase and azoreductase activities	[9]
Suppression of pathogenic bacteria growth	[58]

ABTS: 2,2-azino-bis(3-ethylbenzothiazoline-6-sulfonic acid); APX: ascorbate peroxidase; ARE: antioxidant response element; CAT: catalase; CINC-1: cytokine-induced neutrophil chemoattractant 1; CKs: cytokins; COX-2: cyclooxygenase 2; DPPH: 2,2-diphenyl 1-picrylhydrazyl; FRAP: ferric reducing antioxidant power; GPx: glutathione peroxidase; GSH: reduced glutathione; HSV: herpes simplex virus; ICAM-1: intercellular adhesion molecule; IL: interleukin; MCP-1: cytokine-induced neutrophil chemoattractant 1 monocyte chemoattractant protein-1; MPO: myeloperoxidase; NFκB: nuclear factor kappa B; NO: nitric oxide; Nrf2: nuclear factor-E2-related factor 2; ORAC: oxygen radical absorbance capacity; PARP: poly(ADP-ribose) polymerase; PGE2: prostaglandin E2; ROS: reactive oxygen species; SOD: superoxide dismutase; TBARS: thiobarbituric acid reactive substances; TNF-α: tumor necrosis factor alpha.

3.1. Antioxidant Activity

Thanks to their high content in polyphenols, almonds and their by-products exhibit a potent antioxidant activity, developed by different mechanisms, such as free-radical scavenging activity, antioxidant enzymes induction, modulation of genetic antioxidant response, and oxidative stress and lipid peroxidation biomarker regulation. Considering almond products, the whole seed, skins, husks, and blanch water have been mostly evaluated in terms of antioxidant activity by in vitro and biochemical assays (Table 3).

The determination of radical scavenging activity (RSA) of almond extracts have been carried out by 2,2-diphenyl 1-picrylhydrazyl (DPPH) and 2,2-azino-bis(3-ethylbenzothiazoline-6-sulfonic acid) (ABTS) free radical assays, together with the Oxygen Radical Absorbance Capacity (ORAC), and Ferric Reducing Antioxidant Power (FRAP) assays. In this sense, seven almond cultivars were subjected to the antioxidant activity determination by the DPPH and FRAP determination of flavonoid-enriched almond skin extracts [37]. Specifically, all cultivars showed high RSA rates, with values up to 90% of DPPH scavenging, whereas the Guara cultivar skins showed the highest FRAP values (556 µmol of Trolox equivalents (TE)/g almond skin). Concerning almond skin processing, drying promotes a significant increase of DPPH scavenging from 40.4 µmol TE/g of non-dried almond skins, because of the eventual Maillard reactions formed after this process [59]. In parallel, four different Italian varieties were subjected to the RSA determination of different almond by-products [38,60]. Thus, whole seeds promoted the highest rates of DPPH and ABTS scavenging activity when extracted using ethanol as solvent, whereas the rest of almond parts achieved the maximum values for hydroethanolic extracts. With respect to almond parts, the hydroethanolic skin and hull extracts of Pizzuta cultivar promoted the highest DPPH values 14.76 µg/mL [38,61], whereas the highest ABTS values were observed for hydroethanolic husk extract of Fascionello cultivar: 1.65 mM TE [38]. In the same way, the results for ORAC assay indicated that the mixture of Spanish varieties (0.5 mmol TE/g) possessed a higher activity than American varieties (0.4 mmol TE/g), being consistent with the chemical composition in terms of proanthocyanidins profile [5]. Concerning individual compounds, in terms of ABTS, DPPH, and FRAP, it was recently determined that chlorogenic acid showed the highest antioxidant activity among the polyphenols found in hydromethanolic hull extracts, whereas isorhamnetin was revealed as the most efficient antioxidant from skin extracts [62].

Besides RSA and reducing power, almond extracts have been assessed in terms of antioxidant enzyme induction, especially aqueous skin extracts, thanks to their high proanthocyanidin content, as demonstrated by the promotion of glutathione peroxidase (GPx), catalase (CAT), and superoxide dismutase (SOD) activities, attributed to proanthocyanidins concentrations of 25–50 µg/mL [40]; additionally, the hydromethanolic leaf extracts of the Mazzetto cultivar were shown to induce ascorbate peroxidase (APX) activity [41]. Furthermore, the same extracts were shown to modulate the in vitro expression of signaling pathways, by boosting the activation of nuclear factor-E2-related factor 2 (Nrf2) and promoting the expression of antioxidant response element (ARE), both involved in the cellular antioxidant system [40].

In addition, blanched almond skins have been reported to modulate plasma biomarkers of oxidative stress, such as GPx activity, glutathione concentration and the ratio reduced glutathione/oxidized glutathione [42]. According to a recent study, the antioxidant properties of polyphenol-enriched hydroethanolic almond hull extracts as inhibitors of the toxicity caused by the induced oxidative stress in Caco-2 cancer cell line was reported as a result of reactive oxygen species (ROS) scavenging, and the regulation of cell redox status [43]. Indeed, these oxidative stress-alleviating properties of almonds have been proved by in vivo human trials, indicating that almond consumption prevents the oxidative DNA damage and lipid peroxidation in male smokers [63]. Besides DNA oxidation prevention, ethanolic hull extracts prevented protein oxidation [42]. Moreover, the inhibition of lipid peroxidation of almond peels has been also assessed in vitro [42,64], together with the methanolic almond fruit extracts of different Portuguese varieties [17], by means of the inhibition of peroxidation damage in biomembranes, through the thiobarbituric acid reactive substances (TBARS) formation, and the inhibition of the induced oxidative hemolysis in erythrocytes, mostly developed by Duro Italiano cultivar. Overall, the positive health effects, in terms of antioxidant activity, associated with the consumption of almonds could be partly due to the almond skin phenolic composition, predominantly containing proanthocyanidins and flavonols, that could be used as an added-value by-product to be exploited as dietary antioxidant ingredients [3].

3.2. Anticancer Activity

As observed for antioxidant activity, the anticancer properties associated with almond and its by-products has been assessed by both in vitro and in vivo studies (Table 3), as a consequence of the plethora of bioactive compounds found on these matrices, especially polyphenols; acid-soluble polysaccharides; triterpenoid acids (such as ursolic, oleanolic, and betulinic acids); and UFAs proceeding from different almond seed-associated products, including kernel, skins, hulls, and oil [9]. Nevertheless, the anticancer effects attributed to almond are closely related to those of antioxidant activity, since oxidative stress is considered one of the major process involved in the early stages of carcinogenesis [49].

Taking this into account, the acetonic almond seed extracts, with high concentrations of phenolic compounds, essentially containing phenolic acids and flavonoids, showed high antiproliferative effects on two different human breast cancer cell lines, MCF-7 and MDA-MB-468, exhibiting effective cytocidal concentrations at 10 µg/mL and >20 µg/mL, respectively [44]. In the same way, the aqueous bitter almond kernel extracts showed an impressive cytotoxicity in a dose-dependent manner against MCF-7 and human colon carcinoma cancer (HCT-116) cell lines, whereas the methanolic extracts of the same matrix promoted the highest cytotoxicity against human hepatocellular cancer cell line (HepG2), showing growth inhibitory concentrations (GI_{50}) of 29.5 µg/mL, 24.5 µg/mL, and 10.1 µg/mL, respectively [45]. As a matter of fact, the cytotoxicity attributed to polyphenols from methanolic almond kernel extracts against MFC-7 and HepG2 cancer cell lines was proved to be due to the induction of cell cycle arrest at the G2/M phase, associated with preG1 apoptosis induction, via the coordinated upregulation of cyclin-dependent kinase inhibitor 2A (CDKN2A) and the inhibition of cyclin-dependent kinase 4 (CDK4) genes [50]. Besides the kernel, the almond hull has also been reported as a polyphenol-enriched by-product, whose the hydroacetonic extracts showed a potent cytotoxicity against human osteosarcoma cell line (Saos-2), exhibiting GI_{50} values = 123.7 µg/mL, mostly affecting cancer cell cycle progression, which was arrested at G2/M phase, impairing the mitochondrial function, inducing caspase apoptotic activity, and inhibiting tumoral cell migration [46].

In addition to almond-derived polyphenols, UFAs detected in the almond seed oil, mostly oleic and linoleic acids, also promoted an intense antiproliferative effect on two colon carcinoma cell lines, either primary (Colo-320) or metastatic (Colo-741), in a dose- and time-dependent manner [47]. Moreover, in this particular study, almond seed oil was reported to inhibit cancer cell invasiveness, by the downregulation of bone morphogenetic protein 2 (BMP-2) and β-catenin pathways, and growth, by the inhibition of Ki-67 (a marker of cell proliferation) expression. In the same way, different polysaccharidic fractions from ethanolic almond skin extracts, mainly composed of arabinose, galactose, and mannose, promoted a cytotoxic effect on both human colon carcinoma Caco-2 and murine melanoma B-16 cell lines, with proliferation inhibition ratios of 88.74% and 90.96%, respectively [48]. Finally, terpenoids from ethylacetate Sicilian almond hull extracts were assessed in terms of their antiproliferative properties, specifically betulinic acid, which exhibited an excellent cytotoxic activity toward MCF-7 (GI_{50} = 0.27 µM), even higher than that of the anticancer drug 5-fluorouracil [65,66]. On these bases, both almond and its by-products can be considered as a new source of pharmaceuticals in the management of cancerous tumors of different origins.

3.3. Anti-Inflammatory Activity

Due to the close relationship of inflammation on oxidative stress and carcinogenesis, the determination of anti-inflammatory effects on almond and by-products is essential to provide an added value to these food products. Consequently, the high concentrations of bioactive compounds found on these matrices, mostly polyphenols, UFAs, and protein hydrolysates, are considered the major responsible for this bioactivity. Thus, there is a large amount of scientific evidence about the multifaceted anti-inflammatory effects of almond and by-products, as reported by in vitro, in vivo, and interventional studies in humans, acting as inhibitors of inflammatory enzymes, pro-inflammatory cytokine (CK) production, oxidative stress, and inflammatory marker-lowering agents (Table 3).

In this sense, the in vitro anti-inflammatory properties of almonds have been mostly recorded in lipolysaccharide (LPS)-induced RAW264.7 macrophages cell line. The oleic acid-enriched oily almond extracts promoted the inhibition of the inducible nitric oxide synthase (iNOS), and cyclooxygenase-2 (COX-2), by reducing the levels of the inflammatory mediator tumor necrosis factor alpha (TNF-α) [52]. Furthermore, the same extracts promoted the reduction of pro-inflammatory CKs levels, mostly intereleukin-1β (IL-1β) and IL-6, as well as the mediator nitric oxide (NO) levels. Such effects were in accordance with those observed for pepsin hydrolysates obtained from almond flour [67] and those provided by acetonic dried almond skin extracts from the Corrente/Tuono mixture variety on LPS-induced intestinal IEC-6 cell line, together with the inhibition of oxidative stress, via ROS levels reduction [68].

In parallel, the in vivo almond-associated anti-inflammatory effects have been assessed using different rodent models. In particular, the polyphenol-enriched Marcona almond blanch water was applied to in vivo rat colitis models, causing a reduction in inflammatory cell infiltrate, together with the inhibition of myeloperoxidase (MPO) activity and other related enzymes, such as iNOS [51]. In addition, the same study promoted an inhibition of oxidative stress, together with a reduction in pro-inflammatory CKs, including IL-1β, cytokine-induced neutrophil chemoattractant 1 (CINC-1), and monocyte chemoattractant protein-1 (MCP-1). Accordingly, another colitis-induced mice model fed natural almond skin powder exhibited similar results on the prevention of intestinal inflammation, modulating the associated nuclear factor κB (NFκB) and c-Jun N-terminal kinase (JNK) signaling pathways, inhibiting enzymes such as iNOS and poly(ADP-ribose) polymerase (PARP), and decreasing the levels of leukocyte-activating markers intercellular adhesion molecule (ICAM-1) and P-selectin [55]. Besides such colitis in vivo models, the ethanolic almond seed extracts were applied to renal cell carcinoma-induced rat models, demonstrating an inhibition of the cancer-related inflammatory process by decreasing pro-inflammatory CKs (IL-1β, IL-6, and TNF-α) levels, together with those of several inflammatory mediators (prostaglandin E2, PGE2, and NFκB) in a dose-dependent manner [54].

Finally, the effectiveness of almond and its by-product consumption as anti-inflammatory agents has been reported by different interventional human trials, as indicated by the reduction of circulating levels of C-reactive protein (CRP) and E-selectin inflammatory mediators, mostly guided by the high concentration of UFAs as part of almond-containing diets [69]. Consequently, these results facilitate the characterization of almond as a natural source of anti-inflammatory compounds, although further studies regarding the characterization of the mechanism of action of such compounds are required.

3.4. Antimicrobial Activity

Recent studies have pointed at the polyphenols from almond and its by-products as the major antimicrobial agents found on these matrices, which has been reported for different bacterial, fungal, and viral species (Table 3) [9]. In fact, the antimicrobial activity of almond has been assessed, indicating its high diversification, especially attributed to skins, but also hulls and blanch water [39].

The antibacterial effects of almond skin extracts have been largely assessed for a wide range of both Gram-positive and Gram-negative bacteria. Thus, polyphenol-enriched aqueous methanolic extracts from almond skins and hulls were subjected to antibacterial assays, revealing that skin extracts exhibited significantly higher antibacterial activity rates, even greater than those of antibiotics such as gentamicin, against *Pseudomonas aeruginosa*, *Staphylococcus aureus*, *Enterococcus faecalis*, and *Listeria monocytogenes* [10]. The same authors suggest the high concentrations of phenolic compounds as the cause of this antimicrobial activity, mostly due to naringenin, (-)-epicatechin, protocatechuic acid, catechin, and isorhamnetin-3-O-glucoside. Equally, the methanolic skin extracts of Pizzuta variety showed that blanching promotes a significant loss of the antimicrobial activity, which partially remains in the blanch water [64]. On these bases, these results were in line with other works, in which the bacteriostatic effects of methanolic peel extracts were reported for *Salmonella enterica* var. Typhimurium, *L. monocytogenes*, and *Staphylococcus aureus*, due to the high concentrations of epicatechin, protocatechuic acid, and naringenin [35]. Moreover, the same extracts have been seen to exert a potent inhibition in the proliferation of *Helicobacter pylori* from different clinical isolates, referring to protocatechuic acid as the major responsible for this effect [53]. Due to the wide evidence regarding the antibacterial properties of almond by-products, recent reports have proposed the application of almond gum for the design of antimicrobial zinc oxide nanoparticles with a potent antimicrobial activity against *S. aureus*, *Escherichia coli*, and *Salmonella paratyphi*, as well as antifungal activity against *Candida albicans* [70].

Besides polyphenols, polysaccharides from almond tree gum, mostly containing galactose, arabinose, xylose, mannose, rhamnose, and glucuronic acid as monosaccharide residuals, were explored in the basis of antimicrobial activity, being effective bacteriostatic compounds toward *S. aureus*, *P. aeruginosa*, *S.* Typhimurium, *E. faecalis*, and *E. coli* [71]. Moreover, hemicelluloses from the same almond gum have been seen to promote a higher antibacterial activity than polysaccharides in the case of both Gram-positive species, such as *Bacillus subtilis*, *Actinomyces* sp., and *S. aureus*, and Gram-negative species, such as *Klebsiella pneumoniae* and *Salmonella typhi* [72]. Such results provide insight about the consideration of almond by-products for their revalorization, not only associated with almond fruits, but also the products obtained from almond trees.

In addition to the antibacterial and antifungal properties of almond by-products, their polyphenol-enriched derived extracts have been investigated to determine their antiviral activity. For this purpose, methanolic peel extracts were applied to herpes simplex virus-1 (HSV-1)-infected Vero cell lines, promoting an antiadhesive activity of HSV-1, thus preventing its introduction into the cells, suppressing the synthesis of early viral proteins and the accumulation of viral DNA [39], and increasing the virus exposition to the action of polyphenols, mostly due to the high concentration of flavonones [73]. In parallel, the aqueous peel extracts developed a potent antiviral activity against HSV-2 at extract concentrations of 60 µg/mL, mainly guided by kaempferol glycosides, which were responsible for boosting the immune response in peripheral blood mononuclear cells by modulating the synthesis of different CKs, such as interferon gamma (IFN-γ), IL-4, IL-10, and TNF-α [74]. Due to the multifaceted antimicrobial properties of almond skins, further studies should be aimed at characterizing the mechanisms of action of these natural antimicrobials, as well as the potential synergism between molecules on the treatment of pathogenic bacterial, fungal, and viral infections.

3.5. Prebiotic Activity

Almond and its by-products have been valorized as prebiotic products, as well, especially almond skins, promoting the enhancement of intestinal microbiota diversity and improving the overall gastrointestinal function. In this sense, almond consumption including skins has been reported to increase β-galactosidase activity, considered as a marker of beneficial colonic bacteria, such as bifidobacterial and lactobacilli, thus improving carbohydrate metabolism of chronic ailments, such as Crohn's disease and ulcerative colitis (Table 3) [9].

The prebiotic activity of almond peels was reported by an in vitro digestion model, showing that the dietary fiber derived from this by-product enhanced the bifidobacterial population, including *Clostridium coccoides* and *Eubacterium rectale* after a 24-h incubation period [75]. In addition, the same group demonstrated that both bacterial groups were not negatively affected by the presence of polyphenols in this matrix, and that butyrate production played a critical role in the enhancement of their growth [75], as a consequence of the metabolization by the gut microbiota of UFAs present in almond and its by-products [56]. Such butyrate-mediated bifidobacterial promotion was assessed by Rocchetti et al. (2019), suggesting that polyphenols from almond seeds are catabolized by colonic bacteria during fecal fermentation [57]. Thus, skins constitute a prominent by-product causing the prebiotic effects of almonds, thanks to their high content of dietary fiber, representing the 45% of their weight [75], and a source of bioactive prebiotic molecules, as it is the case of xylooligosaccharides (XOS), polysaccharides, and hemicelluloses [76,77].

Besides the in vitro assessment of prebiotic activity, different clinical interventions in humans have shed light about the effect of almond skin consumption on the gut function, revealing that both the *Bifidobacterium* spp. and *Lactobacillus* spp. increased their populations in fecal samples, whereas it suppressed the multiplication of the pathogenic *Clostridium perfringens* [58]. Moreover, the almond roasting process slightly decreased the prebiotic effects of almonds in comparison with the natural ones, although it improved the metabolic effects at the intestinal tract [78]. In another randomized controlled trial, a positive correlation between almond consumption and gut microbiota promotion was reported, indicating that chopped almonds intensified *Lachnospira*, *Roseburia*, and *Oscillospira* growth, whereas whole almonds increased the *Dialister* populations [56]. As a result, the prebiotic effects of almond skins have generated an increasing interest in the inclusion of this by-product in functional foods, as it is the case of functional biscuits, whose nutritional properties were improved in terms of fiber and phenolic compound content [79], revealing a promising applicability of almond by-products in the inclusion of functional ingredients for the food industry.

3.6. Other Activities

Besides the above-mentioned bioactivities attributed to almond by-products, which have been largely assessed, there are additional functionalities reported on these matrices that open new perspectives for their valorization as source of functional ingredients. In this regard, different authors have highlighted the effectiveness of almond by-products on a range of chronic diseases, including dyslipidemia, diabetes, and cardiovascular diseases (CVD), as well as their role as neuroprotective and hepatoprotective agents (Table 4).

Table 4. Health-promoting effects of almond and its by-products on chronic diseases.

Cholesterol-Lowering and Obesity-Preventing Effects	
Reduction of TC, LDL-C, ApoB levels	[9]
Improvement of lipoprotein profile and inhibition of LDL-C oxidation	[42]
Reduction of body adiposity, body mass index, and body weight	[77]
Cardioprotective effects	
Reduction of atherogenic index	[80]
Reduction of blood pressure	[81]
Antidiabetic effects	
Reduction of blood glucose level via GLP-1 production	[82]
Reduction of carbohydrate absorption	[83]
Reduction of insulin resistance in diabetic patients	[84]
Hepatoprotective effects	
Reduction of serum ALT, AST, and GGT levels	[40,85]
Induction of liver antioxidant enzymes: SOD, GPx, CAT	[40]
Neuroprotective effects	
Alzheimer-preventing mechanisms: anxiolytic, sedative, and memory-enhancing properties	[9,86]

ALT: alanine aminotransferase; AST: aspartate aminotransferase; CAT: catalase; GGT: gamma-glutamyl transferase; GLP-1: glycoprotein 1; GPx: glutathione peroxidase; LDL-C: low-density lypoprotein cholesterol; SOD: superoxide dismutase; TC: total cholesterol.

In the case of dyslipidemia, a great variety of interventional studies have pointed at a beneficial impact of the consumption of almond and by-products on lipid metabolism. Indeed, the daily consumption of almonds has been seen to promote a reduction of total cholesterol (TC), low-density-lipoprotein cholesterol (LDL-C), and apolipoprotein B, without interfering with high-density-lipoprotein cholesterol (HDL-C) [9]. This improvement of lipoprotein profile is accompanied by a reduction in TAGs levels, mostly motivated by the presence of UFAs, minerals, vitamins and phytosterols in almond and by-products [80]. In addition, polyphenol-enriched skin extracts, used as fortifier ingredients of milk, were reported to delay LDL-C oxidation in healthy adults, by promoting the increase of plasma catechin and naringenin levels [42]. Consequently, these effects promote a reduction in the central body adiposity, together with body weight and body mass index reduction, which have been linked to the amelioration of metabolic syndrome, obesity [77,83], and CVD by the reduction of atherogenic lipids levels and blood pressure [80,81].

Concerning the antidiabetic effects of almond, a number of interventional studies have assessed its positive effects, being considered a food with low glycemic index, although the identification of the constituents involved in such effects still remains a challenge [77]. In this sense, several authors have suggested that UFAs and fiber may be the major responsible molecules, developing a glycemic control by reducing blood glucose level through the stimulation of glycoprotein 1 (GLP-1) production [9,82] and carbohydrate absorption, as well as improving insulin sensitivity in type-2 diabetic patients [83,84].

Besides the cardioprotective effects, almond by-products have been also characterized by their effectiveness as protectors of hepatic and neurological functions. The hepatoprotective effects of almond skins have been demonstrated in hepatotoxicity-induced in vivo models, showing that procyanidins-enriched extracts decreased the serum levels of both hepatic alanine aminotransferase (ALT) and aspartate aminotransferase (AST), as well as improving the levels of antioxidant enzymes in the liver, including GPx, SOD, and CAT [40]. Such effects found in vivo were recently assessed in a randomized controlled clinical trial, showing that almond consumption provoked a significant reduction in serum ALT, AST, and gamma-glutamyl transferase (GGT) after 12 weeks, thus promoting an effective liver protection in patients with coronary artery disease [85].

4. Current Trends and Future Perspectives

Currently, consumers' habits are more selective, which has switched the tendency of productive systems into the implementation of more efficient and environmentally friendly production practices. As mentioned above, almond is one of the most important nuts both in terms of surface cultivation area and production worldwide. The agri-food by-products can be used as an alternative source of natural ingredients that may be applied for the development of high added-value products for various industries, facing the production of food, new materials, and energy. As a result, valorization is key factor to provide a proper waste management, leading to their minimization and the consequent establishment of a more integrated and sustainable industrial system, based on the implementation of the circular economy model, including both food-related and non-related sectors. This approach has been regarded as a high-throughput productive system for being able of transforming wastes into profitable products, hence providing both environmental and economic benefits.

4.1. Almond and Its By-Products in the Food Industry

Due to its health benefits, in the last recent decades, almond consumption has increased, as it is a food rich in nutrients associated with health benefits. Therefore, the increment of the almond market size has been accompanied of a larger generation of related by-products [87]. Among the multiple potential applications of almond by-products, the most common ones developed have been mostly focused on biomass generation for their further use in the food and feed industries.

As a consequence of the heterogeneous products derived from almond production systems, almond market is segmented according to different factors: (i) by type—the almond market is divided into "Shelled Type", in which shells are removed from almond fruits, and "In-shell Type" [88], and (ii) by application—the almond market is divided into direct edible use, food processing, and kitchen ingredients [89]. Besides the already mentioned health-promoting of almond kernels, several derived products are gaining a significant relevance in the food industry, especially almond milk and almond oil. Almond milk constitutes an excellent alternative to cow's milk, mostly motivated by the increasing emergence of milk allergy and lactose intolerance among the general population. In this sense, this almond-based product experienced a fast-growing presence in both North American and European markets due to its high content in MUFAs, which assist in body weight management and lowering LDL-C levels [90]. In the same way, almond oil, characterized by their high MUFA content, has been also reported because of its beneficial impact on cardiovascular disease prevention, acting as an enhancer of blood lipoprotein profiles [91]. Moreover, the use of partially delipidified almond flour (PDAF) obtained from the extraction of almond oil is an example of the use of this by-product as an ingredient for the manufacture of biscuits under the name of "almendrados" [1].

As a result, almond by-products represent valuable residues, as they represent a promising source of sustainable and natural ingredients that may have further applications in the food industry.

4.2. Other Uses for Almond By-Products

Besides the importance of almonds and derived products for food applications, these resources have been explored for novel approaches, in order to implement a solid circular economy system around this productive agri-food sector. Among them, almond shells play a central role in this purpose due to their lignocellulosic nature, which provides a wide range of applications associated with the exploitation of this material.

One of the alternative applications that are still in progress is the use of almond shells as heavy metal adsorbents in the wastewater, mainly from the textile industry. The presence of heavy metals in wastewaters is a great environmental problem because of their pollution effects to either superficial or ground waters [3]. Nowadays, the search of different adsorbents to these pollutants is focused on the use of various natural and low-cost materials recovered from by-products. Scientific literature provides a vast variety of cost-effective materials that has been used as adsorbents, such as tannin-rich materials and lignin obtained from pine bark, dead biomass, or sawdust; peat moss; chitin and chitosan from fly ash; modified wool and cotton; rice husks; or animal bones [3]. Almond shells and other similar agricultural by-products, such as sugar cane bagasse or walnut shells, have been pointed out as an economic and rich source of biomass for the preparation of activated carbon. Most of the materials used for activated carbon preparations are not self-renewing. Therefore, the use of almond by-products shows a double environmental benefit. The high added-value ingredient recovered as activated carbon is aimed for adsorption, separation, and purification processes to perform in aqueous or gaseous solution systems. Activated carbon can also be used in catalytic processes, acting itself as a catalyst and thus playing an important role in different chemical, pharmaceutical, and food industries [92,93]. Few studies have analyzed the adsorption capacity of almond shells as an environmentally friendly and low-cost adsorbent on several textile dyes. Different approaches that considered the type of shell, pH, or various activated carbon formulae have been developed for removing various textile dyes (Direct Red 80, methylene blue or crystal violet) from wastewaters. In general terms, the application of almond shells results in a very efficient dye removal that may reach values of 97% that imply dye retentions of nearly 50 mg/L for Direct red 8, 148–833 mg/g for methylene blue, or 625 mg/g for crystal violet [92–94]. Therefore, although the generation of industrial effluents containing textile dyes is almost inevitable, both previous and future experimental studies may provide suitable remediation plans for the minimization of their unfavorable impacts by the elimination of secondary materials from these wastewaters [92].

In addition, the lignocellulosic composition of almond shells has motivated their application for the production of wood-based composites, but also prompted the fortification of plastic materials [95,96]. Concerning other industrial uses with promising results, the high lignin content of shells was correlated with a high heating value, indicating the possible use of this almond by-product as fuel [97]. Moreover, with respect to the agricultural sector, almond shells have been identified by as natural growing media for soilless crop culture, thus being considered as a potent solution to develop their use as an ecological and environmentally friendly substrate, as already demonstrated for tomato culture [98].

On the other hand, almond milk and oil have been reported for their potential use on cosmetics, being valuable skin and hair-care products administered topically with the aim of promoting healthy effects, such as soothing; pain-relief; circulation-enhancement; moisturizing; and in the case of oil, sunlight protection [99]. Consequently, the multifaceted nature of almond and derived products facilitate their exploitation on different economical and industrial sectors, simplifying the establishment of a circular economy system, involving not only the food industry, but also animal feed, cosmetics, pharmaceutical, and material industries.

5. Conclusions

The current increasing trend of almond nut production has been accompanied with an increment in the waste generation, which reveals the importance of seeking solutions to minimize and reutilize them. Traditionally, studies have mainly focused on the phytochemical characterization of almond nuts. However, currently, numerous studies analyze the bioactivities associated with different biomolecules recovered from the almond by-products: shells, skins, hulls, and blanch water. They are considered to have a high content of phenolic compounds, but in very variable quantities depending on various factors, such as the crop selected, the ripening time, the extraction protocol, and the quantification method used. In almond skins, high levels of phenolic acids, including both hydroxybenzoic and hydroxycinnamic acids and flavonoids, are mostly represented by naringenin and catechins. Other compounds described as characteristics of almond by-products are tannins, which can be found both as hydrolysable and condensed forms. As for the total fatty acid content, previous studies have shown that 99% of the fatty acids in almonds are composed of oleic, linoleic, palmitic, and stearic acids, revealing that MUFAs are the most prevalent compounds. Additionally, almonds and their by-products can be an effective source of vegetable proteins, as they have a high percentage (20–25%) in both the kernel and the almond cake. However, almond shells and skins have lower protein levels with 10–13% and 5–7% values, respectively.

The high polyphenol content, both in almonds and their by-products, confer to these natural matrices a potent antioxidant activity related with positive health effects. Regarding by-products, almond shells contain high levels of proanthocyanidins and flavonols with strong antioxidant capacity associated. Besides polyphenols, the presence of other bioactive compounds such as UFAs and high-quality proteins reinforces the antioxidant capacity of almond by-products and prompts their ability as anti-inflammatory and even anticarcinogenic agents. In parallel, the presence of polyphenols in by-products has been described to be capable of acting as antimicrobial agents, being effective against both Gram-positive and Gram-negative bacteria, fungi, and viruses. On the other hand, almond shells can provide prebiotic activity as they may promote the diversification of gut microbiota and improve the gastrointestinal function. In addition to those mentioned above, there are additional functionalities of these matrices that serve as prospects for their valorization as a source of functional ingredients. Indeed, the efficacy of some almond by-products has been proved in several interventional studies to be effective on the prevention of chronic diseases, such as dyslipidemia, diabetes, and cardiovascular diseases, as well as their role as neuroprotective and hepatoprotective agents. Moreover, almond by-products have been described to be applicable to industrial fields such as wastewater treatments, and the production of active carbon, fuel, and heating energy.

In summary, the application of the circular economy model to the exploitation of almond by-products may allow the recovery of natural ingredients to prompt the formulation of new nutritional, cosmetic, and pharmaceutical products, a new productive strategy that would meet future consumer expectations on environmental impact and human health.

Author Contributions: Conceptualization, M.B.-M., M.F.-C., and M.A.P.; methodology, M.B.-M., M.F.-C., and P.G.-P.; software, M.F.-C. and P.G.-P.; validation, M.F.-C. and P.G.-P.; formal analysis, M.B.-M., M.F.-C., and P.G.-P.; investigation, M.B.-M., M.F.-C., and P.G.-P.; writing—original draft preparation, M.B.-M., M.F.-C., and P.G.-P.; writing—review and editing, M.F.-C. and P.G.-P.; visualization, M.A.P. and J.S.-G.; supervision, M.A.P. and J.S.-G. All authors have read and agreed to the published version of the manuscript.

Funding: The program Grupos de Referencia Competitiva (GRUPO AA1-GRC 2018) that supports the work of M.B.-M. The authors are grateful to Ibero-American Program on Science and Technology (CYTED—AQUA-CIBUS, P317RT0003) and to the Bio Based Industries Joint Undertaking (JU) under grant agreement no. 888003 UP4HEALTH Project (H2020-BBI-JTI-2019) that supports the work of P.G.-P. The JU receives support from the European Union's Horizon 2020 research and innovation program and the Bio Based Industries Consortium. The project SYSTEMIC Knowledge Hub on Nutrition and Food Security has received funding from national research funding parties in Belgium (FWO), France (INRA), Germany (BLE), Italy (MIPAAF), Latvia (IZM), Norway (RCN), Portugal (FCT), and Spain (AEI) in a joint action of JPI HDHL, JPI-OCEANS, and FACCE-JPI, launched in 2019 under the ERA-NET ERA-HDHL (no. 696295).

Institutional Review Board Statement: Not applicable.

Informed Consent Statement: Not applicable.

Data Availability Statement: Data sharing not applicable.

Acknowledgments: We would like to thank MICINN for supporting the Ramón y Cajal's grant for M.A. Prieto (RYC-2017-22891), and Xunta de Galicia for supporting the program EXCELENCIA-ED431F 2020/12, the post-doctoral grant of M. Fraga-Corral (ED481B-2019/096).

Conflicts of Interest: The authors declare no conflict of interest.

References

1. Barreira, J.C.M.; Nunes, M.A.; da Silva, B.V.; Pimentel, F.B.; Costa, A.S.G.; Alvarez-Ortí, M.; Pardo, J.E.; Oliveira, M.B.P.P. Almond cold-pressed oil by-product as ingredient for cookies with potential health benefits: Chemical and sensory evaluation. *Food Sci. Hum. Wellness* **2019**, *8*, 292–298. [CrossRef]
2. Alasalvar, C.; Salvadó, J.S.; Ros, E. Bioactives and health benefits of nuts and dried fruits. *Food Chem.* **2020**, *314*, 126192. [CrossRef]
3. Esfahlan, A.J.; Jamei, R.; Esfahlan, R.J. The importance of almond (*Prunus amygdalus* L.) and its by-products. *Food Chem.* **2010**, *120*, 349–360. [CrossRef]
4. Bartolomé, B.; Monagas, M.; Garrido, I.; Gómez-Cordovés, C.; Martín-Álvarez, P.J.; Lebrón-Aguilar, R.; Urpí-Sardà, M.; Llorach, R.; Andrés-Lacueva, C. Almond (*Prunus dulcis* (Mill.) D.A. Webb) polyphenols: From chemical characterization to targeted analysis of phenolic metabolites in humans. *Arch. Biochem. Biophys.* **2010**, *501*, 124–133. [CrossRef]
5. Monagas, M.; Garrido, I.; Lebrón-Aguilar, R.; Bartolome, B.; Gómez-Cordovés, C. Almond (*Prunus dulcis* (Mill.) D.A. Webb) skins as a potential source of bioactive polyphenols. *J. Agric. Food Chem.* **2007**, *55*, 8498–8507. [CrossRef]
6. Prgomet, I.; Goncalves, B.; Domínguez-Perles, R.; Pascual-Seva, N.; Barros, A.I.R.N.A. Valorization challenges to almond residues: Phytochemical composition and functional application. *Molecules* **2017**, *22*, 1774. [CrossRef] [PubMed]
7. Fornés Comas, J.; Alonso Segura, J.M.; Rafel Socias i Company. Shell hardness in almond: Cracking load and kernel percentage. *Sci. Hortic.* **2019**, *245*, 7–11. [CrossRef]
8. Lammi, C.; Bellumori, M.; Cecchi, L.; Bartolomei, M.; Bollati, C.; Clodoveo, M.L.; Corbo, F.; Arnoldi, A.; Mulinacci, N. Extra virgin olive oil phenol extracts exert hypocholesterolemic effects through the modulation of the LDLR pathway: In vitro and cellular mechanism of action elucidation. *Nutrients* **2020**, *12*, 1723. [CrossRef]
9. Karimi, Z.; Firouzi, M.; Dadmehr, M.; Javad-Mousavi, S.A.; Bagheriani, N.; Sadeghpour, O. Almond as a nutraceutical and therapeutic agent in Persian medicine and modern phytotherapy: A narrative review. *Phytother. Res.* **2021**, *35*. [CrossRef] [PubMed]
10. Prgomet, I.; Gonçalves, B.; Domínguez-Perles, R.; Santos, R.; Saavedra, M.J.; Aires, A.; Pascual-Seva, N.; Barros, A. Irrigation deficit turns almond by-products into a valuable source of antimicrobial (poly)phenols. *Ind. Crops Prod.* **2019**, *132*, 186–196. [CrossRef]
11. Moldero, D.; López-Bernal, Á.; Testi, L.; Lorite, I.J.; Fereres, E.; Orgaz, F. Long-term almond yield response to deficit irrigation. *Irrig. Sci.* **2021**. [CrossRef]
12. Aktas, T.; Thy, P.; Williams, R.B.; McCaffrey, Z.; Khatami, R.; Jenkins, B.M. Characterization of almond processing residues from the Central Valley of California for thermal conversion. *Fuel Process. Technol.* **2015**, *140*, 132–147. [CrossRef]
13. Urruzola, I.; Robles, E.; Serrano, L.; Labidi, J. Nanopaper from almond (*Prunus dulcis*) shell. *Cellulose* **2014**, *21*, 1619–1629. [CrossRef]
14. Xie, L.; Bolling, B.W. Characterisation of stilbenes in California almonds (*Prunus dulcis*) by UHPLC–MS. *Food Chem.* **2014**, *148*, 300–306. [CrossRef]
15. Summo, C.; Palasciano, M.; De Angelis, D.; Paradiso, V.M.; Caponio, F.; Pasqualone, A. Evaluation of the chemical and nutritional characteristics of almonds (*Prunus dulcis* (Mill). D.A. Webb) as influenced by harvest time and cultivar. *J. Sci. Food Agric.* **2018**, *98*, 5647–5655. [CrossRef] [PubMed]
16. Bolling, B.W.; Dolnikowski, G.; Blumberg, J.B.; Chen, C.-Y.O. Polyphenol content and antioxidant activity of California almonds depend on cultivar and harvest year. *Food Chem.* **2010**, *122*, 819–825. [CrossRef]

17. Barreira, J.C.M.; Ferreira, I.C.F.R.; Oliveira, M.B.P.P.; Pereira, J.A. Antioxidant activity and bioactive compounds of ten Portuguese regional and commercial almond cultivars. *Food Chem. Toxicol.* **2008**, *46*, 2230–2235. [CrossRef]
18. Čolić, S.D.; Fotirić Akšić, M.M.; Lazarević, K.B.; Zec, G.N.; Gašić, U.M.; Dabić Zagorac, D.Č.; Natić, M.M. Fatty acid and phenolic profiles of almond grown in Serbia. *Food Chem.* **2017**, *234*, 455–463. [CrossRef]
19. Xie, L.; Roto, A.V.; Bolling, B.W. Characterization of ellagitannins, gallotannins, and bound proanthocyanidins from California almond (*Prunus dulcis*) varieties. *J. Agric. Food Chem.* **2012**, *60*, 12151–12156. [CrossRef]
20. Milbury, P.E.; Chen, C.-Y.; Dolnikowski, G.G.; Blumberg, J.B. Determination of flavonoids and phenolics and their distribution in almonds. *J. Agric. Food Chem.* **2006**, *54*, 5027–5033. [CrossRef] [PubMed]
21. Beltrán Sanahuja, A.; Maestre Pérez, S.E.; Grané Teruel, N.; Valdés García, A.; Prats Moya, M.S. Variability of chemical profile in almonds (*Prunus dulcis*) of different cultivars and origins. *Foods* **2021**, *10*, 153. [CrossRef]
22. López-Ortiz, C.M.; Prats-Moya, S.; Sanahuja, A.B.; Maestre-Pérez, S.E.; Grané-Teruel, N.; Martín-Carratalá, M.L. Comparative study of tocopherol homologue content in four almond oil cultivars during two consecutive years. *J. Food Compos. Anal.* **2008**, *21*, 144–151. [CrossRef]
23. Barreira, J.C.M.; Casal, S.; Ferreira, I.C.F.R.; Peres, A.M.; Pereira, J.A.; Oliveira, M.B.P.P. Supervised chemical pattern recognition in almond (*Prunus dulcis*) Portuguese PDO cultivars: PCA- and LDA-based triennial study. *J. Agric. Food Chem.* **2012**, *60*, 9697–9704. [CrossRef] [PubMed]
24. EFSA Panel on Dietetic Products Nutrition and Allergies (NDA). *Scientific Opinion on the Substantiation of Health Claims Related to Almonds and Maintenance of Normal Blood LDL Cholesterol Concentrations (ID 1131) and Maintenance of Normal Erectile Function (ID 2482) Pursuant to Article 13 (1) of Regulation (EC) No 19*; Wiley Online Library: Hoboken, NJ, USA, 2011; Volume 9.
25. Prats-Moya, M.S.; Grané-Teruel, N.; Berenguer-Navarro, V.; Martín-Carratalá, M.L. A chemometric study of genotypic variation in triacylglycerol composition among selected almond cultivars. *J. Am. Oil Chem. Soc.* **1999**, *76*, 267–272. [CrossRef]
26. Cherif, A.; Sebei, K.; Boukhchina, S.; Kallel, H.; Belkacemi, K.; Arul, J. Kernel fatty acid and triacylglycerol composition for three almond cultivars during maturation. *J. Am. Oil Chem. Soc.* **2004**, *81*, 901–905. [CrossRef]
27. Zhu, Y.; Wilkinson, K.L.; Wirthensohn, M. Changes in fatty acid and tocopherol content during almond (*Prunus dulcis*, cv. Nonpareil) kernel development. *Sci. Hortic.* **2017**, *225*, 150–155. [CrossRef]
28. Nawade, B.; Yahyaa, M.; Reuveny, H.; Shaltiel-Harpaz, L.; Eisenbach, O.; Faigenboim, A.; Bar-Yaakov, I.; Holland, D.; Ibdah, M. Profiling of volatile terpenes from almond (*Prunus dulcis*) young fruits and characterization of seven terpene synthase genes. *Plant Sci.* **2019**, *287*, 110187. [CrossRef]
29. Xiao, L.; Lee, J.; Zhang, G.; Ebeler, S.E.; Wickramasinghe, N.; Seiber, J.; Mitchell, A.E. HS-SPME GC/MS characterization of volatiles in raw and dry-roasted almonds (*Prunus dulcis*). *Food Chem.* **2014**, *151*, 31–39. [CrossRef]
30. Oliveira, I.; Malheiro, R.; Meyer, A.S.; Pereira, J.A.; Gonçalves, B. Application of chemometric tools for the comparison of volatile profile from raw and roasted regional and foreign almond cultivars (Prunus dulcis). *J. Food Sci. Technol.* **2019**, *56*, 3764–3776. [CrossRef]
31. Sathe, S.K.; Wolf, W.J.; Roux, K.H.; Teuber, S.S.; Venkatachalam, M.; Sze-Tao, K.W.C. Biochemical characterization of amandin, the major storage protein in almond (*Prunus dulcis* L.). *J. Agric. Food Chem.* **2002**, *50*, 4333–4341. [CrossRef] [PubMed]
32. Ahrens, S.; Venkatachalam, M.; Mistry, A.M.; Lapsley, K.; Sathe, S.K. Almond (*Prunus dulcis* L.) protein quality. *Plant Foods Hum. Nutr.* **2005**, *60*, 123–128. [CrossRef] [PubMed]
33. Roncero, J.M.; Álvarez-Ortí, M.; Pardo-Giménez, A.; Rabadán, A.; Pardo, J.E. Review about non-lipid components and minor fat-soluble bioactive compounds of almond kernel. *Foods* **2020**, *9*, 1646. [CrossRef] [PubMed]
34. De Souza, T.S.P.; Dias, F.F.G.; Oliveira, J.P.S.; de Moura Bell, J.M.L.N.; Koblitz, M.G.B. Biological properties of almond proteins produced by aqueous and enzyme-assisted aqueous extraction processes from almond cake. *Sci. Rep.* **2020**, *10*, 10873. [CrossRef] [PubMed]
35. Mandalari, G.; Faulks, R.M.; Bisignano, C.; Waldron, K.W.; Narbad, A.; Wickham, M.S.J. In vitro evaluation of the prebiotic properties of almond skins (*Amygdalus communis* L.). *FEMS Microbiol. Lett.* **2010**, *304*, 116–122. [CrossRef]
36. Aguilar, A.A.; Smith, N.E.; Baldwin, R.L. Nutritional value of almond hulls for dairy cows. *J. Dairy Sci.* **1984**, *67*, 97–103. [CrossRef]
37. Valdés, A.; Vidal, L.; Beltrán, A.; Canals, A.; Garrigós, M.C. Microwave-assisted extraction of phenolic compounds from almond skin byproducts (*Prunus amygdalus*): A multivariate analysis approach. *J. Agric. Food Chem.* **2015**, *63*, 5395–5402. [CrossRef]
38. Bottone, A.; Masullo, M.; Montoro, P.; Pizza, C.; Piacente, S. HR-LC-ESI-Orbitrap-MS based metabolite profiling of *Prunus dulcis* Mill. (Italian cultivars Toritto and Avola) husks and evaluation of antioxidant activity. *Phytochem. Anal.* **2019**, *30*, 415–423. [CrossRef] [PubMed]
39. Musarra-Pizzo, M.; Ginestra, G.; Smeriglio, A.; Pennisi, R.; Sciortino, M.T.; Mandalari, G. The antimicrobial and antiviral activity of polyphenols from almond (*Prunus dulcis* L.) skin. *Nutrients* **2019**, *11*, 2355. [CrossRef]
40. Truong, V.L.; Bak, M.J.; Jun, M.; Kong, A.N.T.; Ho, C.T.; Jeong, W.S. Antioxidant defense and hepatoprotection by procyanidins from almond (*Prunus amygdalus*) skins. *J. Agric. Food Chem.* **2014**, *62*, 8668–8678. [CrossRef]
41. Zrig, A.; Mohamed, H.B.; Tounekti, T.; Ahmed, S.O.; Khemira, H. Differential responses of antioxidant enzymes in salt-stressed almond tree grown under sun and shade conditions. *J. Plant Sci. Res.* **2015**, *2*, 117–127.
42. Chen, C.Y.O.; Milbury, P.E.; Blumberg, J.B. Polyphenols in almond skins after blanching modulate plasma biomarkers of oxidative stress in healthy humans. *Antioxidants* **2019**, *8*, 95. [CrossRef]

43. An, J.; Liu, J.; Liang, Y.; Ma, Y.; Chen, C.; Cheng, Y.; Peng, P.; Zhou, N.; Zhang, R.; Addy, M.; et al. Characterization, bioavailability and protective effects of phenolic-rich extracts from almond hulls against pro-oxidant induced toxicity in Caco-2 cells. *Food Chem.* **2020**, *322*, 126742. [CrossRef] [PubMed]
44. Dhingra, N.; Kar, A.; Sharma, R.; Bhasin, S. In-vitro antioxidative potential of different fractions from *Prunus dulcis* seeds: Vis a vis antiproliferative and antibacterial activities of active compounds. *S. Afr. J. Bot.* **2017**, *108*, 184–192. [CrossRef]
45. Gomaa, E.Z. In vitro antioxidant, antimicrobial, and antitumor activities of bitter almond and sweet apricot (*Prunus armeniaca* L.) kernels. *Food Sci. Biotechnol.* **2013**, *22*, 455–463. [CrossRef]
46. Khanl, A.; Meshkini, A. Anti-proliferative activity and mitochondria-dependent apoptosis induced by almond and walnut by-product in bone tumor cells. *Waste Biomass Valoriz.* **2021**, *12*, 1405–1416. [CrossRef]
47. Mericli, F.; Becer, E.; Kabadayi, H.; Hanoglu, A.; Hanoglu, D.Y.; Yavuz, D.O.; Ozek, T.; Vatansever, S. Fatty acid composition and anticancer activity in colon carcinoma cell lines of *Prunus dulcis* seed oil. *Pharm. Biol.* **2017**, *55*, 1239–1248. [CrossRef]
48. Dammak, M.I.; Chakroun, I.; Mzoughi, Z.; Amamou, S.; Mansour, H.B.; Le Cerf, D.; Majdoub, H. Characterization of polysaccharides from *Prunus amygdalus* peels: Antioxidant and antiproliferative activities. *Int. J. Biol. Macromol.* **2018**, *119*, 198–206. [CrossRef]
49. García-Pérez, P.; Barreal, M.E.; Rojo-De Dios, L.; Cameselle-Teijeiro, J.F.; Gallego, P.P. Bioactive natural products from the genus Kalanchoe as cancer chemopreventive agents: A review. In *Studies in Natural Products Chemistry*; Rahman, A., Ed.; Elsevier: Amsterdam, The Netherlands, 2019; Volume 61, pp. 49–84.
50. Hikal, D.M.; Awad, N.S.; Abdein, M.A. The anticancer activity of cashew (*Anacardium occidentale*) and almond (*Prunus dulcis*) kernels. *Adv. Environ. Biol.* **2017**, *11*, 31–41.
51. Zorrilla, P.; Rodriguez-Nogales, A.; Algieri, F.; Garrido-Mesa, N.; Olivares, M.; Rondón, D.; Zarzuelo, A.; Utrilla, M.P.; Galvez, J.; Rodriguez-Cabezas, M.E. Intestinal anti-inflammatory activity of the polyphenolic-enriched extract Amanda® in the trinitrobenzenesulphonic acid model of rat colitis. *J. Funct. Foods* **2014**, *11*, 449–459. [CrossRef]
52. Müller, A.K.; Schmölz, L.; Wallert, M.; Schubert, M.; Schlörmann, W.; Glei, M.; Lorkowski, S. In vitro digested nut oils attenuate the lipopolysaccharide-induced inflammatory response in macrophages. *Nutrients* **2019**, *11*, 503. [CrossRef] [PubMed]
53. Bisignano, C.; Filocamo, A.; La Camera, E.; Zummo, S.; Fera, M.T.; Mandalari, G. Antibacterial activities of almond skins on cagA-positive and-negative clinical isolates of Helicobacter pylori. *BMC Microbiol.* **2013**, *13*. [CrossRef]
54. Pandey, P.; Bhatt, P.C.; Rahman, M.; Patel, D.K.; Anwar, F.; Al-Abbasi, F.; Verma, A.; Kumar, V. Preclinical renal chemo-protective potential of *Prunus amygdalus* Batsch seed coat via alteration of multiple molecular pathways. *Arch. Physiol. Biochem.* **2018**, *124*, 88–96. [CrossRef]
55. Mandalari, G.; Bisignano, C.; Genovese, T.; Mazzon, E.; Wickham, M.S.J.; Paterniti, I.; Cuzzocrea, S. Natural almond skin reduced oxidative stress and inflammation in an experimental model of inflammatory bowel disease. *Int. Immunopharmacol.* **2011**, *11*, 915–924. [CrossRef] [PubMed]
56. Holscher, H.D.; Taylor, A.M.; Swanson, K.S.; Novotny, J.A.; Baer, D.J. Almond consumption and processing affects the composition of the gastrointestinal microbiota of healthy adult men and women: A randomized controlled trial. *Nutrients* **2018**, *10*, 126. [CrossRef]
57. Rocchetti, G.; Bhumireddy, S.R.; Giuberti, G.; Mandal, R.; Lucini, L.; Wishart, D.S. Edible nuts deliver polyphenols and their transformation products to the large intestine: An in vitro fermentation model combining targeted/untargeted metabolomics. *Food Res. Int.* **2019**, *116*, 786–794. [CrossRef]
58. Liu, Z.; Lin, X.; Huang, G.; Zhang, W.; Rao, P.; Ni, L. Prebiotic effects of almonds and almond skins on intestinal microbiota in healthy adult humans. *Anaerobe* **2014**, *26*, 1–6. [CrossRef]
59. Pasqualone, A.; Laddomada, B.; Spina, A.; Todaro, A.; Guzmàn, C.; Summo, C.; Mita, G.; Giannone, V. Almond by-products: Extraction and characterization of phenolic compounds and evaluation of their potential use in composite dough with wheat flour. *LWT Food Sci. Technol.* **2018**, *89*, 299–306. [CrossRef]
60. Bottone, A.; Montoro, P.; Masullo, M.; Pizza, C.; Piacente, S. Metabolite profiling and antioxidant activity of the polar fraction of Italian almonds (Toritto and Avola): Analysis of seeds, skins, and blanching water. *J. Pharm. Biomed. Anal.* **2020**, *190*, 113518. [CrossRef] [PubMed]
61. Kahlaoui, M.; Vecchia, S.B.D.; Giovine, F.; Kbaier, H.B.H.; Bouzouita, N.; Pereira, L.B.; Zeppa, G. Characterization of polyphenolic compounds extracted from different varieties of almond hulls (*Prunus dulcis* L.). *Antioxidants* **2019**, *8*, 647. [CrossRef] [PubMed]
62. Smeriglio, A.; Mandalari, G.; Bisignano, C.; Filocamo, A.; Barreca, D.; Bellocco, E.; Trombetta, D.; Karimi, Z.; Firouzi, M.; Dadmehr, M.; et al. Polyphenolic content and biological properties of Avola almond (Prunus dulcis Mill. D.A. Webb) skin and its industrial byproducts. *Phytother. Res.* **2019**, *51*, 283–293. [CrossRef]
63. Li, N.; Jia, X.; Chen, C.Y.O.; Blumberg, J.B.; Song, Y.; Zhang, W.; Zhang, X.; Ma, G.; Chen, J. Almond consumption reduces oxidative DNA damage and lipid peroxidation in male smokers. *J. Nutr.* **2007**, *137*, 2717–2722. [CrossRef]
64. Smeriglio, A.; Barreca, D.; Bellocco, E.; Trombetta, D. Proanthocyanidins and hydrolysable tannins: Occurrence, dietary intake and pharmacological effects. *Br. J. Pharmacol.* **2017**, *174*, 1244–1262. [CrossRef]
65. Amico, V.; Barresi, V.; Condorelli, D.; Spatafora, C.; Tringali, C. Antiproliferative terpenoids from almond hulls (*Prunus dulcis*): Identification and structure-activity relationships. *J. Agric. Food Chem.* **2006**, *54*, 810–814. [CrossRef]
66. Lim, T.K. (Ed.) Prunus dulcis. In *Edible Medicinal and Non-Medicinal Plants: Volume 4, Fruits*; Springer Science & Business Media: Berlin/Heidelberg, Germany, 2012; Volume 4, p. 491. ISBN 9789400740532.

67. Udenigwe, C.C.; Je, J.Y.; Cho, Y.S.; Yada, R.Y. Almond protein hydrolysate fraction modulates the expression of proinflammatory cytokines and enzymes in activated macrophages. *Food Funct.* 2013, *4*, 777–783. [CrossRef] [PubMed]
68. Lauro, M.R.; Marzocco, S.; Rapa, S.F.; Musumeci, T.; Giannone, V.; Picerno, P.; Aquino, R.P.; Puglisi, G. Recycling of almond by-products for intestinal inflammation: Improvement of physical-chemical, technological and biological characteristics of a dried almond skins extract. *Pharmaceutics* 2020, *12*, 884. [CrossRef]
69. Barreca, D.; Nabavi, S.M.; Sureda, A.; Rasekhian, M.; Raciti, R.; Silva, A.S.; Annunziata, G.; Arnone, A.; Tenore, G.C.; Süntar, İ.; et al. Almonds (*Prunus dulcis* Mill. D. A. webb): A source of nutrients and health-promoting compounds. *Nutrients* 2020, *12*, 672. [CrossRef] [PubMed]
70. Theophil Anand, G.; Renuka, D.; Ramesh, R.; Anandaraj, L.; John Sundaram, S.; Ramalingam, G.; Magdalane, C.M.; Bashir, A.K.H.; Maaza, M.; Kaviyarasu, K. Green synthesis of ZnO nanoparticle using *Prunus dulcis* (Almond Gum) for antimicrobial and supercapacitor applications. *Surf. Interfaces* 2019, *17*, 100376. [CrossRef]
71. Bouaziz, F.; Koubaa, M.; Helbert, C.B.; Kallel, F.; Driss, D.; Kacem, I.; Ghorbel, R.; Chaabouni, S.E. Purification, structural data and biological properties of polysaccharide from *Prunus amygdalus* gum. *Int. J. Food Sci. Technol.* 2015, *50*, 578–584. [CrossRef]
72. Bouaziz, F.; Koubaa, M.; Ellouz Ghorbel, R.; Ellouz Chaabouni, S. Biological properties of water-soluble polysaccharides and hemicelluloses from almond gum. *Int. J. Biol. Macromol.* 2017, *95*, 667–674. [CrossRef] [PubMed]
73. Bisignano, C.; Mandalari, G.; Smeriglio, A.; Trombetta, D.; Pizzo, M.M.; Pennisi, R.; Sciortino, M.T. Almond skin extracts abrogate HSV-1 replication by blocking virus binding to the cell. *Viruses* 2017, *9*, 178. [CrossRef]
74. Arena, A.; Bisignano, C.; Stassi, G.; Mandalari, G.; Wickham, M.S.J.; Bisignano, G. Immunomodulatory and antiviral activity of almond skins. *Immunol. Lett.* 2010, *132*, 18–23. [CrossRef]
75. Mandalari, G.; Bisignano, C.; D'Arrigo, M.; Ginestra, G.; Arena, A.; Tomaino, A.; Wickham, M.S.J. Antimicrobial potential of polyphenols extracted from almond skins. *Lett. Appl. Microbiol.* 2010, *51*, 83–89. [CrossRef]
76. Singh, R.D.; Nadar, C.G.; Muir, J.; Arora, A. Green and clean process to obtain low degree of polymerisation xylooligosaccharides from almond shell. *J. Clean. Prod.* 2019, *241*, 118237. [CrossRef]
77. Martins, I.M.; Chen, Q.; Oliver Chen, C.Y. Emerging functional foods derived from almonds. In *Wild Plants, Mushrooms and Nuts: Functional Food Properties and Applications*; John Wiley & Sons: Hoboken, NJ, USA, 2016; pp. 445–469. ISBN 9781118944653.
78. Liu, Z.; Wang, W.; Huang, G.; Zhang, W.; Ni, L. In vitro and in vivo evaluation of the prebiotic effect of raw and roasted almonds (Prunus amygdalus). *J. Sci. Food Agric.* 2016, *96*, 1836–1843. [CrossRef]
79. Pasqualone, A.; Laddomada, B.; Boukid, F.; de Angelis, D.; Summo, C. Use of almond skins to improve nutritional and functional properties of biscuits: An example of upcycling. *Foods* 2020, *9*, 1705. [CrossRef]
80. Kalita, S.; Khandelwal, S.; Madan, J.; Pandya, H.; Sesikeran, B.; Krishnaswamy, K. Almonds and cardiovascular health: A review. *Nutrients* 2018, *10*, 468. [CrossRef]
81. Eslampour, E.; Asbaghi, O.; Hadi, A.; Abedi, S.; Ghaedi, E.; Lazaridi, A.V.; Miraghajani, M. The effect of almond intake on blood pressure: A systematic review and meta-analysis of randomized controlled trials. *Complement. Ther. Med.* 2020, *50*, 102399. [CrossRef] [PubMed]
82. Ren, M.; Zhang, H.; Qi, J.; Hu, A.; Jiang, Q.; Hou, Y.; Feng, Q.; Ojo, O.; Wang, X. An almond-based low carbohydrate diet improves depression and glycometabolism in patients with type 2 diabetes through modulating gut microbiota and GLP-1: A randomized controlled trial. *Nutrients* 2020, *12*, 3036. [CrossRef]
83. Gupta, A.; Sharma, R.; Sharma, S. Almond. In *Antioxidants in Vegetables and Nuts—Properties and Health Benefits*; Nayik, G.A., Gull, A., Eds.; Springer: New York, NY, USA, 2020; pp. 423–452.
84. Li, S.C.; Liu, Y.H.; Liu, J.F.; Chang, W.H.; Chen, C.M.; Chen, C.Y.O. Almond consumption improved glycemic control and lipid profiles in patients with type 2 diabetes mellitus. *Metabolism* 2011, *60*, 474–479. [CrossRef] [PubMed]
85. Jamshed, H.; Arslan, J.; Sultan, F.T.; Siddiqi, H.S.; Qasim, M.; Gilani, A.-u.G. Almond protects the liver in coronary artery disease—A randomized controlled clinical trial. *J. Pak. Med. Assoc.* 2020, 1–15. [CrossRef]
86. Gorji, N.; Moeini, R.; Memariani, Z. Almond, hazelnut and walnut, three nuts for neuroprotection in Alzheimer's disease: A neuropharmacological review of their bioactive constituents. *Pharmacol. Res.* 2018, *129*, 115–127. [CrossRef] [PubMed]
87. Tungmunnithum, D.; Elamrani, A.; Abid, M.; Drouet, S.; Kiani, R.; Garros, L.; Kabra, A.; Addi, M.; Hano, C. A quick, green and simple ultrasound-assisted extraction for the valorization of antioxidant phenolic acids from Moroccan almond cold-pressed oil residues. *Appl. Sci.* 2020, *10*, 3313. [CrossRef]
88. Reisman, E. Superfood as spatial fix: The ascent of the almond. *Agric. Hum. Values* 2020, *37*, 337–351. [CrossRef]
89. Ryan, N.T. World almond market. In *Almonds: Botany, Production and Uses*; Gradziel, T.M., Ed.; CABI: Wallingford, UK, 2017; pp. 449–459.
90. Vanga, S.K.; Wang, J.; Orsat, V.; Raghavan, V. Effect of pulsed ultrasound, a green food processing technique, on the secondary structure and in-vitro digestibility of almond milk protein. *Food Res. Int.* 2020, *137*, 109523. [CrossRef] [PubMed]
91. Čolić, S.D.; Zec, G.; Nati, M.; Fotirić Akšić, M.M. Almond (*Prunus dulcis*) oil. In *Fruit Oils: Chemistry and Functionality*; Ramadan, M.F., Ed.; Springer Nature Switzerland AG: Cham, Switzerland, 2019; pp. 149–180.
92. Doulati Ardejani, F.; Badii, K.; Limaee, N.Y.; Shafaei, S.Z.; Mirhabibi, A.R. Adsorption of Direct Red 80 dye from aqueous solution onto almond shells: Effect of pH, initial concentration and shell type. *J. Hazard. Mater.* 2008, *151*, 730–737. [CrossRef]
93. İzgi, M.S.; Saka, C.; Baytar, O.; Saraçoğlu, G.; Şahin, Ö. Preparation and characterization of activated carbon from microwave and conventional heated almond shells using phosphoric acid activation. *Anal. Lett.* 2019, *52*, 772–789. [CrossRef]

94. Ait Ahsaine, H.; Zbair, M.; Anfar, Z.; Naciri, Y.; El haouti, R.; El Alem, N.; Ezahri, M. Cationic dyes adsorption onto high surface area 'almond shell' activated carbon: Kinetics, equilibrium isotherms and surface statistical modeling. *Mater. Today Chem.* **2018**, *8*, 121–132. [CrossRef]
95. Pirayesh, H.; Khazaeian, A. Using almond (Prunus amygdalus L.) shell as a bio-waste resource in wood based composite. *Compos. Part B Eng.* **2012**, *43*, 1475–1479. [CrossRef]
96. Sabbatini, A.; Lanari, S.; Santulli, C.; Pettinari, C. Use of almond shells and rice husk as fillers of poly(methyl methacrylate) (PMMA) composites. *Materials* **2017**, *10*, 872. [CrossRef]
97. Demirbaş, A. Fuel characteristics of olive husk and walnut, hazelnut, sunflower, and almond shells. *Energy Sources* **2002**, *24*, 215–221. [CrossRef]
98. Urrestarazu, M.; Martínez, G.A.; Salas, M.D.C. Almond shell waste: Possible local rockwool substitute in soilless crop culture. *Sci. Hortic.* **2005**, *103*, 453–460. [CrossRef]
99. Krist, S. Almond oil. In *Vegetable Fats and Oils*; Puri, S., Ed.; Springer Nature Switzerland AG: Cham, Switzerland, 2021; pp. 77–79. ISBN 9783030303143.

Article

Physical Chemical and Textural Characteristics and Sensory Evaluation of Cookies Formulated with Date Seed Powder

Zein Najjar [1], Maitha Alkaabi [1], Khulood Alketbi [1], Constantinos Stathopoulos [2,*] and Meththa Ranasinghe [1]

[1] Department of Food Science, College of Agriculture and Veterinary Medicine, United Arab Emirates University, Al Ain P.O. Box 15551, United Arab Emirates; zeinrnajjar@gmail.com (Z.N.); 201401011@uaeu.ac.ae (M.A.); 201401097@uaeu.ac.ae (K.A.); 201990013@uaeu.ac.ae (M.R.)

[2] Faculty of Agrobiology, Food and Natural Resources, Czech University of Life Sciences Prague, 165 00 Prague, Czech Republic

* Correspondence: stathopoulos@af.czu.cz

Abstract: Date seeds are a major waste product that can be utilised as a valuable and nutritional material in the food industry. The aim of the present study was to improve cookies quality in terms of functional and textural value and assess the effect of date seed powder flour substitution on the physical and chemical characteristics of cookies. Three substitution levels (2.5, 5 and 7.5%) of flour by fine date seed powder from six varieties locally named *Khalas*, *Khinaizi*, *Sukkary*, *Shaham*, *Zahidi* and *Fardh* were prepared. Two types of flour were used (white flour and whole wheat) at two different baking temperatures: 180 and 200 °C. The incorporation of date seed had no or slight effect on moisture, ash, fat and protein content of the baked cookies. On the other hand, incorporation significantly affected the lightness and hardness of cookies; the higher level of addition, the darker and crispier the resulting cookies. The sensory analysis indicated that the produced cookies were acceptable in terms of smell, taste, texture and overall acceptability. The results indicate that the most acceptable cookies across all evaluated parameters were produced using whole wheat flour with 7.5% levels of date seed powder using *Khalas* and *Zahidi* varieties. Overall, the analysis indicated that cookies with acceptable physical characteristics and an improved nutritional profile could be produced with partial replacement of the white/whole wheat flour by date seed powder.

Keywords: waste utilisation; date seed powder; cookies; sensory analysis

1. Introduction

The arid and semiarid regions of countries in North Africa and the Middle East are good habitats for the date palm plant [1–4]. Date palm is of economic and social importance for people in date-producing countries [5]. The pericarp of the date fruit is the edible part, and the pit is a waste product [3]. Date pits, on average, make up 10–15% of the total date fruit's mass, depending on the variety [6]. Dates of low quality and a hard texture, as well as contaminated ones, are rejected and sometimes used for feed in the animal and poultry industries [7]; up to 800,000 tonnes/year of seeds could be disposed of [2]. Date seeds are composed of 2.10–7.10% moisture, 2.3–6.4% protein, 5–13.2% fat, 0.9–1.8% ash and 72.5–80.2% dietary fibre [8].

Cookies hold an important position as snack foods due to the variety of taste, texture and digestibility [9–12]; they are also one of the most popular bakery products consumed by almost all levels of society due to their long shelf-life and low cost [13,14]. To improve the nutritive value of cookies, they can be prepared with fortified or composite flour [10–12,15,16]. Incorporating date seed powder into baked goods has been tried previously [17–20] with promising results. Such incorporation would not only improve the nutritional value of the produced cookies, for instance, by enhancing dietary fibre content and antioxidant properties, but also contribute to the utilisation of a waste product of regional significance, date seeds.

In most of the date-processing plants, including date confectionery and date syrup, the seed is considered as the major waste product which amounts to approximately 10% of the fruit mass. Thus, the utilisation of date waste is important for date cultivation and to increase the income in date-processing units. Currently, seeds are used mainly for animal feed for various livestock and sometimes as a soil additive [2,21]. The prevailing global trends aim for a more circular economy. In this context, recently developed food recovery hierarchies in Europe and US highlight the idea that priority should be given to formulating human diets rather than animal feed [22,23]. According to this point of view, there is an increased interest in sustainable and healthier diets along with building up a circular economy. Therefore, the main focus is the valorisation of by-products by incorporating them into food formulations, hence enabling their re-entry into the food chain as new products [24,25].

At the same time, baked food items such as cookies, biscuits, muffins and cakes are very popular among consumers due to their taste and availability. When considering the health aspects, they are usually high in sugar, whereas they are low in antioxidants, minerals and fibre content [26]. Hence, incorporating a by-product, such as date seeds, into the baked foods can be recognised as a perfect solution to improve the food quality as well as to reduce the impact of food processing waste.

Therefore, the purpose of this study was to determine the physicochemical and sensory attributes of date seed powder (DSP)-substituted cookies in order to enable the development and production of better cookies in terms of quality and nutritional value while reducing waste.

2. Materials and Methods

2.1. Materials

Date seed powder of six fully ripe Emirati date varieties locally known as *Khalas*, *Khinaizi*, *Sukkari*, *Shaham*, *Zahidi* and *Fardh* were used. The seeds were collected from local farms located in Al Ain, United Arab Emirates. Flour (whole wheat flour—12% protein; white flour—13% protein), sugar, salt, palm oil, food-grade ammonium bicarbonate (NH_4HCO_3) and food-grade sodium bicarbonate ($NaHCO_3$) were purchased from a local market (Al Ain, United Arab Emirates).

2.2. Samples Preparation

2.2.1. Date Seed Powder

The received seeds were cleaned, washed with water, air-dried, ground into powder according to [17] and sieved to obtain fine particles less than 250 μm in diameter. Date seed powder was packed in zip-lock plastic bags and stored at −20 °C until use.

2.2.2. Cookies Preparation

The recipe of cookies described in the American Association of Cereal Chemists (AACC) method 1054 [27] was used with modifications. The original formula was as follows: 80 g white/whole wheat flour, 30 g palm oil (raised to 40 g when using whole wheat flour), 35 g sucrose, 0.8 g $NaHCO_3$, 1.0 g salt, 0.4 g NH_4HCO_3 and 17.6 g water. The flour was replaced with preparations of date seed powder to a level of 2.5%, 5.0% and 7.5% in the cookie recipe, and cookies with no substitution were produced as control samples. Two types of flour—white (WF) and whole wheat flour (WW)—and two baking temperatures of 180 °C and 200 °C (for 10 and 8 min, respectively) were used, making twelve different combinations for each variety of date seeds. The dough was kneaded and sheeted to a uniform thickness using a pasta machine set at 3 mm. Then, the dough was cut with a stainless-steel cutter into circular shapes of a diameter of 4.5 cm. The cookies were baked in a wall oven (Miele & Cie. KG, Bielefeld, Germany) and were packed in sealed plastic bags and stored at −20 °C until further analysis.

2.3. Proximate Composition

Moisture percentage was determined by the oven dry method (Association of Official Analytical Chemists (AOAC) method 934.01) [28] using an oven (Carbolite Gero Limited, Sheffield, UK) at 85 °C for 4 h and ash by direct analysis (AOAC method 940.26) using a furnace (CWF 1100, Carbolite Gero Limited, Derbyshire, UK). Protein was determined by Kjeldahl nitrogen (AOAC method 920.152) [29] using Kjeldahl apparatus (AutoKjeldal unit K-370, BUCHI, Flawil, Switzerland), and protein was calculated using the general factor (6.25). The percentage of crude fat was determined by the Soxtec automated extraction method (AOAC, 2003.05) [28] using the Soxtec Auto fat extraction system (FOSS Analytical, Hillerød, Denmark). All assessments were conducted in triplicate. Carbohydrate content was calculated by difference: 100 − (moisture + ash + protein + fat).

2.4. Physical Analysis

2.4.1. Colour Analysis

The colour of the date seed powder of six varieties and cookie samples was determined using a Hunter Colourimeter fitted with an optical sensor (HunterLab ColourFlex EZ spectrophotometer. Hunter Associates Laboratory, Inc., Reston, VA, USA) on the basis of the L*, a*, b* colour system. L* is the lightness component that measures black (0) to white (100), a* parameter goes from green to red, and b* parameter from blue to yellow. Throughout this manuscript, the colour analysis was performed only on the basis of lightness (L*). Six cookies were evaluated for each formulation.

2.4.2. Texture Analysis

Measuring hardness was determined using Texture Analyzer (BROOKFIELD CT3 Texture Analyzer, Brookfield Engineering Labs, Inc., Middleborough, MA, USA) equipped with a 3-Point Bending Rig in a compression mode with a sharp blade-cutting probe (TA7). Pre-test and test speeds were 1 and 0.5 mm/s, respectively, at a trigger load of 3.0 g. Hardness (maximum peak force) was measured with at least 3 cookies for each sample. The peak force, when the cookie began to break, was reported in Newton (N).

2.4.3. Sensory Evaluation

Cookies were subjected to a sensory evaluation using 30 untrained panellists recruited from within the university community. Sensory properties (taste, aroma, colour, crispiness and overall acceptability) were evaluated. The ratings were on a 9-point hedonic scale ranging from 9 (like extremely) to 1 (dislike extremely).

2.5. Statistical Analysis

The data were shown as mean ± standard deviation (SD). Data were assessed by Tukey test using (Minitab 19, USA) statistical software package at $p < 0.05$ to determine the level of significance.

3. Results and Discussion

3.1. Proximate Analysis

3.1.1. Date Seed Powder Chemical Composition

The chemical composition for DSP varieties used in making cookies is shown in Table 1. Regarding moisture, *Sukkari* seed powder was the least moist, followed by *Zahidi Fardh*, *Khinaizi*, *Shaham* and *Khalas* seeds. Differences in cultivars, as well as cultivation under divergent climatic conditions, make it possible to have discrete moisture levels. Previous studies on date seed flour also demonstrated different moisture contents in different varieties [30,31]. For ash content, *Sukkari* seed powder had the lowest, while seeds of the *Fardh* variety had the highest. In terms of crude fat, *Zahidi* seeds had a high amount of fat, opposite to the *Khinaizi* seed variety. No significant difference was observed in protein content between the varieties used. The lowest carbohydrate content was in *Khalas* powder; it was significantly different from *Sukkari*, *Shaham* and *Zahidi* varieties. In previous findings for *Khalas*, *Sukkari* and *Fardh*, the results are quite different from this study [30–32]. These differences of proximate compositions, moisture and ash may be due to the climatic conditions or the variability of the cultivars used, as postulated also by [33]. Data about the chemical compositions of the remaining date seeds varieties—*Khinaizi*, *Shaham* and *Zahidi*—could not be found.

Table 1. Chemical composition of date seed powder [1].

Variety	Moisture	Ash	Crude Fat	Crude Protein	Carbohydrate
Khalas	5.08 ± 0.10 [a]	3.31 ± 0.26 [bc]	9.81 ± 0.37 [ab]	8.13 ± 2.45 [a]	73.68 ± 2.20 [c]
Khinaizi	4.06 ± 0.32 [ab]	3.52 ± 0.26 [b]	8.26 ± 38.47 [c]	6.84 ± 0.86 [a]	77.31 ± 38.44 [abc]
Sukkari	1.24 ± 0.18 [d]	2.04 ± 0.16 [d]	9.94 ± 2.98 [ab]	8.30 ± 1.11 [a]	78.48 ± 3.61 [ab]
Shaham	4.08 ± 0.58 [ab]	3.37 ± 0.12 [bc]	7.70 ± 0.12 [bc]	6.14 ± 0.38 [a]	78.70 ± 0.79 [ab]
Zahidi	2.53 ± 0.87 [c]	2.65 ± 0.04 [cd]	12.68 ± 0.08 [a]	6.92 ± 0.46 [a]	75.22 ± 1.26 [ab]
Fardh	3.72 ± 0.21 [bc]	6.90 ± 0.63 [a]	8.03 ± 0.16 [bc]	7.29 ± 1.12 [a]	74.06 ± 1.16 [c]

[1] Means ± SD are presented. Data expressed as g/100 g on a fresh weight basis, and different lowercase superscript letters in a column denote significant differences, $p < 0.05$.

3.1.2. Proximate Composition of Cookies

The percentage of moisture, ash content and crude fat content of different cookie samples are shown in Tables 2–4, respectively.

Table 2. Moisture content of cookies [1].

Baking Temp, Flour Type, Addition Level		Khalas	Khinaizi	Sukkari	Shaham	Zahidi	Fardh
180 °C White Flour	0	3.25 ± 0.19 [bc]	3.25 ± 0.19 [a]	3.25 ± 0.19 [bc]	3.25 ± 0.19 [c]	3.25 ± 0.19 [b]	3.25 ± 0.19 [a]
	2.5	3.14 ± 0.44 [cB]	3.03 ± 0.07 [aB]	4.01 ± 0.27 [aAB]	4.41 ± 0.13 [aA]	4.00 ± 0.10 [aAB]	3.80 ± 0.46 [aAB]
	5.0	3.85 ± 0.08 [abA]	3.31 ± 0.28 [aAB]	3.4 ± 0.36 [abAB]	3.89 ± 0.24 [bA]	2.78 ± 0.14 [cB]	3.55 ± 0.28 [aA]
	7.5	3.95 ± 0.14 [aA]	2.90 ± 0.06 [aBC]	2.66 ± 0.06 [cC]	3.38 ± 0.06 [cAB]	2.76 ± 0.18 [cBC]	3.18 ± 0.47 [aBC]
180 °C Whole Wheat Flour	0	2.70 ± 0.09 [b]	2.70 ± 0.09 [a]	2.70 ± 0.09 [a]	2.70 ± 0.09 [a]	2.70 ± 0.09 [a]	2.72 ± 0.09 [a]
	2.5	3.42 ± 0.05 [aA]	2.34 ± 0.13 [aB]	2.07 ± 0.13 [bBC]	1.69 ± 0.24 [bC]	2.09 ± 0.23 [bBC]	1.07 ± 0.22 [bD]
	5.0	3.09 ± 0.17 [abA]	2.20 ± 0.09 [aB]	1.96 ± 0.13 [bBC]	2.06 ± 0.05 [bBC]	1.82 ± 0.09 [bC]	0.33 ± 0.09 [bcD]
	7.5	2.98 ± 0.31 [bA]	1.14 ± 0.46 [bBC]	1.19 ± 0.08 [cBC]	1.19 ± 0.13 [cBC]	1.76 ± 0.19 [bB]	0.50 ± 0.33 [cC]
200 °C White Flour	0	2.87 ± 0.11 [bc]	2.87 ± 0.11 [a]	2.87 ± 0.11 [b]	2.87 ± 0.11 [b]	2.87 ± 0.11 [a]	2.87 ± 0.11 [b]
	2.5	3.12 ± 0.08 [abB]	2.72 ± 0.16 [aBC]	3.19 ± 0.05 [aB]	3.89 ± 0.33 [aA]	2.53 ± 0.03 [bC]	4.07 ± 0.32 [aA]
	5.0	2.81 ± 0.04 [cB]	2.91 ± 0.02 [aB]	2.67 ± 0.05 [cB]	3.55 ± 0.24 [aA]	2.05 ± 0.26 [cC]	2.45 ± 0.21 [bcBC]
	7.5	3.30 ± 0.16 [aA]	2.88 ± 0.07 [aB]	1.69 ± 0.06 [dD]	2.89 ± 0.13 [bB]	2.02 ± 0.04 [cCD]	2.28 ± 0.22 [cC]
200 °C Whole Wheat Flour	0	2.06 ± 0.02 [ab]	2.06 ± 0.02 [a]	2.06 ± 0.02 [a]	2.06 ± 0.02 [a]	2.06 ± 0.02 [a]	2.06 ± 0.02 [a]
	2.5	2.52 ± 0.15 [aA]	2.24 ± 0.28 [aAB]	1.21 ± 0.04 [bC]	1.68 ± 0.11 [bBC]	1.35 ± 0.13 [bC]	0.62 ± 0.35 [bD]
	5.0	2.31 ± 0.06 [abA]	2.20 ± 0.03 [aA]	1.17 ± 0.05 [bC]	2.19 ± 0.01 [aA]	1.40 ± 0.19 [bB]	0.41 ± 0.00 [bD]
	7.5	1.95 ± 0.32b [bA]	1.44 ± 0.38 [bAB]	0.84 ± 0.17 [cBC]	1.19 ± 0.27 [cAB]	1.42 ± 0.23 [bAB]	0.26 ± 0.38 [bC]

[1] Means ± SD are presented. Data are expressed as g/100 g on a fresh weight basis; different lowercase superscript letters in a column and different uppercase superscript letters in a row denote significant differences, $p < 0.05$.

Table 3. Ash content of cookies [1].

Baking Temp, Flour Type, Addition Level		Khalas	Khinaizi	Sukkari	Shaham	Zahidi	Fardh
180 °C White Flour	0	1.50 ± 0.07 [a]	1.50 ± 0.07 [a]	1.50 ± 0.07 [a]	1.50 ± 0.07 [a]	1.50 ± 0.07 [a]	1.50 ± 0.07 [a]
	2.5	1.68 ± 0.18 [aA]	1.42 ± 0.07 [aA]	1.32 ± 0.11 [aA]	1.87 ± 0.44 [aA]	1.96 ± 0.04 [aA]	1.08 ± 0.81 [aA]
	5.0	1.37 ± 0.43 [aB]	1.40 ± 0.05 [aB]	1.42 ± 0.03 [aAB]	1.54 ± 0.02 [aAB]	1.95 ± 0.04 [aA]	1.33 ± 0.23 [aB]
	7.5	1.46 ± 0.05 [aBC]	1.53 ± 0.05 [aB]	1.36 ± 0.04 [aB]	1.56 ± 0.05 [aC]	2.07 ± 0.06 [bA]	0.93 ± 0.01 [aD]

Table 3. Cont.

Baking Temp, Flour Type, Addition Level		Khalas	Khinaizi	Sukkari	Shaham	Zahidi	Fardh
180 °C Whole Wheat Flour	0	1.63 ± 0.04 ^{a}	1.63 ± 0.04 ^{a}	1.63 ± 0.04 ^{a}	1.63 ± 0.04 ^{a}	1.63 ± 0.04 ^{a}	1.63 ± 0.04 ^{a}
	2.5	1.65 ± 0.04 ^{bA}	1.61 ± 0.02 ^{aA}	1.65 ± 0.05 ^{aA}	1.74 ± 0.12 ^{aA}	1.63 ± 0.1 ^{aA}	1.62 ± 0.08 ^{aA}
	5.0	1.66 ± 0.04 ^{bB}	1.66 ± 0.01 ^{aB}	1.63 ± 0.06 ^{aB}	1.55 ± 0.03 ^{aB}	2.08 ± 0.23 ^{bA}	1.59 ± 0.09 ^{aB}
	7.5	1.81 ± 0.04 ^{bB}	1.66 ± 0.03 ^{aBCD}	1.53 ± 0.04 ^{aCD}	1.60 ± 0.08 ^{aD}	2.08 ± 0.12 ^{bA}	1.76 ± 0.06 ^{aBC}
200 °C White Flour	0	1.32 ± 0.04 ^{a}	1.32 ± 0.04 ^{a}	1.32 ± 0.04 ^{a}	1.32 ± 0.04 ^{a}	1.32 ± 0.04 ^{a}	1.32 ± 0.04 ^{a}
	2.5	1.50 ± 0.12 ^{aB}	0.79 ± 0.17 ^{bC}	1.22 ± 0.07 ^{aB}	1.31 ± 0.26 ^{aB}	1.96 ± 0.04 ^{abA}	1.55 ± 0.03 ^{aB}
	5.0	1.53 ± 0.02 ^{aB}	0.84 ± 0.12 ^{bD}	1.32 ± 0.1 ^{aBC}	1.42 0.04 ^{aC}	1.8 ± 0.04 ^{bA}	1.49 ± 0.05 ^{bBC}
	7.5	1.44 ± 0.14 ^{aB}	0.88 ± 0.07 ^{bC}	1.33 ± 0.02 ^{aB}	1.50 ± 0.11 ^{aB}	2.08 ± 0.18 ^{cA}	0.79 ± 0.03 ^{cC}
200 °C Whole Wheat Flour	0	1.60 ± 0.01 ^{a}	1.60 ± 0.01 ^{a}	1.60 ± 0.01 ^{a}	1.60 ± 0.01 ^{a}	1.60 ± 0.01 ^{a}	1.60 ± 0.01 ^{a}
	2.5	1.53 ± 0.09 ^{aAB}	1.12 ± 0.12 ^{abC}	1.55 ± 0.04 ^{aABC}	1.44 ± 0.03 ^{aAB}	1.26 ± 0.26 ^{abC}	1.67 ± 0.08 ^{abA}
	5.0	1.59 ± 0.01 ^{aAB}	0.99 ± 0.16 ^{abC}	1.58 ± 0.08 ^{aBC}	1.25 ± 0.17 ^{abAB}	1.48 ± 0.05 ^{aAB}	1.76 ± 0.18 ^{abA}
	7.5	1.67 ± 0.08 ^{aA}	0.55 ± 0.42 ^{bB}	1.54 ± 0.02 ^{aA}	1.55 ± 0.02 ^{bA}	1.5 ± 0.03 ^{aA}	1.88 ± 0.07 ^{bA}

[1] Means ± SD are presented. Data expressed as g/100 g on a fresh weight basis; different lowercase superscript letters in a column and different uppercase superscript letters in a row denote significant differences, $p < 0.05$.

Table 4. Crude fat content of cookies [1].

Baking Temp Flour Type and Addition Level		Khalas	Khinaizi	Sukkari	Shaham	Zahidi	Fardh
180 °C White Flour	0	19.34 ± 0.97 ^{a}	19.34 ± 0.97 ^{a}	19.34 ± 0.97 ^{a}	19.34 ± 0.97 ^{a}	19.34 ± 0.97 ^{a}	19.34 ± 0.97 ^{a}
	2.5	19.15 ± 0.49 ^{aA}	19.34 + 1.95 ^{aA}	20.71 ± 0.68 ^{aA}	19.37 ± 0.15 ^{aA}	20.07 ± 0.48 ^{aA}	19.65 ± 0.28 ^{aA}
	5.0	19.44 ± 0.60 ^{aA}	20.60 ± 0.10 ^{aA}	19.67 ± 2.60 ^{aA}	20.16 ± 0.57 ^{aA}	19.62 ± 0.21 ^{aA}	19.52 ± 0.48 ^{aA}
	7.5	20.46 ± 0.35 ^{aA}	20.58 ± 0.64 ^{aA}	20.87 ± 0.25 ^{aA}	20.91 ± 0.29 ^{aA}	20.11 ± 0.68 ^{aA}	20.60 ± 0.43 ^{aA}
180 °C Whole Wheat Flour	0	25.69 ± 0.31 ^{a}	25.69 ± 0.31 ^{a}	25.69 ± 0.31 ^{a}	25.69 ± 0.31 ^{a}	25.69 ± 0.31 ^{a}	25.69 ± 0.31 ^{a}
	2.5	26.04 ± 0.78 ^{aA}	24.81 ± 0.10 ^{aA}	25.59 ± 0.99 ^{aA}	26.27 ± 0.92 ^{aA}	25.84 ± 0.49 ^{aA}	24.58 ± 0.57 ^{bA}
	5.0	26.64 ± 0.32 ^{aA}	25.72 ± 0.46 ^{aAB}	25.5 ± 1.18 ^{aAB}	26.52 ± 0.29 ^{aA}	26.12 ± 0.14 ^{aAB}	24.65 ± 0.29 ^{bB}
	7.5	26.21 ± 0.56 ^{aA}	25.17 ± 1.82 ^{aA}	26.79 ± 0.31 ^{aA}	25.84 ± 0.4 ^{aA}	26.44 ± 0.31 ^{aA}	25.24 ± 0.35 ^{abA}
200 °C White Flour	0	20.26 ± 0.76 ^{a}	20.26 ± 0.76 ^{a}	20.26 ± 0.76 ^{a}	20.26 ± 0.76 ^{a}	20.26 ± 0.76 ^{a}	20.26 ± 0.76 ^{a}
	2.5	20.79 ± 1.54 ^{aA}	21.31 ± 0.27 ^{aA}	21.29 ± 0.82 ^{aA}	20.54 ± 0.26 ^{aA}	20.36 ± 0.87 ^{aA}	17.05 ± 1.33 ^{bB}
	5.0	20.20 ± 0.52 ^{aA}	20.43 ± 0.42 ^{aA}	20.30 ± 0.15 ^{aA}	20.81 ± 0.32 ^{aA}	20.39 ± 0.65 ^{aA}	21.30 ± 1.16 ^{aA}
	7.5	19.95 ± 0.15 ^{aB}	21.52 ± 0.41 ^{aAB}	22.68 ± 1.54 ^{aA}	20.89 ± 0.2 ^{aAB}	20.36 ± 0.79 ^{aB}	22.83 ± 0.53 ^{aA}
200 °C Whole Wheat Flour	0	26.38 ± 0.63 ^{a}	26.38 ± 0.63 ^{ab}	26.38 ± 0.63 ^{a}	26.38 ± 0.63 ^{a}	26.38 ± 0.63 ^{a}	26.38 ± 0.63 ^{a}
	2.5	25.54 ± 0.53 ^{aA}	26.27 ± 0.58 ^{abA}	25.74 ± 0.08 ^{aA}	25.64 ± 0.23 ^{aA}	25.76 ± 0.50 ^{aA}	25.21 ± 0.54 ^{abA}
	5.0	25.55 ± 0.29 ^{aAB}	25.65 ± 0.42 ^{bA}	26.34 ± 0.44 ^{aA}	25.66 ± 1.13 ^{aA}	27.01 ± 0.02 ^{aA}	23.63 ± 0.60 ^{bB}
	7.5	26.13 ± 0.33 ^{aA}	26.97 ± 0.29 ^{aA}	26.24 ± 0.68 ^{aA}	25.84 ± 1.28 ^{aA}	27.25 ± 0.74 ^{aA}	26.06 ± 0.19 ^{aA}

[1] Means ± SD are presented. Data expressed as g/100 g on a fresh weight basis; different lowercase superscript letters in a column and different uppercase superscript in a row denote significant differences, $p < 0.05$.

In Table 2, when considering the differences between white flour and whole wheat flour cookies: at 180 °C, the WF cookies consistently had higher moisture content. At 200 °C, the same trend was observed with WF cookies having higher moisture content than the WW composite cookies. The control samples showed the same pattern. We believe that the higher proportion of gluten in WF cookies leads to this effect. The only exception in these observations was the *Khalas* cookies at 2.5% addition, where the WW cookies had higher moisture than the WF ones. The reasons for this difference are not clear.

When considering the differences between baking temperatures (for the same flour type) overall, as expected, the cookies baked at 200 °C had lower moisture content than those baked at 180 °C. This observation was true for the controls as well as the composite cookies. The exceptions were the *Shaham* cookies that exhibited no significant differences between the two baking temperatures.

When considering the differences among varieties, for the 180 °C WF cookies, at 2.5 and 5% addition, *Shaham* had the highest values, while at 7.5%, the highest was *Khalas*. For the 180 °C WW cookies, at all levels of addition, the Khalas variety cookies exhibited the highest values. For WF cookies baked at 200 °C, there was no clear pattern emerging on the basis of date seed variety. For WW cookies baked at 200 °C, *Khalas* showed the highest moisture values at all addition levels.

In previous studies, improved water holding capacities resulted in bread formulated with defatted date seed powder [20] and extracted polysaccharides of date seeds [34]. Therefore, the results in this study could be attributed to the improved water holding capacity [35] due to the high amount of fibre content in date seeds [17–19]. When it comes to preservation, convenience in packaging, storage and transport, the moisture content is an important quality factor. Differences in moisture contents among date seed varieties could also be responsible for the discrete levels of moisture in composite cookies.

Ash content differences between cookies are listed in Table 3. Regarding the effect of flour type, in the control samples, WF cookies had lower ash content than WW ones. In composite cookies, however, there was no clear trend observed. With regards to the effect of cooking temperature (for the same flour type), there were no significant differences in the control samples (no DSP addition), while no trend was apparent for composite cookies. Assessing the results by DSP variety, for WF at 180 °C, *Zahidi* cookies had the highest values at all levels of addition. For WW cookies at 180 °C, again, the *Zahidi* variety had the highest values at the higher levels of addition (5% and 7.5%). For WF cookies prepared at 200 °C, the *Zahidi* variety had significantly higher values than the rest, although no significant differences were observed between levels of addition. For WW cookies baked at 200 °C, *Fardh* cookies had the highest values at all levels of addition. Overall, the results regarding the ash content are somewhat surprising as they did not correspond to the ash content of the DSP.

There was no significant improvement in ash content of some of the composite cookies compared to control samples, although a considerable amount of ash was observed in date seed powder, especially *Fardh*. This may be due to the addition of lower amounts of date seed powder in composite cookies, leading to no significant difference in ash content between the control and composite samples. High ash content is an indication of high mineral content. Date seeds are found to be rich in minerals [3,36], and improved mineral content was observed in composite bakery products in a previous study [37]. Therefore, the addition of higher amounts of date seed powder may increase the ash content and hence mineral content in composite cookies.

The fat content results in Table 4, when examined with regards to the flour type used, indicate that for both controls and composite cookies, WF cookies had lower fat content than WW cookies at both temperatures. The addition of DSP did not result in significant differences between controls and composite cookies. When looking at the effect of baking temperature (for the same flour type), no significant differences were observed for both controls and composite cookies. Regarding the effect of DSP variety, no significant differences were observed at any flour-temperature combination.

The different stages of cookie preparation change the physical and chemical properties of the various flour constituents, including fat, present in the cookie dough formula [38]. Moreover, the fat content in different date seed varieties would result in different levels of fat in the final product. Fat provides a number of benefits to improve the physical and textural quality of cookies [38].

The incorporation of date pits in the form of powder at different levels into cookies resulted in similar fat content, and this can be attributed to the high amount of fat in the oil and flour used (accounting for about 96% of the total fat content of cookies) that masks the fat content coming from date seed powder. The same explains the insignificant difference in the protein content of the cookies, where the protein content came mainly from the high amount of protein in whole wheat and white wheat flour.

However, using even higher date seed powder addition levels, as shown by Platat et al. [17], may make a significant difference, although the consumer acceptability of such formulations would need to be confirmed. The chemical composition of date seed pits is known to vary depending on the variety and can be attributed to the use of fertilisers, harvest time and post-harvest treatments [30]. Thus, it is quite expected to notice a difference in results among the varieties.

In terms of protein content, the incorporation of date seed powder of any variety in making cookies did not affect the protein content when compared to cookies with zero incorporation, which agrees with Platat et al. [17]. The date seed powders used contained similar levels of protein to the wheat flours used, and, considering the levels of addition, this resulted in non-significant differences in the protein content of the composite cookies.

3.2. Physical Properties of Cookies

3.2.1. Colour Analysis

The colour measurements of date seed powder varieties and the cookies substituted with various date seed powder are listed in Table 5. The lightness of the six varieties used was significantly different from each other; *Sukkari* date seed powder was the lightest in colour, and *Khalas* was the darkest; however, the differences in the level of lightness between the date seed powder did not affect the final colour of the cookies. White flour cookies with no addition of date seed powder were obviously lighter than cookies of whole wheat flour, and the lightness of the composite cookies displayed a decreasing trend with the increasing level of addition; the higher level of addition, the darker the cookie. These results are in accordance with Gómez and Martinez [16], Aksoylu et al. [39] and Ashoush and Gadallah [15], who reported the colour alterations in biscuits due to incorporation of by-products such as seeds. Therefore, the colour of cookies was significantly affected by the DSP addition; the composite cookies were darker, and the dark colour is caused by the natural dark pigmentation of date seeds regardless of the variety.

Table 5. Lightness of date powder and cookie [1] samples.

Lightness		White Flour 93.27 [A]		Whole Wheat Flour 87.00 [B]			
		Khalas 37.72 [F]	Khinaizi 42.78 [B]	Sukkari 45.13 [A]	Shaham 42.40 [C]	Zahidi 37.90 [E]	Fardh 41.99 [D]
Baking temp flour type and addition level							
180 °C White Flour	0	70.20 ± 2.28 [a]	70.20 ± 2.28 [a]	70.20 ± 2.28 [a]	70.20 ± 2.28 [a]	70.20 ± 2.28 [a]	70.20 ± 2.28 [a]
	2.5	48.32 ± 0.76 [bD]	50.27 ± 1.27 [bBC]	50.83 ± 0.10 [bB]	48.28 ± 0.72 [bD]	53.13 ± 0.09 [bA]	48.81 ± 0.33 [bCD]
	5.0	42.54 ± 1.55 [cBC]	41.78 ± 0.39 [cC]	44.71 ± 0.69 [cAB]	41.79 ± 0.84 [cC]	45.77 ± 0.10 [cA]	41.78 ± 1.37 [cC]
	7.5	40.55 ± 1.56 [cA]	39.39 ± 0.33 [cA]	41.39 ± 0.06 [cA]	38.73 ± 0.47 [cA]	39.61 ± 1.24 [dA]	38.70 ± 0.52 [cA]
180 °C Whole Wheat Flour	0	60.25 ± 0.25 [a]	60.25 ± 0.25 [a]	60.25 ± 0.25 [a]	60.25 ± 0.25 [a]	60.25 ± 0.25 [a]	60.25 ± 0.25 [a]
	2.5	49.70 ± 0.55 [bA]	47.21 ± 0.38 [bB]	47.71 ± 0.79 [bB]	47.20 ± 1.01 [bB]	47.60 ± 0.31 [bB]	47.52 ± 0.41 [bB]
	5.0	42.36 ± 1.31 [cAB]	41.99 ± 0.32 [cAB]	43.32 ± 0.56 [cA]	41.26 ± 0.28 [cB]	42.99 ± 0.30 [cA]	42.52 ± 0.10 [bAB]
	7.5	40.06 ± 0.70 [dAB]	38.05 ± 0.43 [dAB]	38.62 ± 0.52 [dAB]	38.81 ± 0.36 [dAB]	41.79 ± 1.07 [cA]	35.03 ± 4.30 [cB]
200 °C White Flour	0	71.14 ± 1.40 [a]	71.14 ± 1.40 [a]	71.14 ± 1.40 [a]	71.14 ± 1.40 [a]	71.14 ± 1.40 [a]	71.14 ± 1.40 [a]
	2.5	46.44 ± 0.33 [bD]	50.02 ± 0.56 [bB]	48.86 ± 0.60 [bBC]	48.60 ± 0.58 [bC]	52.11 ± 0.16 [bA]	47.91 ± 0.52 [bC]
	5.0	40.66 ± 1.32 [cB]	42.38 ± 0.44 [cB]	45.02 ± 0.40 [cA]	42.22 ± 0.12 [cB]	46.28 ± 0.60 [cA]	41.48 ± 0.20 [cB]
	7.5	39.96 ± 1.14 [cAB]	39.70 ± 0.44 [dABC]	40.29 ± 0.48 [dA]	37.98 ± 0.84 [dC]	38.37 ± 0.49 [dABC]	38.29 ± 0.52 [dBC]
200 °C Whole Wheat Flour	0	58.10 ± 1.79 [a]	58.10 ± 1.79 [a]	58.10 ± 1.79 [a]	58.10 ± 1.79 [a]	58.10 ± 1.79 [a]	58.10 ± 1.79 [a]
	2.5	47.89 ± 1.04 [bA]	47.59 ± 0.27 [bA]	46.44 ± 0.32 [bA]	47.02 ± 1.11 [bA]	47.19 ± 1.36 [bA]	46.14 ± 0.97 [bA]
	5.0	41.30 ± 0.25 [cB]	42.10 ± 0.30 [cB]	43.25 ± 0.04 [cA]	42.02 ± 0.52 [cB]	43.46 ± 0.65 [cA]	41.32 ± 0.44 [cB]
	7.5	38.07 ± 2.71 [cA]	38.25 ± 0.71 [dA]	37.27 ± 0.62 [dA]	38.96 ± 0.21 [dA]	39.48 ± 0.40 [dA]	37.05 ± 0.37 [dA]

[1] Means ± SD are presented. Different lowercase superscript letters in a column and different uppercase superscript in a row denote significant differences, $p < 0.05$.

3.2.2. Texture Analysis

Cookies' hardness assessments are listed in Table 6. The texture of all cookies made with *Shaham* date seed powder was not affected by the addition level and was almost similar to the texture of *Zahidi* cookies, which were softer when prepared with white flour at 200 °C at an addition level of 7.5%. The effect of the addition level of either *Fardh* or *Sukkari* seed powder on the texture of cookies was different; the more powder added, the softer the cookies. Moreover, the case was the same in *Khalas* cookies, except for whole wheat flour *Khalas* cookies baked at 180 °C, where softer cookies were obtained. While in whole wheat flour *Khinaizi* cookies at any baking temperature, the softness was not affected, differing from white flour *Khinaizi* cookies, where a decrease in hardness with the addition level was recognisable.

Table 6. Hardness assessments of cookies (N).

Baking Temp Flour Type and Addition Level		*Khalas*	*Khinaizi*	*Sukkari*	*Shaham*	*Zahidi*	*Fardh*
180 °C White Flour	0	22.82 ± 1.05 [a]	22.82 ± 1.05 [a]	22.82 ± 1.05 [a]	22.82 ± 1.05 [a]	22.82 ± 1.05 [a]	22.82 ± 1.05 [a]
	2.5	19.94 ± 2.02 [abABC]	18.40 ± 0.16 [bBC]	16.73 ± 1.33 [bC]	22.10 ± 0.59 [aAB]	23.02 ± 2.01 [aA]	19.90 ± 1.20 [bABC]
	5.0	17.72 ± 0.19 [bA]	17.41 ± 1.78 [bA]	16.28 ± 1.57 [bA]	19.50 ± 2.56 [aA]	17.59 ± 2.43 [aA]	18.82 ± 0.68 [bA]
	7.5	17.81 ± 0.22 [bAB]	18.02 ± 0.97 [bAB]	14.71 ± 0.14 [bB]	19.31 ± 0.49 [aA]	18.33 ± 3.12 [aAB]	15.11 ± 0.85 [cB]
180 °C Whole Wheat Flour	0	11.04 ± 1.20 [a]	11.04 ± 1.20 [a]	11.04 ± 1.20 [a]	11.04 ± 1.20 [a]	11.04 ± 1.20 [a]	11.04 ± 1.20 [a]
	2.5	10.05 ± 0.89 [aA]	10.62 ± 0.62 [aA]	10.81 ± 0.60 [abA]	10.83 ± 0.72 [aA]	9.36 ± 1.38 [aA]	9.43 ± 0.46 [bA]
	5.0	9.70 ± 0.36 [aA]	10.40 ± 0.35 [aA]	9.21 ± 0.45 [abA]	9.90 ± 0.41 [aA]	9.60 ± 1.67 [aA]	9.23 ± 0.38 [bA]
	7.5	9.06 ± 0.39 [aA]	9.74 ± 0.49 [aA]	8.53 ± 1.11 [bA]	9.69 ± 0.31 [aA]	8.91 ± 1.61 [aA]	8.14 ± 0.50 [cA]
200 °C White Flour	0	21.90 ± 1.54 [a]	21.90 ± 1.54 [a]	21.90 ± 1.54 [a]	21.90 ± 1.54 [a]	21.90 ± 1.54 [a]	21.90 ± 1.54 [a]
	2.5	19.83 ± 1.76 [abAB]	17.22 ± 0.44 [bB]	18.02 ± 1.63 [bB]	23.61 ± 3.07 [aA]	21.44 ± 1.97 [aAB]	19.55 ± 0.85 [bAB]
	5.0	17.34 ± 0.27 [bAB]	17.40 ± 1.29 [bAB]	15.91 ± 0.63 [bcB]	20.02 ± 0.45 [aA]	17.20 ± 2.79 [aAB]	16.33 ± 0.87 [bB]
	7.5	18.21 ± 0.90 [bAB]	16.05 ± 0.89 [bBC]	13.73 ± 0.72 [cC]	19.70 ± 1.14 [aA]	17.05 ± 1.47 [bAB]	13.88 ± 1.01 [bC]
200 °C Whole Wheat Flour	0	10.94 ± 0.79 [a]	10.94 ± 0.79 [a]	10.94 ± 0.79 [a]	10.94 ± 0.79 [a]	10.94 ± 0.79 [a]	10.94 ± 0.79 [a]
	2.5	10.86 ± 0.28 [abA]	11.07 ± 0.70 [aA]	10.22 ± 0.61 [abA]	10.74 ± 0.63 [aA]	9.62 ± 1.30 [aA]	9.85 ± 0.90 [aA]
	5.0	9.42 ± 0.37 [bcA]	10.11 ± 0.73 [aA]	7.97 ± 0.37 [bcA]	9.94 ± 0.54 [aA]	9.85 ± 1.38 [aA]	9.06 ± 0.35 [aA]
	7.5	9.03 ± 0.68 [cAB]	9.34 ± 0.55 [aA]	8.31 ± 0.91 [cAB]	9.62 ± 0.64 [aA]	9.33 ± 1.43 [aA]	6.84 ± 1.06 [bB]

Different lowercase superscript letters in a column and different uppercase superscript in a row denote significant differences, $p < 0.05$.

With the addition of date seed flour, cookies become softer when the control is white flour, except for *Shahm* and *Zahidi*. In the case of whole wheat flour, a significant difference was noticeable in *Khalas* and *Sukkari* at higher addition levels and the 7.5% addition level in *Fardh* at 200 °C, whereas at 180 °C, it was only in *Fardh* and 7.5% addition level in *Sukkari*.

The most pronounced decrease in the hardness of the obtained composite cookies was observed for the 7.5% addition level of white flour *Sukkari* cookie samples at 200 °C, which showed a prominent significant difference from the control. However, it was not significantly different from *Khinaizi* and *Fardh* at the same treatment conditions. In comparison among varieties, the increase in the softness was quite underlined in *Fardh* cookies in all the treatment conditions at 7.5% inclusion level except white flour *Fardh* cookies at 200 °C. In that treatment condition, all the inclusion levels of 2.5%, 5% and 7.5% were also significantly different from the control, although it did not show significant differences between the addition levels.

Composite cookies were softer in accordance with similar studies on seed incorporation, such as amaranth flour composite cookies [10,40]. According to Chauhan et al. [10], the decrease in hardness was due to the replacement of wheat flour with the seed flour, which results in a gluten content reduction in the cookie dough which, in turn, contributed to the substantial decrease in hardness. This phenomenon is applicable to the observations in our study. Moreover, several studies on bakery items, such as bread and biscuits, demonstrated that hardness is mainly due to the interactions between gluten and fibre, where dietary fibre leads to higher water absorption and interferes with gluten development time [41,42]. Bouaziz et al. [34] tried incorporating the extracted dietary fibre from date seeds, which resulted in a decrease in bread hardness. Since date seeds are rich in dietary fibre [17–19,34,43], this can be a reason for the reduction in hardness with increasing levels of date seeds.

3.2.3. Sensory Evaluation

The evaluation of sensory properties of cookies is illustrated in Figures 1–5, colour, smell, texture, taste and overall acceptability for all cookies made with different varieties of date seed powder were acceptable and were rated between 5 and 7 on a hedonic scale.

Figure 1. Smell evaluation for date seed composite cookies.

Figure 2. Colour evaluation for date seed composite cookies.

Figure 3. Texture evaluation for date seed composite cookies.

Figure 4. Taste evaluation for date seed composite cookies.

The consumers' evaluation in terms of colour, texture and taste showed that the produced cookies were highly acceptable. A darker cookie colour correlates with other physical, chemical and sensorial indicators of product quality. Colour is considered a fundamental physical property of foods and agricultural products, and it affects the assessment of external quality in both the food industry and food engineering research [44,45]. When analysing the results for colour, we cannot say the panellists are only interested in the light colour products, as some scores are higher for the darker ones, such as in the varieties *Khalas* and *Zahidi*. It is noteworthy that in some instances, date seed composite cookies with high substitutions are the ones which are more preferred compared to zero or low-level samples. When the texture is considered, consumer acceptability is high in the samples with high date seed flour substitutions. This may be due to the softness of biscuits. According to the results of hardness, date seed composite cookies have lower hardness compared to the control samples. Hardness is a textural property that plays a major role in the evaluation of baked goods, as it is associated with the human perception of freshness—the lower this parameter is, the more desirable the product [46]. However, the overall acceptability of the date seed composite cookies is significant depending on the variety and the heat treatment.

Figure 5. Overall acceptability of date seed composite cookies.

4. Conclusions

This study concluded that the by-products or the waste of date-processing units, such as date seeds, can be used to enhance the quality of cookies in terms of their physical, chemical and textural properties. There were no significant differences in the protein content of cookies with date seed flour incorporated since the composite cookies had a protein content similar to the control samples. For cookies baked at 180 °C, whole wheat flour formulations had lower moisture content than those made with white flour. A similar trend was observed at 200 °C, with whole wheat flour formulations having lower moisture content than white flour for each date variety. Different varieties with different addition levels showed a variation in moisture levels, but at all addition levels, *Khalas'* moisture was the highest. It appears that the moisture content level is primarily affected by the baking temperature and the type of flour, while varietal differences seem to be less pronounced. When considering the ash content, *Zahidi* composite cookies at the 7.5% addition level in both flour types had the highest values. The colour of cookies was significantly affected by the incorporation of date seed flour, making them darker. This may be the reason for the somewhat reduced score for the colour attribute in sensory, even though some addition levels show higher consumer acceptance. However, the sensory analysis results showed that the overall acceptance of the composite cookies was higher than the others with whole wheat flour at both temperature levels. Therefore, based on the sensory analysis, Khalas and Zahidi composite cookies at 7.5% addition level with whole wheat flour can be recommended to obtain the most preferred final product. As expected, an increased softness was observed in date seed composite cookies, except for the *Shaham* and *Zahidi* varieties (where the increment was not significant), leading the path to formulate high-quality cookies in terms of texture.

Author Contributions: Conceptualisation, C.S.; Methodology, C.S. and Z.N.; formal analysis, Z.N., M.R. and C.S.; investigation, Z.N., M.A. and K.A.; writing-original draft preparation, Z.N.; writing-review and editing, M.R. and C.S.; visualisation, C.S.; supervision, C.S.; project administration, C.S.; funding acquisition, C.S. All authors have read and agreed to the published version of the manuscript.

Funding: This research was funded by United Arab Emirates University start up grant (G00002958).

Institutional Review Board Statement: Not applicable.

Informed Consent Statement: Not applicable.

Data Availability Statement: Not applicable.

Conflicts of Interest: The authors declare no conflict of interest.

References

1. Al Juhaimi, F.; Ghafoor, K.; Ozcan, M.M. Physical and chemical properties, antioxidant activity, total phenol and mineral profile seeds of seven different date fruit (*Phoenix dactylifera*) varieties. *Int. J. Food Sci. Nutr.* **2012**, *63*, 84–89. [CrossRef]
2. Al-Farsi, M.A.; Lee, C.Y. Optimization of phenolics and dietary fibre extraction from date seeds. *Food Chem.* **2008**, *108*, 977–985. [CrossRef] [PubMed]
3. Besbes, S.; Blecker, C.; Deroanne, C.; Drira, N.E.; Attia, H. Date seeds: Chemical composition and characteristic profiles of the lipid fraction. *Food Chem.* **2004**, *84*, 577–584. [CrossRef]
4. Golshan, T.A.; Solaimani, D.N.; Yasini Ardakani, S.A. Physicochemical properties and applications of date seed and its oil. *Int. Food Res. J.* **2017**, *24*, 1399–1406.
5. Basuni, A.M.M.; Al-Marzooq, M.A. Production of mayonnaise from date pit oil. *Food Nutr. Sci.* **2011**, *2*, 938–943. [CrossRef]
6. Ghnimi, S.; Umer, S.; Karim, A.; Kamal-Eldin, A. Date fruit (*Phoenix dactylifera* L.): An underutilized food seeking industrial valorization. *NFS* **2017**, *6*, 1–10. [CrossRef]
7. Al Juhaimi, F.; Ozcan, M.M.; Adiamo, O.Q.; Alsawmahi, O.N.; Ghafoor, K.; Babiker, E.E. Effect of date varieties on physico-chemical properties, fatty acid composition, tocopherol contents, and phenolic compounds of some date seed and oils. *J. Food Process. Preserv.* **2018**, *42*, e13584. [CrossRef]
8. Amany, M.M.B.; Shaker, M.A.; Abeer, A.K. Antioxidant activities of date pits in a model meat system. *Int. Food Res.* **2012**, *19*, 223–227.
9. Jemziya, M.B.F.; Mahendran, T. Quality characteristics and sensory evaluation of cookies produced from composite blends of sweet potato (*Ipomoea batatas* L.) and wheat (*Triticum aestivum* L.) flour. *SLJFA* **2015**, *1*, 23–30. [CrossRef]
10. Chauhan, A.; Saxena, D.C.; Singh, S. Physical, textural, and sensory characteristics of wheat and amaranth flour blend cookies. *Cogent Food Agric.* **2016**, *2*, 1125773. [CrossRef]
11. Baumgartner, B.; Özkaya, B.; Saka, I.; Özkaya, H. Functional and physical properties of cookies enriched with dephytinized oat bran. *J. Cereal Sci.* **2018**, *80*, 24–30. [CrossRef]
12. Khouryieh, H.; Aramouni, F. Physical and sensory characteristics of cookies prepared with flaxseed flour. *J. Sci. Food Agric.* **2012**, *92*, 2366–2372. [CrossRef] [PubMed]
13. Hrušková, M.; Švec, I. Cookie making potential of composite flour containing wheat, barley and hemp. *Czech. J. Food Sci.* **2015**, *33*, 545–555. [CrossRef]
14. Jan, R.; Saxena, D.C.; Singh, S. Physico-chemical, textural, sensory and antioxidant characteristics of gluten–Free cookies made from raw and germinated Chenopodium (*Chenopodium album*) flour. *LWT-Food Sci. Technol.* **2016**, *71*, 281–287. [CrossRef]
15. Ashoush, I.S.; Gadallah, M.G. Utilization of mango peels and seed kernels powders as sources of phytochemicals in biscuit. *World J. Dairy Food Sci.* **2011**, *6*, 35–42.
16. Gómez, M.; Martinez, M.M. Fruit and vegetable by-products as novel ingredients to improve the nutritional quality of baked goods. *Critical Rev. Food Sci. Nutri.* **2018**, *58*, 2119–2135. [CrossRef]
17. Platat, C.; Habib, H.M.; Hashim, I.B.; Kamal, H.; AlMaqbali, F.; Souka, U.; Ibrahim, W.H. Production of functional pita bread using date seed powder. *J. Food Sci. Technol.* **2015**, *10*, 6375–6384. [CrossRef]
18. Ambigaipalan, P.; Shahidi, F. Date seed flour and hydrolysates affect physicochemical properties of muffin. *Food Biosci.* **2015**, *12*, 54–60. [CrossRef]
19. Almana, H.A.; Mahmoud, R.M. Palm date seeds as an alternative source of dietary fibre in Saudi bread. *Ecol. Food Nutri.* **1994**, *32*, 261–270. [CrossRef]
20. Bouaziz, M.A.; Amara, W.B.; Attia, H.; Blecker, C.; Besbes, S. Effect of the addition of defatted date seeds on wheat dough performance and bread quality. *J. Texture Stud.* **2010**, *41*, 511–531. [CrossRef]
21. Suresh, S.; Guizani, N.; Al-Ruzeiki, M.; Al-Hadhrami, A.; Al-Dohani, H.; Al-Kindi, I.; Rahman, M.S. Thermal characteristics, chemical composition and polyphenol contents of date-pits powder. *J. Food Eng.* **2013**, *119*, 668–679. [CrossRef]
22. Storup, K.; Mattfolk, K.; Voinea, D.; Jakobsen, B.; Bain, M.; Reverté Casas, M.; Oliveira, P. Combating Food Waste: An Opportunity for the EU to Improve the Resource-Efficiency of the Food Supply Chain. *Eur. Court. Audit. Eur. Union* **2016**, *34*, 6–7.

23. US Environmental Protection Agency. Food Recovery Hierarchy. In *Sustainable Management of Food*; US Environmental Protection Agency: Washington, DC, USA, 2015.
24. Grasso, S.; Liu, S.; Methven, L. Quality of muffins enriched with upcycled defatted sunflower seed flour. *LWT* **2020**, *119*, 108893. [CrossRef]
25. Grasso, S.; Pintado, T.; Pérez-Jiménez, J.; Ruiz-Capillas, C.; Herrero, A.M. Characterisation of Muffins with Upcycled Sunflower Flour. *Foods.* **2021**, *10*, 426. [CrossRef]
26. Heo, Y.; Kim, M.J.; Lee, J.W.; Moon, B. Muffins enriched with dietary fibre from kimchi by-product: Baking properties, physical–chemical properties, and consumer acceptance. *Food Sci. Nutr.* **2019**, *7*, 1778–1785. [CrossRef]
27. Approved Methods of Analysis. AACC. Method 10-54.01. *Baking Quality of Cookie Flour-Micro Wire-Cut Formulation*, 11th ed.; Cereals & Grains Association: St. Paul, MN, USA, 1999; ISBN 978-1-891127-68-2.
28. The Association of Official Analytical Chemists. *AOAC (2003) Official Methods of Analysis*; AOAC International: Washington, DC, USA, 2003.
29. The Association of Official Analytical Chemists. *AOAC (1995) Official Methods of Analysis*, 15th ed.; AOAC International: Washington, DC, USA, 1995.
30. Habib, H.M.; Ibrahim, W.H. Nutritional quality evaluation of eighteen date pit varieties. *Int. J. Food Sci. Nutr.* **2009**, *60*, 99–111. [CrossRef]
31. Ismail, B.; Haffar, I.; Baalbaki, R.; Mechref, Y.; Henry, J. Physico-chemical characteristics and total quality of five date varieties grown in the United Arab Emirates. *Int. J. Food Sci. Technol.* **2006**, *41*, 919–926. [CrossRef]
32. Platat, C.; Habib, H.M.; Al Maqbali, F.D.; Jaber, N.N.; Ibrahim, W.H. Identification of date seeds varieties patterns to optimize nutritional benefits of date seeds. *J. Nutr. Food Sci. S* **2014**, *8*. [CrossRef]
33. Saafi, E.B.; Trigui, M.; Thabet, R.; Hammami, M.; Achour, L. Common date palm in Tunisia: Chemical composition of pulp and pits. *Int. J. Food Sci. Technol.* **2008**, *43*, 2033–2037. [CrossRef]
34. Bouaziz, F.; Ben Abdeddayem, A.; Koubaa, M.; Ellouz Ghorbel, R.; Ellouz Chaabouni, S. Date seeds as a natural source of dietary fibres to improve texture and sensory properties of wheat bread. *Foods.* **2020**, *9*, 737. [CrossRef]
35. Bölek, S. Effects of waste fig seed powder on quality as an innovative ingredient in biscuit formulation. *J. Food Sci.* **2021**, *86*, 55–60. [CrossRef] [PubMed]
36. Nehdi, I.; Omri, S.; Khalil, M.I.; Al-Resayes, S.I. Characteristics and chemical composition of date palm (*Phoenix canariensis*) seeds and seed oil. *Ind. Crops Prod.* **2010**, *32*, 360–365. [CrossRef]
37. Ahfaiter, H.; Zeitoun, A.; Abdallah, A.E. Physicochemical Properties and Nutritional Value of Egyptian Date Seeds and Its Applications in Some Bakery Products. *J. Adv. Agr. Res.* **2018**, *23*, 260–279.
38. Pareyt, B.; Delcour, J.A. The role of wheat flour constituents, sugar, and fat in low moisture cereal based products: A review on sugar-snap cookies. *Crit. Rev. Food Sci. Nutr.* **2008**, *48*, 824–839. [CrossRef] [PubMed]
39. Aksoylu, Z.; Çağindi, Ö.; Köse, E. Effects of blueberry, grape seed powder and poppy seed incorporation on physicochemical and sensory properties of biscuit. *J. Food Qual.* **2015**, *38*, 164–174. [CrossRef]
40. Sindhuja, A.; Sudha, M.L.; Rahim, A. Effect of incorporation of amaranth flour on the quality of cookies. *Eur. Food Res. Technol.* **2005**, *221*, 597–601. [CrossRef]
41. Saeed, S.M.; Urooj, S.; Ali, S.A.; Ali, R.; Mobin, L.; Ahmed, R.; Sayeed, S.A. Impact of the incorporation of date pit flour an underutilized biowaste in dough and its functional role as a fat replacer in biscuits. *J. Food Process. Preserv.* **2021**, *45*, e15218. [CrossRef]
42. Gómez, M.; Ronda, F.; Blanco, C.A.; Caballero, P.A.; Apesteguía, A. Effect of dietary fibre on dough rheology and bread quality. *Eur. Food Res. Technol.* **2003**, *216*, 51–56. [CrossRef]
43. Al-Farsi, M.; Alasalvar, C.; Al-Abid, M.; Al-Shoaily, K.; Al-Amry, M.; Al-Rawahy, F. Compositional and functional characteristics of dates, syrups, and their by-products. *Food Chem.* **2007**, *104*, 943–947. [CrossRef]
44. Segnini, S.; Dejmek, P.; Öste, R. A low cost video technique for colour measurement of potato chips. *Lebensm-Wiss. U.-Technol.* **1999**, *32*, 216–222. [CrossRef]
45. Abdullah, M.Z.; Guan, L.C.; Lim, K.C.; Karim, A.A. The applications of computer vision system and tomographic radar imaging for assessing physical properties of food. *J. Food Eng.* **2004**, *61*, 125–135. [CrossRef]
46. Assis, L.M.; Zavareze, E.; Radünz, A.; Dias, Á.; Gutkoski, L.; Elias, M. Nutritional, technological and sensory properties of cookies with substitution of wheat flour by oat flour or parboiled rice flour. *Alim. Nutr. Araraquara* **2009**, *20*, 15–24.

Communication

Characterisation of Muffins with Upcycled Sunflower Flour

Simona Grasso [1,*], Tatiana Pintado [2], Jara Pérez-Jiménez [2], Claudia Ruiz-Capillas [2] and Ana Maria Herrero [2]

[1] Institute of Food, Nutrition and Health, School of Agriculture, Policy and Development, University of Reading, Reading RG6 6AH, UK

[2] Institute of Food Science, Technology and Nutrition (ICTAN-CSIC), 28040 Madrid, Spain; tatianap@ictan.csic.es (T.P.); jara.perez@ictan.csic.es (J.P.-J.); claudia@ictan.csic.es (C.R.-C.); ana.herrero@ictan.csic.es (A.M.H.)

* Correspondence: simona.grasso@ucdconnect.ie; Tel.: +44-118-3786-576

Abstract: There is an increased interest and need to make our economy more circular and our diets healthier and more sustainable. One way to achieve this is to develop upcycled foods that contain food industry by-products in their formulation. In this context, the aim of this study was to develop muffins containing upcycled sunflower flour (a by-product from the sunflower oil industry) and assess the effects of sunflower flour addition on the fibre, protein, amino acid, mineral content, and antioxidant activity measured by a Ferric Reducing Antioxidant Power (FRAP) assay and Photo chemiluminescence (PCL) assay. Results show that the sunflower flour inclusion significantly improved all the parameters analysed as part of this study. A more balanced muffin amino acid profile was achieved, thanks to the increased levels of lysine, threonine, and methionine, the limiting essential amino acids of wheat flour. We can conclude that upcycled ingredients, such as sunflower flour, could be used for the nutritional improvement of baked goods, such as muffins. Their addition can result in several nutritional advantages that could be communicated on packaging through the use of the appropriate EU nutrition claims, such as those on protein, fibre, and mineral content.

Keywords: muffins; by-product; valorisation; sunflower flour; amino acid profile; antioxidant activity; mineral content; fibre content; FRAP; PCL assay

Citation: Grasso, S.; Pintado, T.; Pérez-Jiménez, J.; Ruiz-Capillas, C.; Herrero, A.M. Characterisation of Muffins with Upcycled Sunflower Flour. *Foods* **2021**, *10*, 426. https://doi.org/10.3390/foods10020426

Academic Editor: Stefania Masci

Received: 25 January 2021
Accepted: 12 February 2021
Published: 15 February 2021

Publisher's Note: MDPI stays neutral with regard to jurisdictional claims in published maps and institutional affiliations.

Copyright: © 2021 by the authors. Licensee MDPI, Basel, Switzerland. This article is an open access article distributed under the terms and conditions of the Creative Commons Attribution (CC BY) license (https://creativecommons.org/licenses/by/4.0/).

1. Introduction

In order to make our economies more circular and our diets more sustainable, there is an increased need to valorise food industry by-products into ingredients that can re-enter the food chain as part of new foods.

Sunflower cake is a by-product of the sunflower oil industry which has been traditionally used as animal feed [1]. Rich in protein, fibre, essential amino acids and minerals [1], it has been reported to have a high antioxidant potential [2]. Recently, sunflower cake has been upcycled into a functional flour by a US start-up through the patented use of novel technologies, such as extrusion and steam explosion [3], which have opened up this under-valorised ingredient to a whole new range of food applications. Researchers have so far used it on both baked goods [4,5] and meat product applications [6], reporting promising results.

Circular economy principles should push us to valorise food industry by-products as ingredients for human diets, rather than just as animal feed, as explained in the food recovery hierarchies developed in the EU and in US [7,8]. This is especially relevant if we consider that food industry by-products contain several nutrients of interest, such as protein, fibre, minerals and vitamins. We know that the demand for proteins will continue to increase in the future [9]; therefore, valorising food industry by-products rich in proteins could be a positive step towards sustainable protein production. Similarly, in 2015, the Scientific Advisory Committee on Nutrition brought the recommended daily intake of fibre to 30 g, while the average intake in adults is around 18 g of fibre daily [10]; therefore, fibre-rich by-products could play a key role in meeting this nutritional need when suitably incorporated into new foods.

Popular baked goods, such as muffins, cakes, or biscuits, are usually high in sugar and fat but low in fibre, antioxidants, and minerals [11], so they could represent ideal foods to be reformulated to be healthier through the use of upcycled ingredients. Efforts to include ingredients such as pecan nut meal, spent coffee, and several fruit and vegetable pomaces [4,12–14] in muffins have recently been reported, showing an increased research interest in this area.

The aim of this short communication was to partially replace wheat flour with sunflower flour (at 15% or 30%) in muffins and evaluate the effects of this replacement on their fibre, mineral, and amino acid content, as well as antioxidant activity. No attempts have been made so far to investigate these parameters in the development of muffins with upcycled sunflower flour, while an investigation on the proximate composition (moisture, ash, protein, and fat), physical analyses, and sensory quality of muffins with sunflower flour can be found in a study by Grasso, Liu, and Methven [4].

2. Materials and Methods
2.1. Muffin Manufacture

Muffins were manufactured according to the recipe and procedure shown by Ateş and Elmacı [15] and the ingredients reported by Grasso, Liu, and Methven [4]. Muffins with sunflower flour were prepared by replacing wheat flour with sunflower flour at either 15% or 30% [4]. The following ingredients made up 100 g of control dough: 28.2 g sugar, 24.4 g wheat flour, 20.7 g whole egg, 15.8 g sunflower oil, 8.6 g water, 1.2 g skimmed milk powder, 0.9 g baking powder, and 0.2 g salt. Experimental muffins were prepared by replacing wheat flour with sunflower flour at either 15% (3.7 g/100 g) or 30% (7.3 g/100 g) [4]. Briefly, egg and sugar were mixed for 1 min with a Kenwood Hand Mixer (HM520, Reading, UK) at low speed. Then, oil, milk powder in water, flour only (for control muffins) or flour and sunflower flour (for 15% and 30%), baking powder, and salt were added. The ingredients were mixed for 3 min at high speed. Batter portions of 40 g ± 0.5 g were baked in paper muffin cases placed onto muffin trays in batches of 12 units in a pre-heated, ventilated oven (Kwick_Co, Salva, Gipuzkoa, Spain) for 20 min at 190 °C. After 1 h of cooling time, the muffins were kept in sealed plastic bags to prevent moisture loss.

2.2. Dietary Fibre Content

Dietary fibre content was evaluated according to Grasso, Pintado, Pérez-Jiménez, Ruiz-Capillas, and Herrero [6]. Duplicate measurements were carried out for each sample, and results were expressed as g/100 g of sample.

2.3. Protein Content

Protein content was measured in duplicate with a Nitrogen Determinator LECO FP-2000 (Leco Corporation, St Joseph, MI, USA). The factor used to convert nitrogen content to protein was 6.25, and the results were expressed as g/100 g of the sample.

2.4. Mineral Content

For mineral content determination, freeze-dried samples (Lyophilizer Telstar-Cryodos Equipment, Tarrasa, Spain) were prepared by acid digestion with nitric acid in a microwave digestion system (ETHOS 1, Milestone, Srl, Sorisole, Italy). The minerals were quantified on a ContrAA 700 High-Resolution Continuum Source spectrophotometer (Analytik Jena AG, Jena, Germany) equipped with a Xenon short-arc lamp (GLE, Berlin, Germany). Three determinations were carried out per sample to measure Calcium (Ca), Magnesium (Mg), Sodium (Na), Potassium (K), Phosphorus (P), Iron (Fe), Zinc (Zn), Copper (Cu), and Manganese (Mn). The determinations were made in duplicate, and the results were expressed as mg/100 g of the sample. More information on the mineral content analysis can be found in a study by Sánchez-Faure et al. [16].

2.5. Amino Acid Content

Amino acid content was determined and measured using ninhydrin derivative reagent and separated by means of cation-exchange chromatography, using a Biochron 20 automatic amino-acid analyser (Amersham Pharmacia Biotech. Biocom, Uppsala, Sweden) where we injected the extract of samples that was dried and hydrolysed in vacuum-sealed glass tubes at 110 °C for 22 h in the presence of 6 N HCl containing 0.1% phenol and nor leucine (Sigma Aldrich, Inc.) as the internal standard. After hydrolysis, samples were again vacuum-dried, dissolved in application buffer, and injected onto a Biochrom 20 amino-acid analyser (Pharmacia, Barcelona, Spain). A mixture of amino acids was used as the standard (Sigma Aldrich, Inc., Madrid, Spain). The determinations were made in duplicate, and the results expressed as mg/g of the sample.

2.6. Antioxidant Activity

For the determination of antioxidant capacity, an aqueous-organic extraction was carried out in duplicate following the methodology of Jiménez et al. [17].

2.6.1. Ferric-Reducing Antioxidant Power (FRAP) Assay

FRAP reagent, freshly prepared and warmed to 37 °C, was mixed (150 µL) with distilled water (15 µL) and the test sample, Trolox, or appropriate blank solvent (5 µL). Readings at 595 nm in a Synergy MX (BioTek, Madrid, Spain) spectrophotometer after 30 min were selected to calculate the FRAP values. Results were expressed as µg eq Trolox/mg after interpolating in the calibration curve.

2.6.2. Photo Chemiluminescence (PCL) Assay

This assay was used to determine antioxidant capacity using an automated photo chemiluminescent system (Photochem, Analytik Jena Model AG; Analytic Jena USA, The Woodlands, TX, USA), which measures the capacity to quench free radicals. This method is based on controlled photochemical generation of radicals, part of which is quenched by the antioxidant, and the remaining radicals are quantified by a sensitive chemiluminescence-detection reaction. Results were expressed as µg eq Trolox/mg sample (liposoluble fraction) and µg eq ascorbic acid/mg (hydro soluble fraction).

2.7. Statistical Analysis

The baking experiment was repeated twice on two different days. One-way analysis of variance (ANOVA) was carried out to evaluate differences between formulations using the SPSS program (v.22, IBM SPSS Inc., Chicago, IL, USA). To compare mean values between formulations, least squares differences and Tukey's HSD tests were used to identify significant differences ($p < 0.05$) between formulations.

3. Results and Discussion

The fibre and protein content of the sunflower flour and the three muffins are shown in Table 1. Soluble dietary fibre was below the limit of detection, and this was expected, due to the mainly insoluble nature of the fibre reported in sunflower by-products [2]. The insoluble dietary fibre and protein content increased with increasing sunflower flour inclusion. Both muffins with sunflower flour provide at least 3% fibre; therefore, they would represent "a source of fibre", according to the current EU regulations [18]. This is a positive result, as food industry by-products could be used as ingredients to enhance the nutritional content of baked goods, such as muffins, as recently shown with spent coffee grounds [13] and grape pomace [19].

Table 1. Dietary fibre and protein content (g/100 g of sample) of sunflower flour and muffins.

	Sunflower Flour	Control	15%	30%
Soluble dietary fibre	1.84 ± 0.15	<LOD	<LOD	<LOD
Insoluble dietary fibre	24.55 ± 1.67	2.27 ± 0.17 [c]	3.60 ± 0.10 [b]	4.58 ± 0.17 [a]
Protein	30.99 ± 0.16	7.08 ± 0.11 [c]	8.33 ± 0.12 [b]	9.52 ± 0.07 [a]

LOD: limit of detection. Data are expressed as means ± standard deviation ($n = 4$). Different letters indicate significant differences ($p < 0.05$) for the same analysis.

The mineral content of the sunflower flour and the muffins is shown in (Table 2). Sunflower oil cake, on a dry basis, contains 0.48 g/100 g calcium, 0.84 g/100 g phosphorus, 0.44 g/100 g magnesium, and 3.49 g/100 g potassium [20]. As a result of the sunflower flour inclusion, all minerals subject to analysis significantly increased (except for sodium), and both 15% and 30% muffins can be considered "a source of" or "high in" several minerals, according to the current EU regulations [21]. The 15% and 30% muffins could be considered a source of magnesium (>56.2 mg/100 g) and manganese (>0.4 mg/100 g), as well as high in phosphorous (>210 mg/100 g). The 30% muffins could be considered a source of potassium (>300 mg/100 g), iron (>2.2 mg/100 g), and zinc (>1.6 mg/100 g). Finally, the 15% muffins can be considered a source of copper (>0.2 mg/100 g), and the 30% muffins can be considered high in copper (>0.4 mg/100 g). Mehta et al. [22] also reported a significant mineral content increase with the addition of tomato pomace in bread and muffins, so food industry by-products could be used as ingredients to increase the micronutrient value of appropriately reformulated baked goods.

Table 2. Mineral content (mg/100 g sample) of sunflower flour and muffins.

Mineral	Sunflower Flour	Control	15%	30%
Calcium	54.00 ± 1.16	65.45 ± 1.63 [c]	78.50 ± 2.62 [b]	97.21 ± 3.11 [a]
Magnesium	64.17 ± 1.91	16.44 ± 1.24 [c]	68.34 ± 7.67 [b]	106.4 ± 4.93 [a]
Sodium	2.04 ± 0.04	400.0 ± 24.9 [a]	397.5 ± 14.8 [a]	407.7 ± 18.8 [a]
Potassium	213.0 ± 14.5	137.6 ± 2.85 [c]	203.1 ± 7.64 [b]	303.1 ± 6.51 [a]
Phosphorus	70.39 ± 3.99	195.4 ± 10.6 [c]	223.7 ± 10.2 [b]	254.7 ± 9.87 [a]
Iron	1.48 ± 0.02	1.01 ± 0.12 [c]	1.56 ± 0.19 [b]	2.49 ± 0.28 [a]
Zinc	1.41 ± 0.00	0.74 ± 0.03 [c]	1.25 ± 0.02 [b]	1.80 ± 0.02 [a]
Copper	0.31 ± 0.00	0.17 ± 0.01 [b]	0.23 ± 0.01 [b]	0.43 ± 0.01 [a]
Manganese	0.38 ± 0.00	0.24 ± 0.01 [c]	0.49 ± 0.02 [b]	0.78 ± 0.03 [a]

Data are expressed as means ± standard deviation ($n = 4$). Different letters indicate significant differences ($p < 0.05$) for the same mineral among muffins.

Table 3 shows the results of the amino acid analysis performed on the sunflower flour and the muffins. For four non-essential amino acids (aspartic acid, glycine, alanine, arginine) and three essential amino acids (valine, methionine, leucine), the addition of sunflower flour resulted in a significant amino acid increase, compared to the control (30–60% increase between control and 30% muffins). Additionally, for these amino acids, the 30% muffins showed significantly higher content than the 15% muffins. For four amino acids (essential threonine, isoleucine, phenylalanine, and non-essential tyrosine), there was a significant amino acid increase only between the control and 30% muffins (increase in the range 27–39%), but the amino acid content was similar between the control and 15% muffins. For the non-essential amino acids glutamic acid and proline, there was no significant difference in terms of content across the three muffins, while cysteine was the only amino acid where a non-significant decrease was recorded in sunflower muffins compared to the control. For the essential amino acids histidine and lysine, the addition of sunflower flour resulted in similar levels in the 15% and 30% muffins, and in lower levels in the control muffins (26–33% increase between control and 30% muffins). Finally, the content of the non-essential amino acid serine was highest in the 30% muffins and lowest in the control muffins, while the 15% muffins had an intermediate serine content and were not significantly different from the control or 30% muffins. The addition of distillers' grain flour was reported to improve the amino acid content of muffins, especially the levels of threonine, serine, glutamic acid, alanine, methionine, leucine, and histidine [23]. As reported by Siddiqi et al. [24], the amino acid composition of wheat is quite unbalanced, lacking the essential amino acids lysine, threonine, and methionine. Since sunflower flour addition increased the content of these amino acids lacking in wheat, the incorporation of sunflower flour could help to achieve a more balanced amino acid profile in muffins.

Table 3. Amino acid content (mg/g sample) of sunflower flour and muffins, and percentage amino acid content change between sunflower flour muffins and control.

	Amino Acid	Sunflower Flour	Control	15%	% Change 15%-Control	30%	% Change 30%-Control
Non-essential amino acids	Aspartic acid	24.66 ± 0.28	6.30 ± 0.37 [c]	7.58 ± 0.13 [b]	+20	8.82 ± 0.17 [a]	+40
	Serine	11.67 ± 0.62	5.57 ± 0.29 [b]	5.79 ± 0.05 [ab]	+4	6.30 ± 0.07 [a]	+13
	Glutamic acid	49.63 ± 2.21	18.43 ± 1.11 [a]	19.51 ± 0.32 [a]	+6	20.74 ± 0.44 [a]	+13
	Proline	17.55 ± 0.81	7.88 ± 0.30 [a]	8.47 ± 0.01 [a]	+7	8.53 ± 0.22 [a]	+8
	Glycine	14.42 ± 0.31	2.76 ± 0.16 [c]	3.78 ± 0.05 [b]	+37	4.49 ± 0.04 [a]	+63
	Alanine	11.55 ± 0.26	3.69 ± 0.19 [c]	4.51 ± 0.03 [b]	+22	5.03 ± 0.02 [a]	+36
	Cysteine	2.28 ± 0.11	0.86 ± 0.11 [a]	0.66 ± 0.02 [a]	−23	0.72 ± 0.01 [a]	−16
	Tyrosine	7.39 ± 0.24	1.36 ± 0.13 [b]	1.36 ± 0.04 [b]	0	1.76 ± 0.11 [a]	+29
	Arginine	16.19 ± 0.45	2.15 ± 0.07 [c]	2.64 ± 0.06 [b]	+23	3.45 ± 0.23 [a]	+60
Essential amino acids	Valine	11.71 ± 0.33	3.65 ± 0.12 [c]	4.06 ± 0.08 [b]	+11	4.74 ± 0.05 [a]	+30
	Methionine	4.67 ± 0.14	0.73 ± 0.09 [c]	1.05 ± 0.11 [b]	+44	1.33 ± 0.10 [a]	+82
	Isoleucine	9.80 ± 0.29	2.35 ± 0.15 [b]	2.69 ± 0.09 [b]	+14	3.27 ± 0.06 [a]	+39
	Leucine	16.86 ± 0.47	4.51 ± 0.03 [c]	4.89 ± 0.14 [b]	+8	5.89 ± 0.16 [a]	+31
	Threonine	10.01 ± 0.28	3.05 ± 0.16 [b]	3.31 ± 0.06 [b]	+9	3.88 ± 0.06 [a]	+27
	Phenylalanine	14.23 ± 0.38	3.05 ± 0.30 [b]	3.30 ± 0.06 [b]	+8	4.12 ± 0.18 [a]	+35
	Histidine	7.24 ± 0.33	2.05 ± 0.12 [b]	2.49 ± 0.02 [a]	+21	2.72 ± 0.07 [a]	+33
	Lysine	10.66 ± 0.41	4.33 ± 0.36 [b]	5.13 ± 0.07 [a]	+18	5.47 ± 0.12 [a]	+26

Data are expressed as means ± standard deviation (n = 4). Different letters indicate significant differences ($p < 0.05$) for the same amino acid among muffins.

Table 4 shows the Food and Agriculture Organization (FAO) adult amino acid requirements [25], the amino acid content of sunflower flour, and the amino acid score of sunflower flour. The amino acid score determines the effectiveness with which absorbed dietary nitrogen can meet the indispensable amino acid requirement at the safe level of protein intake [25]. This is achieved by a comparison of the content of the amino acid in the protein with its content in the requirement pattern [25]. It can be seen that the first limiting amino acid in sunflower flour is lysine, while all the other sunflower amino acids have a score of at least one, and up to almost two.

Table 4. Adult FAO amino acid requirements, amino acid content, and amino acid score of sunflower flour.

Amino Acid	Adult Requirement (FAO) mg/g Protein	Sunflower Flour mg/g Protein	AA Score Sunflower Flour
Methionine + cysteine	22	22.43	1.02
Isoleucine	30	31.62	1.05
Leucine	59	54.40	0.92
Threonine	23	32.30	1.40
Phenylalanine + tyrosine	38	69.76	1.84
Histidine	15	23.36	1.56
Lysine	45	34.40	0.76
Valine	39	37.79	0.97

Table 5 shows the results of the antioxidant capacity tests carried out on the muffins. The addition of sunflower flour resulted in a dose-dependent significant increase in the antioxidant activity of the muffins, with the 30% muffins showing significantly higher antioxidant capacity than the 15% muffins, with the 15% muffins showing higher values than the control muffins. Previous results on biscuits with sunflower flour also showed an increased antioxidant capacity through the 2,2-Diphenyl-1-picrylhydrazyl (DPPH) assay and the cupric reducing antioxidant capacity (CUPRAC) assay [5], which was related to the higher total phenolic content of the sunflower flour compared to wheat flour. It has been shown that sunflower meal is a good source of phenolic compounds with high antioxidant capacity (such as chlorogenic, caffeic, p-hydroxybenzoic, p-coumaric, cinamic, m-hydroxybenzoic, vanillic, syringic, transcinnamic, isoferulic, and sinapic acids [26]), while wheat flour has a very low polyphenol content [27]. An increase in the natural antioxidant content of baked goods could help in terms of shelf life by lowering the oxidation of fats and would help to keep the food as a "clean label" [28].

Table 5. Antioxidant capacity of muffins evaluated by FRAP (Ferric Reducing Antioxidant Power) and PCL (photo chemiluminescence).

	Control	15%	30%
FRAP (µg eq Trolox/mg sample)	1.52 ± 0.16 [c]	2.99 ± 0.17 [b]	4.36 ± 0.36 [a]
PCL—liposoluble (µg eq Trolox/mg sample)	* nd	0.44 ± 0.04 [b]	1.20 ± 0.13 [a]
PCL—hydrosoluble (µg eq ascorbic acid/mg sample)	0.04 ± 0.02 [c]	6.04 ± 0.25 [b]	18.79 ± 1.07 [a]

* nd: not detected. Data are expressed as means ± standard deviation (n = 4). Different letters indicate significant differences ($p < 0.05$) for the same analysis.

4. Conclusions

The use of upcycled ingredients in baked goods, such as sunflower flour in muffins, could result in several nutritional advantages as here shown, such as improved fibre content, mineral content, amino acid profile, and antioxidant activity. The development of baked goods with a balanced amino acid profile through the use of upcycled ingredients is of particular interest and should be explored in further research. Upcycled ingredients could be promoted on the packaging if they are used at sufficient levels to make nutrition claims, such as those on fibre, protein, or mineral content. The sensory quality of muffins with sunflower flour was investigated through a Quantitative Descriptive Analysis by Grasso, Liu, and Methven [4]. Results showed that the 15% muffins were the most similar to the control, and that further reformulation was needed to improve the sunflower samples. Future efforts should also concentrate on developing recipes that are healthier overall (for example, by using less sugar in the batter). A holistic and multi-disciplinary approach should be used in the development of such novel baked goods, considering several aspects at once, such as the nutritional profile, as well as sensory and technological aspects, to create new foods that deliver in taste and that will be well-received by consumers..

Author Contributions: Conceptualization, S.G. and C.R.-C.; methodology, S.G. and C.R.-C.; software, S.G., T.P., J.P.-J., A.M.H.; validation, T.P., J.P.-J., A.M.H.; formal analysis, S.G., T.P., J.P.-J., A.M.H.; investigation, S.G., T.P., J.P.-J., A.M.H. and C.R.-C.; resources, S.G., T.P., J.P.-J., A.M.H. and C.R.-C.; data curation, S.G., T.P., J.P.-J., A.M.H.; writing—original draft preparation, S.G., T.P., J.P.-J., A.M.H. and C.R.-C.; writing—review and editing, S.G.; visualization, S.G., T.P., J.P.-J., A.M.H. and C.R.-C.; supervision, S.G. and C.R.-C.; project administration, S.G. and C.R.-C.; funding acquisition, S.G. and C.R.-C. All authors have read and agreed to the published version of the manuscript.

Funding: The APC for this article was funded by the University of Reading Open Access Fund, OA177790.

Institutional Review Board Statement: Not applicable.

Informed Consent Statement: Not applicable.

Data Availability Statement: Not applicable.

Acknowledgments: The authors are grateful to the company Planetarians for donating the defatted sunflower seed flour.

Conflicts of Interest: The authors declare no conflict of interest.

References

1. Yegorov, B.; Turpurova, T.; Sharabaeva, E.; Bondar, Y. Prospects of using by-products of sunflower oil production in compound feed industry. *J. Food Sci. Technol. Ukr.* **2019**, *13*, 106–113. [CrossRef]
2. Wanjari, N.; Waghmare, J. Phenolic and antioxidant potential of sunflower meal. *Adv. Appl. Sci. Res.* **2015**, *6*, 221–229.
3. Manchuliantsau, A.; Tkacheva, A. Upcycling Solid Food Wastes and by-Products into Human Consumption Products. U.S. Patent 16/370,896, 15 February 2021.
4. Grasso, S.; Liu, S.; Methven, L. Quality of muffins enriched with upcycled defatted sunflower seed flour. *LWT* **2020**, *119*, 108893. [CrossRef]
5. Grasso, S.; Omoarukhe, E.; Wen, X.; Papoutsis, K.; Methven, L. The use of upcycled defatted sunflower seed flour as a functional ingredient in biscuits. *Foods* **2019**, *8*, 305. [CrossRef]
6. Grasso, S.; Pintado, T.; Pérez-Jiménez, J.; Ruiz-Capillas, C.; Herrero, A.M. Potential of a Sunflower Seed By-Product as Animal Fat Replacer in Healthier Frankfurters. *Foods* **2020**, *9*, 445. [CrossRef]
7. Storup, K.; Mattfolk, K.; Voinea, D.; Jakobsen, B.; Bain, M.; REVERTÉ CASAS, M.; Oliveira, P. Combating Food Waste: An Opportunity for the EU to Improve the Resource-Efficiency of the Food Supply Chain. *Eur. Court. Audit. Spec. Ed.* **2016**, *34*, 6–7.
8. US Environmental Protection Agency. Food Recovery Hierarchy. In *Sustainable Management of Food*; US Environmental Protection Agency: Washington, DC, USA, 2015.
9. Henchion, M.; Hayes, M.; Mullen, A.M.; Fenelon, M.; Tiwari, B. Future Protein Supply and Demand: Strategies and Factors Influencing a Sustainable Equilibrium. *Foods* **2017**, *6*, 53. [CrossRef] [PubMed]
10. SACN. *Scientific Advisory Committee on Nutrition-Carbohydrates and Health*; SACN: London, UK, 2015.
11. Heo, Y.; Kim, M.J.; Lee, J.W.; Moon, B. Muffins enriched with dietary fiber from kimchi by-product: Baking properties, physical–chemical properties, and consumer acceptance. *Food Sci. Nutr.* **2019**, *7*, 1778–1785. [CrossRef] [PubMed]
12. Marchetti, L.; Califano, A.; Andrés, S. Partial replacement of wheat flour by pecan nut expeller meal on bakery products. Effect on muffins quality. *LWT Food Sci. Technol.* **2018**, *95*, 85–91. [CrossRef]
13. Severini, C.; Caporizzi, R.; Fiore, A.G.; Ricci, I.; Onur, O.M.; Derossi, A. Reuse of spent espresso coffee as sustainable source of fibre and antioxidants. A map on functional, microstructure and sensory effects of novel enriched muffins. *LWT* **2020**, *119*, 108877. [CrossRef]
14. Gómez, M.; Martinez, M.M. Fruit and vegetable by-products as novel ingredients to improve the nutritional quality of baked goods. *Crit. Rev. Food Sci. Nutr.* **2018**, *58*, 2119–2135. [CrossRef] [PubMed]
15. Ateş, G.; Elmacı, Y. Physical, chemical and sensory characteristics of fiber-enriched cakes prepared with coffee silverskin as wheat flour substitution. *J. Food Meas. Charact.* **2019**, *13*, 755–763. [CrossRef]
16. Sánchez-Faure, A.; Calvo, M.M.; Pérez-Jiménez, J.; Martín-Diana, A.B.; Rico, D.; Montero, M.P.; Gómez-Guillén, M.d.C.; López-Caballero, M.E.; Martínez-Alvarez, O. Exploring the potential of common iceplant, seaside arrowgrass and sea fennel as edible halophytic plants. *Food Res. Int.* **2020**, *137*, 109613. [CrossRef]
17. Jiménez, J.P.; Serrano, J.; Tabernero, M.; Arranz, S.; Díaz-Rubio, M.E.; García-Diz, L.; Goñi, I.; Saura-Calixto, F. Effects of grape antioxidant dietary fiber in cardiovascular disease risk factors. *Nutrition* **2008**, *24*, 646–653. [CrossRef] [PubMed]
18. EFSA. Regulation 1924/2006 on nutrition and health claims made on foods. *Off. J. Eur. Union* **2006**, *L404*, 9–25.
19. Ortega-Heras, M.; Gomez, I.; De Pablos-Alcalde, S.; Gonzalez-Sanjose, M.L. Application of the Just-About-Right Scales in the Development of New Healthy Whole-Wheat Muffins by the Addition of a Product Obtained from White and Red Grape Pomace. *Foods* **2019**, *8*, 419. [CrossRef]
20. Ratcliff, R.K. Nutritional value of sunflower meal for ruminants. Ph.D. Thesis, Texas Tech University, Lubbuck, TX, USA, 1977.
21. EFSA. EU Register on Nutrition and Health Claims. Available online: https://ec.europa.eu/food/safety/labelling_nutrition/claims/nutrition_claims_en (accessed on 15 February 2021).
22. Mehta, D.; Prasad, P.; Sangwan, R.S.; Yadav, S.K. Tomato processing byproduct valorization in bread and muffin: Improvement in physicochemical properties and shelf life stability. *J. Food Sci. Technol.* **2018**, *55*, 2560–2568. [CrossRef]
23. Reddy, N.; Pierson, M.; Cooler, F. Supplementation of wheat muffins with dried distillers grain flour. *J. Food Qual.* **1986**, *9*, 243–249. [CrossRef]
24. Siddiqi, R.A.; Singh, T.P.; Rani, M.; Sogi, D.S.; Bhat, M.A. Diversity in Grain, Flour, Amino Acid Composition, Protein Profiling, and Proportion of Total Flour Proteins of Different Wheat Cultivars of North India. *Front. Nutr.* **2020**, *7*. [CrossRef]
25. Joint FAO; World Health Organization. *Protein and Amino Acid Requirements in Human Nutrition: Report of a Joint FAO/WHO/UNU Expert Consultation*; World Health Organization: Geneva, Switzerland, 2007.
26. Lomascolo, A.; Uzan-Boukhris, E.; Sigoillot, J.-C.; Fine, F. Rapeseed and sunflower meal: A review on biotechnology status and challenges. *Appl. Microbiol. Biotechnol.* **2012**, *95*, 1105–1114. [CrossRef]
27. Vaher, M.; Matso, K.; Levandi, T.; Helmja, K.; Kaljurand, M. Phenolic compounds and the antioxidant activity of the bran, flour and whole grain of different wheat varieties. *Procedia Chem.* **2010**, *2*, 76–82. [CrossRef]
28. Do Nascimento, K.; Paes, S.; Augusta, I.M. A review "Clean Labeling": Applications of natural ingredients in bakery products. *J. Food Nutr. Res.* **2018**, *6*, 285–294. [CrossRef]

Article

Valorization of Mango By-Products to Enhance the Nutritional Content of Maize Complementary Porridges

Juliana Mandha [1,2], Habtu Shumoy [1], Athanasia O. Matemu [2] and Katleen Raes [1,*]

[1] Research Unit VEG-i-TEC, Department of Food Technology, Safety and Health, Ghent University Campus Kortrijk, Sint-Martens-Latemlaan 2B, 8500 Kortrijk, Belgium; Juliana.Mandha@UGent.be (J.M.); Habtu.Shumoy@UGent.be (H.S.)
[2] Department of Food Biotechnology and Nutritional Sciences, Nelson Mandela African Institution of Science and Technology, Arusha 447, Tanzania; athanasia.matemu@nm-aist.ac.tz
* Correspondence: Katleen.Raes@UGent.be; Tel.: +32-56-322008

Citation: Mandha, J.; Shumoy, H.; Matemu, A.O.; Raes, K. Valorization of Mango By-Products to Enhance the Nutritional Content of Maize Complementary Porridges. *Foods* **2021**, *10*, 1635. https://doi.org/10.3390/foods10071635

Academic Editors: Simona Grasso, Konstantinos Papoutsis, Claudia Ruiz-Capillas and Ana Herrero Herranz

Received: 11 June 2021
Accepted: 13 July 2021
Published: 15 July 2021

Publisher's Note: MDPI stays neutral with regard to jurisdictional claims in published maps and institutional affiliations.

Copyright: © 2021 by the authors. Licensee MDPI, Basel, Switzerland. This article is an open access article distributed under the terms and conditions of the Creative Commons Attribution (CC BY) license (https://creativecommons.org/licenses/by/4.0/).

Abstract: Mango by-products are disregarded as waste contributing to greenhouse gas emissions. This study used mango seed and kernel to enhance the nutritional content of maize complementary porridges. Composite maize-based porridges (MBP) were formulated by fortifying maize flour with fine ground mango seed and kernel at different levels (31%, 56%, 81%). The by-products and formulated porridges were characterized for their nutritional composition, mineral content, total phenolic content, and antioxidant capacity. Furthermore, the bioaccessibility of essential minerals during in vitro gastrointestinal digestion of the formulated porridges was determined using inductively coupled plasma optical emission spectrometry. Mango seed had a high fat (12.0 g/100 g dw) and protein content (4.94 g/100 g dw), which subsequently doubled the fat content of the porridges. Mango by-products increased the total phenolic content of maize porridge by more than 40 times and the antioxidant capacity by 500 times. However, fortification with mango by-products significantly decreased the bioaccessibility of minerals, especially manganese, copper, and iron, as the highest percentages of insoluble minerals were recorded in MBP 81 at 78.4%, 71.0%, and 62.1%, respectively. Thus, the results suggest that mango seed and kernel could increase the nutritional value of maize porridge, but fortification should be done at lower levels of about 31–56%.

Keywords: mango by-products; valorization; fortification; value addition; in vitro digestion; maize porridge

1. Introduction

Fruit processing from industry and agriculture generates high amounts of by-products that are often disregarded as waste. This waste is decomposed in open fields or landfills and emits dangerous greenhouse gasses, such as methane, carbon dioxide, and nitrogen dioxide, which cause environmental problems [1]. Strategies to recycle of these fruit by-products back into the food system are essential to protect the environment and to reduce the social and economic burden.

Mango (*Mangifera indica* L.) is among the top 10 fruits of major economic importance cultivated in Africa. It belongs to the family of *Anacardiaceae* and is considered a high-quality fruit as the 'King of fruits' in the Orient [2]. Over 8 million tonnes of mangoes are produced in the region per year [3] and are often consumed either as fresh fruit or processed into various products such as pulp/juice, puree, pickle, jam, powder, and nectar. Mango processing generates approximately 3 million tonnes of the by-products (seeds, peels, kernels) making up 25–40% of fresh fruit. These are rich sources of nutrients and bioactive compounds, i.e., mango seed and kernel are excellent sources of dietary fiber, carotenoids, protein, fat, minerals, and phenolic compounds [4–6]. The bioactive compounds in mango by-products have antioxidant, antibacterial, cardioprotective, anti-inflammatory, and antiproliferative properties [7] that protect against chronic non-communicable diseases. Hence, mango by-products could be transformed into useful products.

One possibility of utilizing mango by-products could be to use them to enrich the nutrient content of maize porridges through fortification. Maize (*Zea mays*) is a staple food that is largely consumed in developing countries (52 g/person/day) [8] and is the major complementary porridge for children. This food crop is mainly composed of carbohydrates (70%) and lacks essential micronutrients [9,10]. Food-to-food fortification is currently viewed as an emerging strategy against micronutrient deficiencies [11] and the use of locally available products is a cost-effective and sustainable strategy. Research on mango as a fortificant is limited and studies have mainly focused on the use mango pulp [12] but not its seed and kernel. Some authors have shown the use of mango pomaces to enrich wheat flour in the production of bakery products, such as cookies and biscuits [13,14] and starch-molded snacks [15], but not in maize porridges. Besides, investigations on how processes such as fortification and cooking alter the interactions of nutrients in the food matrix during in vitro gastrointestinal digestion broaden our understanding on how to utilize these by-products.

Therefore, the objectives of this study were to investigate the nutritional value, total phenolic content, antioxidant capacity, and minerals of mango seed, mango kernel, and maize flour, to formulate a composite maize porridge with the mango seed and kernel and to investigate the bioaccessibility of essential minerals in the porridges during in vitro gastrointestinal digestion.

2. Materials and Methods

2.1. Chemicals

ABTS (2,2'-azino-bis(3-ethylbenzothiazoline-6-sulfonic acid)), Trolox (6-hydroxyl-2,5,7,8-tetramethylchroman-2-carboxylic acid), DPPH (2,2-diphenyl-1-picrylhydrazyl), TFA (trifluoroacetic acid), gallic acid, Folin–Ciocalteu reagent (FC), pepsin from porcine gastric mucosa-lyophilized powder, α-amylase and pancreatin from porcine pancreas, heat-stable α-amylase solution, protease, amyloglucosidase solution, 2-(N-Morpholino) ethane-sulfonic acid (MES), and Tris (hydroxymethyl) aminomethane (TRIS) were purchased from Sigma-Aldrich Co. (St. Louis, MO, USA). Technical grades of methanol (MeOH) (>98.5%), ethanol (C_2H_5OH) (93%), sodium hydroxide (NaOH) (97%), sodium chloride (NaCl) (97%), potassium persulfate ($K_2S_2O_8$) (100%), sodium carbonate (Na_2CO_3) (>99%), nitric acid (HNO_3) (>99%), sulfuric acid (H_2SO_4) (>97%), hydrochloric acid (HCl) (37.2% w/w), petroleum ether (C_6H_{14}) (30–60 °C, A.R), Kjeldahl tablet ($CuH_{10}O_9S$), potassium chloride (KCl) (> 99%), sodium hydrogen carbonate ($NaHCO_3$) (>99%), sodium chloride (NaCl) (97%), magnesium chloride ($MgCl_2$) (99%), ammonium chloride (NH_4Cl) (100%), calcium chloride ($CaCl_2$) (>93%), monopotassium phosphate (KH_2PO_4), and indicators (tashiro and phenolphthalein) were purchased from VWR International (Leuven, Belgium). Inductively coupled plasma (ICP) multi-element standard solution IV was procured from Merck KGak (Darmstadt, Germany).

2.2. Plant Materials

Mature mango fruits (*Mangifera indica*, L. cv Kagoogwa) and maize (*Zea mays*) flour were purchased from Nakasero market, Kampala, Uganda (latitude: 00°18′42.34″ N, longitude: 32°34′46.34″ E). The mango fruits (10,000 g) were physically checked for integrity, insect contamination, and size/color uniformity. The screened samples were then packaged in air-tight boxes and cold transported by air to the Research Unit VEG-i-TEC of Ghent University, Kortrijk, Belgium. On arrival, the mango fruits were again inspected, washed, and the peel and fruit mesocarp was removed. The mango kernel was then manually separated from the mango seed and fine ground mango seed and kernel (wet basis) were obtained using a KitchenAid blender (Joseph, MI, USA). The obtained by-products were rapidly stored in sealed plastic bags at −20 °C until further analyses.

2.3. Sample Preparation

Using mass balance, all the porridge formulations were calculated to provide the recommended dietary intake (RDA) for energy (1200 kcal/day) at 300 g wt. Maize flour was substituted for fine ground mango seed and kernel at different levels—31%, 56%, and 81%—to formulate composite maize-blended porridges (MBP): MBP 31, MBP 56, and MBP 81, respectively. Porridge formulations are shown in Table 1. The maize control porridge (MCP) constituted of only maize flour and water (1:10 m/v). The flours were cooked into porridges using a traditional method, which involved cooking at 80–100 °C for 25 min (Figure 1).

Table 1. Formulations of the composite porridges (g per 300 g wt).

Raw Materials	Composite Porridge Formulation			
	MCP	MBP 31%	MBP 56%	MBP 81%
Maize flour (g)	300	206.2	131.9	57.6
Mango seed (g)	0	66.1	129.3	192.5
Mango kernel (g)	0	27.7	38.8	49.9
Energy (kcal)	1213.7	1200	1200	1200

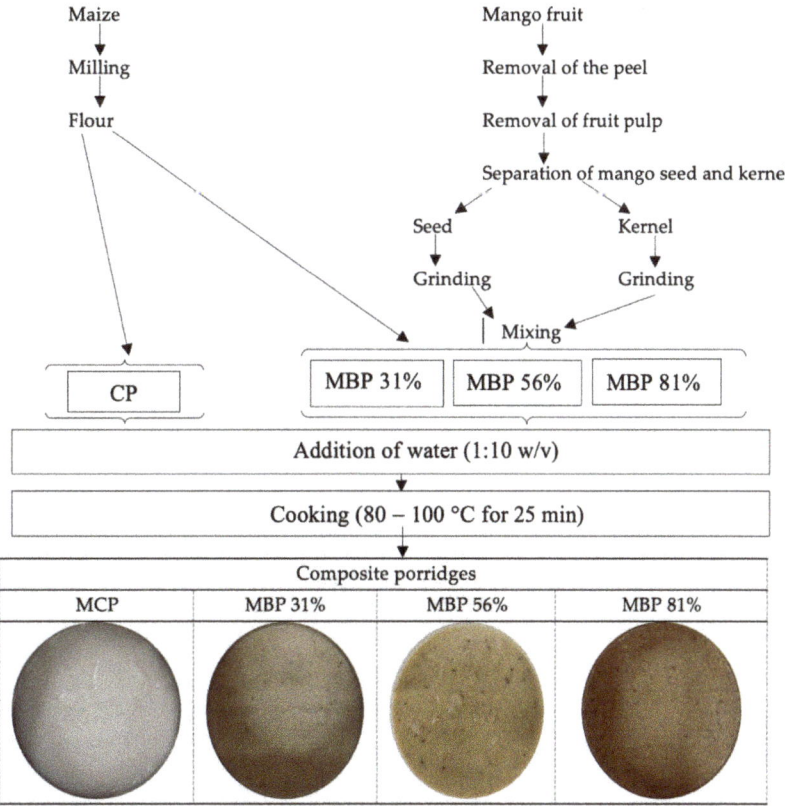

Figure 1. Flowchart showing the preparation of maize composite porridges (wt).

2.4. Determination of Proximate Composition

The amounts of ash, fat, and soluble and insoluble dietary fiber of the raw materials (mango seed, kernel, maize flour) and composite porridges were assessed using AOAC standard methods, namely, 945.46, 920.39, and 991.43, respectively [16]. Moisture and protein concentrations were determined using ISO 1442-1973 and ISO 937-1978, respectively. A factor of 6.25 was used for conversion of nitrogen to crude protein. Digestible carbohydrates were calculated by difference, subtracting the sum of protein, ash, lipid, and total dietary fiber from 100.

2.5. Determination of Mineral Content

The concentrations of essential minerals such as iron (Fe), zinc (Zn), manganese (Mn), copper (Cu), magnesium (Mg), potassium (K), calcium (Ca), and sodium (Na) were determined using the inductively coupled plasma optical emission spectrometry (ICP-OES) (Varian, PTY Ltd., Victoria, Australia) [17]. Briefly, 2 g of mango seed, mango kernel, and maize flour and 5 g of porridge sample were completely carbonized in a high-form porcelain crucible followed by overnight ashing in a muffle oven at 550 °C. The obtained ash was subsequently dissolved in 5 mL of 65% HNO_3, filtered, and its mineral concentration measured using ICP-OES with Thermo iCAP 7200 spectrometer (Thermo Fisher Scientific Inc., Waltham, MA, USA). The ICP-OES was equipped with a peristaltic pump (0.76 mm), cyclonic spray chamber, concentric nebulizer, quartz plasma torch, and 2.0 mm alumina internal diameter injector. The instrumental parameters used were: 1180 W RF power, 12.0 L/min plasma flow rate, 0.5 L/min auxiliary gas flow rate, 0.5 L/min nebulizer flow rate, radial view, 15 min UV exposure time, 5 min VIS exposure time, 10 min warm up, and 40 min wash time. The wavelengths selected for Cu, Fe, Mn, Zn, K, Na, Ca, and Mg were 224.7, 239.6, 260.6, 202.6, 766.5, 589.6, 422.7, and 285.2 nm, respectively. Standard plot analytical curves for each element with a fit factor of above 0.99 were used to calculate the concentration of the elements in the samples in comparison to multi-element stock standard. Results were expressed as mg/100 g dry basis (dw) for the mango seed, kernel, maize flour, and mg/100 g wet basis (wt) for the porridges.

2.6. Determination of Total Phenolic Content

Total phenolic compounds were extracted using 80% methanol as described by Gonzales et al. [18]. Briefly, either 2 g mango seed, 2 g mango kernel, 5 g maize flour, or 5 g porridge sample was added to 15 mL of methanol (MeOH 80%) and homogenized using Ultra Turrax homogenizer (T 18 digital, IKA, Staufen, Germany) at $1422 \times g$ (10,000 rpm) for 45 s and immediately kept on ice for 15 min. The homogenate was then centrifuged (Hermle Z300K, Hermle Labortechnik GmbH., Wehingen, Germany) at $2540 \times g$ (4000 rpm), 4 °C for 15 min and then filtered (filter paper particle retention 5–13 µm, VWR International., Leuven, Belgium) in the dark. The pellets were re-extracted with 10 mL MeOH (80%) using the same procedure. The collected supernatants were then pooled, filled to 25 mL with MeOH, and stored at −20 °C in the dark until further analyses.

Total phenolic content (TPC) was determined using the Folin–Ciocalteu (FC) method according to Huynh et al. [19] and Singleton et al. [20]. In brief, 1 mL methanolic extract was added to 1 mL deionized water and vortex (Vortex-genie 2, Thermo Fisher Scientific Inc., Waltham, MA, USA) mixed with 0.5 mL of 10 times diluted FC reagent in deionized water. After 6 min of standing, 1.5 mL Na_2CO_3 (20% w/v) and 1 mL deionized water were added, vortex mixed, and incubated in the dark for 2 h at room temperature. The absorbance of the mixture was then measured at 760 nm using a Spectrophotometer (Shimadzu UV-1800 spectrophotometer, Kioto, Japan) and the TPC concentration was expressed as mg gallic acid equivalent (GAE)/100 g of dried sample (dw).

2.7. Determination of Antioxidant Capacity

2.7.1. Using ABTS Radical Scavenging Activity

Stock solution of $ABTS^+$ was prepared by mixing equal amounts of 7 mM ABTS radical cation and 2.45 mM of potassium persulfate and allowing them to react for 12−16 h in the dark at room temperature. The working solution was subsequently prepared by diluting the stock solution with MeOH (90%) to an absorbance of 0.70 ± 0.02 at 734 nm equilibrated at 30 °C. Aliquots (20 µL) of each sample extract and standard Trolox solution or MeOH (90%) (blank) were then added to 2 mL of the ABTS$^+$ solution, vortex mixed, and incubated for 5 min in the dark at room temperature. Thereafter, the absorbance of the resulting solution was measured spectrophotometrically (Shimadzu UV-1800 spectrophotometer, Kioto, Japan) at 734 nm, and results were expressed in mg Trolox equivalent (TE)/100 g of dried sample (dw) [21].

2.7.2. Using DPPH Radical Scavenging Activity

The reducing ability of the antioxidants in the samples towards DPPH was measured using the procedure by Brand-Williams et al. [22]. Aliquots (200 µL) of sample extracts/Trolox standards solutions were vortex mixed for 10 s with 4 mL of DPPH solution (prepared by dissolving 3.94 mg DPPH in 100 mL pure MeOH) and incubated for 30 min at room temperature in the dark. The absorbance of the mixture was then measured using a spectrophotometer at 517 nm. Results were expressed in mg TE/100 g of dried sample (dw).

2.8. In Vitro Gastrointestinal Digestion

In vitro digestion of the fortified porridges was done using the international consensus static in vitro digestion method as described by Minekus et al. [23]. Porridge samples were digested in three simulated digestive solutions: salivary, gastric, and intestinal fluids simulating digestion in the mouth, stomach, and intestine, respectively. In the oral phase, 5 g of porridge was mixed thoroughly with 3.5 mL of simulated salivary fluid (KCl–15.1 mmol/L, KH_2PO_4–3.7 mmol/L, $NaHCO_3$–13.6 mmol/L, $MgCl_2$–0.15 mmol/L, $(NH_4)_2CO_3$–0.06 mmol/L, pH 7), 0.5 mL α-amylase (0.15 g/mL–1500 units/mL), and 1000 µL 0.3 M $CaCl_2$. The mixture was thoroughly homogenized, pH was adjusted to 7 using either 1 M NaOH or 1 M HCl, and then the mixture was incubated for 2 min at 37 °C with constant shaking in a warm water bath (Memmert WNB 45, Schwabach, Germany).

This was followed by the gastric phase, whereby the oral bolus was mixed with 7.5 mL of simulated gastric fluid (KCl–6.9 mmol/L, KH_2PO_4–0.9 mmol/L, $NaHCO_3$–25 mmol/L, NaCl–47.2 mmol/L, $MgCl_2$–0.1 mmol/L, $(NH_4)_2CO_3$–0.5 mmol/L, pH 3), 1.6 mL of porcine pepsin (0.00781 g/mL–25,000 units/mL), and 700 µL 0.3 M $CaCl_2$ made in simulated gastric fluid. The pH was adjusted to 3 using 1 M HCl. The mixture was then incubated at 37 °C in a shaking water bath for 90 min. Next, dialysis bags of 15.5 cm (molecular weight cut-off (MWCO), 12–14 kDa) containing 5.5 mL 0.9% NaCl and 5.5 mL 0.5 M $NaHCO_3$ were inserted in the gastric chyme and incubation continued for further 30 min.

The gastric chyme was then thoroughly mixed with 11 mL of simulated intestinal fluid (KCl–6.8 mmol/L, KH_2PO_4–0.8 mmol/L, $NaHCO_3$–85 mmol/L, NaCl–38.4 mmol/L $MgCl_2$–0.33 mmol/L, pH 7), followed by 5.0 mL pancreatin (800 units/mL, 1 g in 250 mL), 2.5 mL of fresh bile prepared, 40 µL 0.3 M $CaCl_2$, 1.31 mL deionized water, and pH adjusted to 7 using either 1 M NaOH or 1 M HCl. The digestion continued at 37 °C for 2 h in a shaking water bath.

After the intestinal digestion phase, the dialysis bags were taken out, rinsed, and dried using a paper cloth and its contents (dialyzed minerals (D)) were transferred into falcon tubes. The remaining digestion solution was centrifuged at 2540× g (4000 rpm) for 15 min at 4 °C and supernatants had the soluble and non-dialyzable (SND) mineral while the pellets had insoluble minerals (P). The fractions were then oven dried, ashed, and solubilized using 1 M HNO_3. Minerals were determined using inductively coupled plasma optical emission spectroscopy (ICP-OES) as previously described. The percentages of bioaccessible minerals were calculated using Equation (1).

$$D\% = (D \text{ minerals}/\text{Total minerals } (D + SND + P)) \times 100 \tag{1}$$

$$SND\% = (SND \text{ minerals}/\text{Total minerals } (D + SND + P)) \times 100 \tag{2}$$

where D = dialyzed minerals, SND = soluble and non-dialyzable minerals, and P = insoluble minerals.

2.9. Determination of Porridge Viscosity

The porridge viscosity of each formulation was determined using a Brookfield DV2T viscometer (USA) with SNLV spindle at 37 °C, which is a suitable temperature considered for eating porridge. The viscometer was first standardized and the rotational speed ranging from 0.1 to 200 rpm was selected. Results with a torque between 10% and 100% were used and apparent viscosity measured as a function of shear rate (s^{-1}) was presented in mPa·s.

2.10. Statistical Analysis

Data followed a normal distribution and variances were homogeneous. The experimental results were analyzed using one-way analysis of variance (ANOVA) using XL-STAT, (version 2020.1, Addinsoft, Paris, France), to check for any variations among treatments and comparison of means was done using multiple range test (Tukey's HSD test). Significance difference was accepted at $p < 0.05$ and values are expressed as mean ± SD of two independent samples. Graphs were obtained using GraphPad Prism (Version 8.0.0 for macOS, San Diego, CA, USA). All the experiments were carried out in duplicates.

3. Results and Discussion

3.1. Nutrient Composition

Mango seed had the highest fat (12.0 g/100 g dw) and protein content (4.94 g/100 g dw) compared to mango kernel and maize flour (Table 2). The obtained protein values in the mango kernel were lower than the results by Mutua et al. [24] that showed an average of 6.74–9.20%. This could be attributed to differences fruit variety, geographical conditions, and climatic growing conditions. The ash in the mango seed was 1.39 g/100 g dw, similar to Bertha et al. [25] showing 1.29 g/100 g dw. Mango kernel had the highest insoluble (34.6 g/100 g dw) and soluble dietary fiber (5.07 g/100 g dw), while maize flour had the lowest total dietary fiber (4.82 g/100 g dw), although this was higher than prior reported results (2.68 g/100 g) [26]. Soluble fibers are known to reduce blood serum cholesterol, hence preventing the development of non-communicable diseases, and insoluble fibers improve regular bowel movements, hence reducing the risk of constipation and colon cancer by maintaining gut health [27]. Maize flour had a high carbohydrate of 81 g/100 g dw, which was consistent with previous analyses [9,10].

Table 2. Nutrient composition (g/100 g dw) of mango by-products and maize flour. The results are expressed as mean ± SD.

	Mango Kernel	Mango Seed	Maize Flour
Moisture	36.6 ± 0.02	51.7 ± 0.01	6.14 ± 0.04
Carbohydrate	18.5 ± 7.82	6.85 ± 3.15	81.26 ± 0.95
Fat	2.92 ± 0.02	12.0 ± 1.43	4.26 ± 0.63
Protein	1.58 ± 0.07	4.94 ± 0.09	3.29 ± 0.14
Ash	0.70 ± 0.07	1.39 ± 0.06	0.23 ± 0.05
Soluble dietary fiber	5.07 ± 0.76	1.83 ± 0.93	2.21 ± 0.06
Insoluble dietary fiber	34.6 ± 7.04	21.6 ± 7.41	2.61 ± 0.06
Total dietary fiber	39.7 ± 7.80	23.4 ± 8.45	4.82 ± 0.00

After fortification of maize flour with mango seed and kernel, the composite maize porridges varied in their nutrient composition at the different levels of formulations as presented in Figure 2. Mango by-products significantly ($p < 0.05$) increased the protein, fat, and ash of the porridges, and this increment was proportional to the percentage of mango by-products. MBP 81 had the highest fat content at 3.29 g/100 g wt, as compared to MBP 31 at 2.30 g/100 g and the control 1.20 g/100 g ($p = 0.001$). Crude ash, which is an indicator of total minerals, ranged from 1.32 g/100 g wt in MBP 81 to 0.19 g/100 g wt in the MCP (*p* = 0.001), and the protein content was highest in MBP 31 and MBP 56 at 0.38 and 0.39 g/100 g (*p* = 0.004), respectively. Moisture content ranged from 83.6 in composite porridges to 89.4 g/100 g wt in the control. Maize starch granules are known to be insoluble in cold water but upon heating, they absorb water and swell due to the presence of amylose and amylopectin [28].

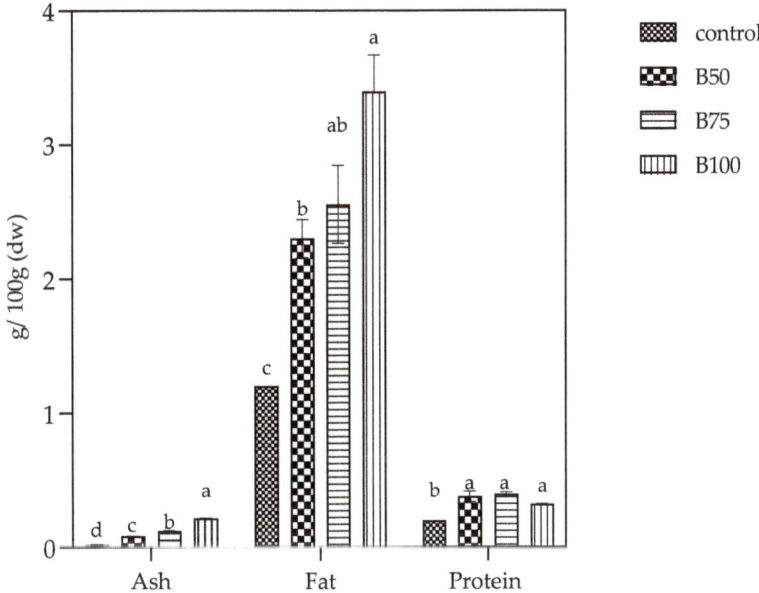

Figure 2. Ash, fat, and protein content of maize porridges. The results are expressed mean ± SD shown as error bars. Different lower-case letters (a–d) represent statistical differences. MCP is maize control porridge; MBP is formulated composite porridges.

3.2. Mineral Content

Minerals are vital nutrients needed by the body to carry out different metabolic functions. The concentration of essential minerals in the mango seed and mango kernel were on average double that of maize flour (Table 3). Minerals such as Fe, Zn, Mn, K, Ca, and Mg are needed for the proper functioning of the body. Iron and Cu were highest in the mango kernel at 4.65 and 5.13 mg/100 g dw, respectively, and lowest in maize flour at 0.69 and 0.57 mg/100 g dw, respectively. Among the abundant elements, K and Ca had the highest values at 405 and 504 mg/100 g dw, respectively, in mango seed.

Table 3. Mineral composition of mango by-products and maize flour (mg/100 g dw). The results are expressed as mean ± SD.

Mineral	Mango Kernel	Mango Seed	Maize Flour
Copper, Cu	5.13 ± 0.27	3.39 ± 0.51	0.57 ± 0.04
Iron, Fe	4.65 ± 0.16	4.29 ± 0.23	0.69 ± 0.06
Manganese, Mn	1.14 ± 0.07	1.42 ± 0.03	0.18 ± 0.04
Zinc, Zn	2.01 ± 0.08	2.01 ± 0.43	0.74 ± 0.09
Potassium, K	163 ± 19	405 ± 28	76.7 ± 4.2
Sodium, Na	103 ± 6	139 ± 10	29.6 ± 0.5
Calcium, Ca	395 ± 24	504 ± 14	82.8 ± 0.4
Magnesium, Mg	76.5 ± 2.8	121 ± 1	19.5 ± 0.2

Table 4 shows the composition of trace (Cu, Fe, Mn, and Zn) and abundant (K, Na, Ca, and Mg) minerals in the porridges. Among the trace minerals, Fe, which plays an important role in the formation of hemoglobin, significantly differed ($p = 0.030$) in the porridges ranging from 0.36 to 0.49 mg/100 wt in the composite porridges compared to the control at 0.29 mg/100 g wt. Additionally, mango seed and kernel significantly ($p < 0.05$) increased the Mn and Zn amounts. Manganese is a scavenger of free radicals in the body and Zn improves the body's immunity, especially in children. Regarding abundant minerals, Mg and Ca had the highest concentrations at 80.3 and 57 mg/100 g wt, respectively, and their amounts did not vary with the fortification of mango fruit by-products. However, there was a significant ($p = 0.001$) increment of K from 7.02 mg/100 g wt in the control to 40.7 mg/100 g wt in MBP 81 porridge.

Table 4. Mineral composition of formulated maize porridges (mg/100 g wt). The results are expressed as mean ± SD. Different lower-case letters represent statistical differences.

Mineral	MCP	MBP 31	MBP 56	MBP 81	p Value
Copper, Cu	0.32 ± 0.02 [a]	0.39 ± 0.04 [a]	0.37 ± 0.03 [a]	0.34 ± 0.02 [a]	0.349
Iron, Fe	0.29 ± 0.04 [b]	0.36 ± 0.02 [ab]	0.49 ± 0.03 [a]	0.47 ± 0.03 [ab]	0.030
Manganese, Mn	0.08 ± 0.01 [c]	0.12 ± 0.01 [bc]	0.15 ± 0.01 [b]	0.22 ± 0.01 [a]	0.001
Zinc, Zn	0.38 ± 0.02 [b]	0.28 ± 0.00 [b]	0.47 ± 0.00 [ab]	1.06 ± 0.21 [a]	0.021
Potassium, K	7.02 ± 0.53 [c]	16.3 ± 0.6 [bc]	25.36 ± 0.93 [b]	40.7 ± 3.7 [a]	0.001
Sodium, Na	22.8 ± 2.5 [a]	24.8 ± 2.6 [a]	19.9 ± 0.2 [a]	23.2 ± 1.3 [a]	0.432
Calcium, Ca	57.4 ± 2.5 [a]	55.1 ± 7.1 [a]	57.0 ± 0.5 [a]	54.8 ± 3.1 [a]	0.956
Magnesium, Mg	63.3 ± 3.8 [a]	68.7 ± 8.3 [a]	71.7 ± 0.1 [a]	80.3 ± 4.0 [a]	0.256

3.3. Total Phenolic Content and Antioxidant Capacity

Phenolic compounds are a major group of phytochemicals commonly distributed in plants. Mango seed was observed as a potentially good source of total phenolic compounds as it had the highest TPC of 3714 mg GAE/100 g dw, followed by mango kernel at 263 mg GAE/100 g dw (Table 5). These results are comparable with previous findings of Bertha el al. [25] and Nguyen el al. [29] wherein the TPC of tropical mango seed was 8.95–12.8 g/100 g dw. The distribution of total phenolic compounds could be affected by the type of crop, as maize flour had the lowest TPC of 18.7 mg GAE/100 g dw. These compounds have anti-inflammatory, antioxidant, and anti-proliferative properties protecting the human body from infections. The antioxidant activity was measured by two different methods: ABTS[+] and DPPH. The mango seed had the highest antioxidant activity according to DPPH at 10,659 mg TE/100 g dw and ABTS[+] at 10,568 mg TE/100 g dw.

Table 5. Total phenolic content and antioxidant capacity of mango by-products, maize flour, and formulated composite porridges. The results are expressed as mean ± SD. Different lower-case letters represent statistical differences.

	TPC (GAE)	Antioxidant Capacity	
		ABTS (TE)	DPPH (TE)
Materials (mg/100 g dw)			
Mango kernel	263 ± 0.59	745 ± 13.45	670 ± 2.93
Mango seed	3714 ± 11.91	10568 ± 73.05	10659 ± 419.69
Maize flour	18.7 ± 0.00	15.66 ± 5.49	4.08 ± 1.30
Formulated Porridges (mg/100 g wt)			
MCP	2.20 ± 0.09 [d]	12.5 ± 0.90 [d]	3.46 ± 0.09 [d]
MBP 31	131 ± 3.46 [c]	853 ± 42.5 [c]	530 ± 1.86 [c]
MBP 56	309 ± 1.87 [b]	1569 ± 80.3 [b]	1245 ± 18.0 [b]
MBP 81	479 ± 11.08 [a]	3846 ± 22.7 [a]	2276 ± 17.9 [a]
p-value	0.000	0.000	0.000

Interestingly, mango by-products increased the TPC of maize porridge by more than 40 times (Table 5). The results showed that higher fortification of maize flour with mango by-products (seed and kernel) resulted in higher TPC values of ($p < 0.05$) from 2.2 mg GAE/100 g wt in MCP to 479 mg GAE/100 g wt in MBP 81. Bertha et al. [25] reported the main antioxidant polyphenol compounds in mango seed as gallic acid > ellagic acid > mangiferin > catechin > gallocatechin > gallocatechin gallate. These compounds could be responsible for the TPC of the porridges. These health-promoting substances are essential in the removal of free radicals as antioxidant agents. Similarly, the antioxidant capacity of the porridges using both $ABTS^+$ and DPPH increased with the fortification of mango by-products. The antioxidant capacity could also be attributed to other compounds besides phenolic compounds in the porridges such as vitamin C and carotenoids, among others. However, since the porridges are cooked at high temperatures (>80 °C), it can be assumed that vitamin C is degraded, and does not play a role in the antioxidant capacity of the porridges.

3.4. Bioaccessibility of Minerals during In Vitro Gastrointestinal Digestion

Mineral content was investigated during in vitro gastrointestinal digestion, which predicts the proportion of nutrients released from food along the gastrointestinal tract making the nutrients available for absorption into the bloodstream for biological functions [23]. Overall, the percentage of bioaccessible, i.e., dialyzable and soluble non-dialyzable minerals, were significantly ($p < 0.05$) higher in MCP as compared to the MBP composite porridges (Figure 3). Among the essential micro minerals, zinc had the highest dialyzable percentage (38.9%), followed by Cu (34.0%) and Fe (32.7%). Fortification with mango by-products significantly ($p < 0.05$) decreased the bioaccessibility of micro minerals, especially Mn, Cu, and Fe, as the highest percentages of insoluble minerals were recorded in MBP 81 in at 78.4%, 71.0%, and 62.1%, respectively. Similarly, there was a gradual decrease in SND% and D% of macro minerals such as K and Mg ($p < 0.05$) in correlation with an increased proportion of fortification with mango seed and kernel.

The observed lower proportions of bioaccessible minerals in the composite porridges than the control may be likely due to the presence of considerable amounts of phenolic compounds, phytates, and tannins in the mango by-products [30] that could have a pronounced effect of binding minerals, thus forming complexes that are insoluble for absorption. Additionally, as previously shown in Table 2, the mango seed and mango kernel had higher amounts of soluble and insoluble dietary fibers than maize flour, which could consequently influence the bioaccessibility of minerals. Dietary fibers decrease the diffusion kinetics of mineral sources by binding and physically entrapping minerals [31,32]. Fernández et al. [33] suggested that bioaccessibility of minerals could be affected by interaction with other components in the digested food. Minerals may also interfere with each other's absorption [34], for example; the presence of Ca impairs Zn absorption due to co-precipitation, Mn affects Fe absorption as the intestine cannot differentiate them, and finally, Zn and Fe compete for intestinal absorption due to similar configuration [35,36].

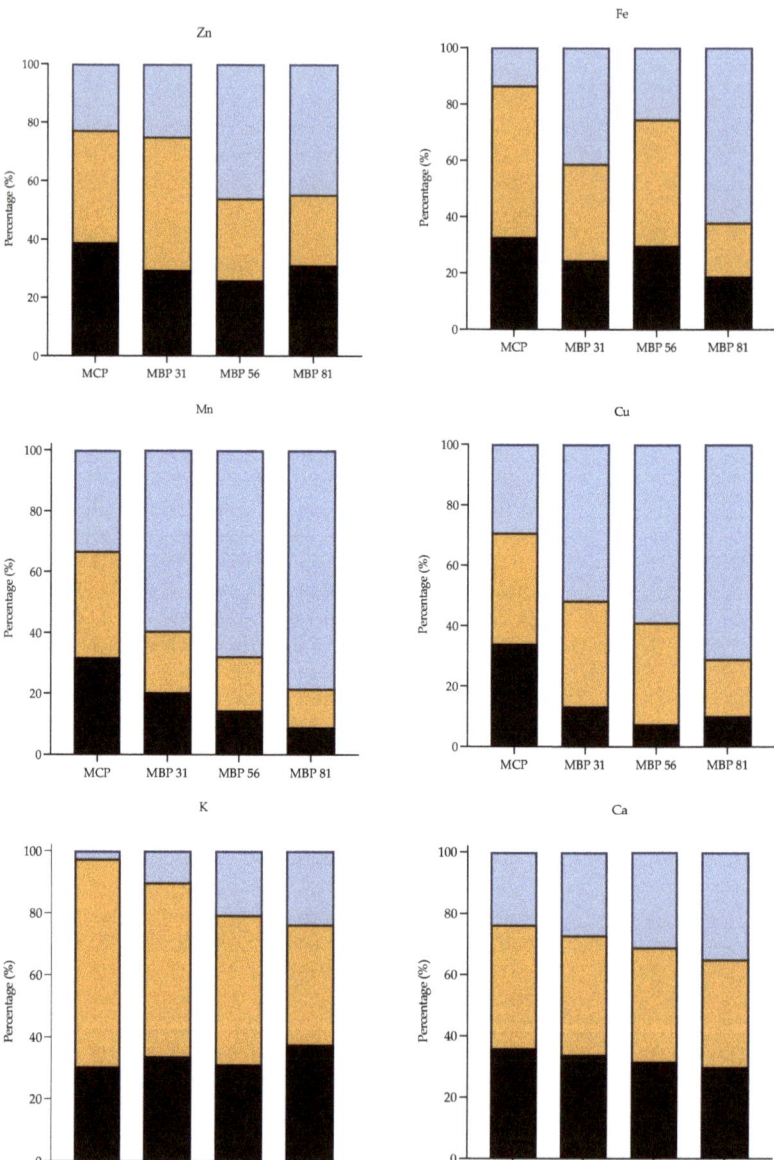

Figure 3. Cont.

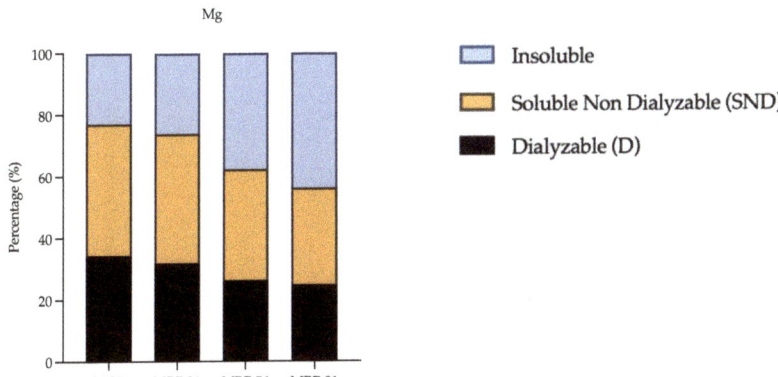

Figure 3. Percentage (%) bioaccessibility of dialyzable (D), soluble non-dialyzable (SND), and insoluble minerals during in vitro digestion in the formulated composite porridges.

3.5. Viscosity of the Porridges

From the data in Figure 4, it can be seen that the apparent viscosity of all the porridges generally decreased with increasing shear rate; hence, it followed non-Newtonian characteristics of pseudoplastic fluids. However, this change was observed to be reversible as the decreasing shear rate had similar apparent viscosity with increasing shear rate. These porridges therefore have thixotropic properties as they revert to their original state on standing [37]. Viscosity is an important quality parameter for the utilization of starch-based foods and is desirable when adequate solid content is maintained and allows for ease of consumption.

It was further observed that at a constant initial shear rate of 0.068 s^{-1}, viscosity differed significantly ($p < 0.05$) across the porridges. The control porridge and MBP 31 had the highest viscosities at 321,300 and 352,500 mPa·s, respectively, unlike MBP 56 and MBP 81, which registered lower viscosities of 70,000 and 15,000 mPa·s, respectively. A highly dense porridge reduces food/energy intake and may not be easily palatable by children. The high viscosities of MCP and MBP 31 porridges could be due to their greater proportion of starch, as starch granules gelatinize during cooking, increasing viscosity. Additionally, soluble fibers in the mango seed and kernel may absorb and retain moisture, increasing viscosity and gel formation [38]. However, fortification of maize flour with mango by-products (seed and kernel) at higher levels (>50%) significantly decreased ($p < 0.05$) porridge viscosity, possibly due to the insoluble solids and fat in these porridges (Table 2), as fat forms insoluble complexes with starch granules and/or forms a fatty layer around starch granules, reducing their water absorption capacity during cooking, and hence reducing viscosity [39]. A porridge with low viscosity but with high energy density is desirable for infants and young children for easy mastication and swallowing.

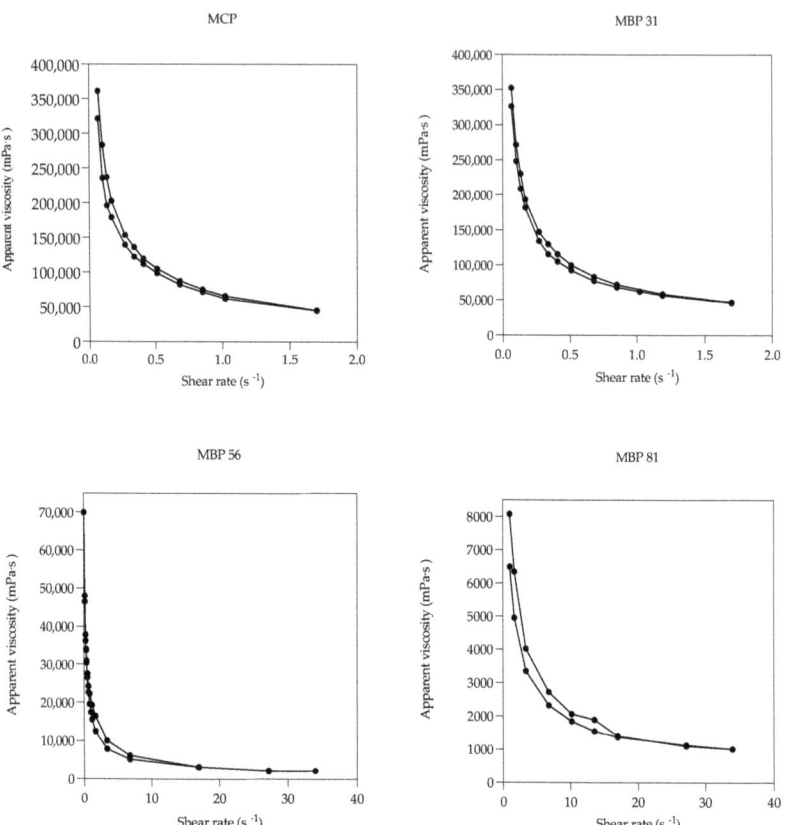

Figure 4. Effect of mango by-products on the apparent viscosity (mPa·s) of maize-based porridges expressed as a factor of shear rate (s^{-1}).

4. Conclusions

This study was conducted to enhance the nutritional content of maize complementary porridges with mango by-products (seed and kernel). Based on the results, mango seed and kernel increased the protein, fat, and mineral (Fe, Mn, Zn) content of maize porridges in proportion to the level of fortification. In addition, the by-products increased the total phenolic content of maize porridges, which subsequently improved their antioxidant capacity. However, the by-products decreased mineral bioaccessibility during in vitro gastrointestinal digestion. Therefore, fortification using mango seed and kernel should be done at lower levels of about 31–56%. Further studies to understand the structures of the mineral inhibitors in mango seed and kernel may be conducted. Overall, this work provides an alternative way of utilizing agro-industrial by-products in the formulation of food products.

Author Contributions: Conceptualization, J.M., A.O.M. and K.R.; methodology, J.M., H.S. and K.R.; validation, J.M., H.S. and K.R.; formal analysis, J.M.; investigation, J.M. and H.S.; writing—original draft preparation, J.M.; writing—review and editing, H.S., A.O.M. and K.R.; visualization, J.M.; supervision, A.O.M. and K.R. All authors have read and agreed to the published version of the manuscript.

Funding: This research was funded by the Bijzonder Onderzoeksfonds UGent, (01W03717), the Centre for Research, Agricultural Advancement, Teaching Excellence, and Sustainability (CREATES) at the Nelson Mandela African Institution of Science and Technology (NM-AIST) and L'OREAL–UNESCO for Women in Science Sub-Saharan regional fellowship.

Data Availability Statement: The authors confirm that the data supporting the findings of this study are available within the article.

Acknowledgments: The authors thank Xenia Verheire, a thesis student, for her practical work that contributed to the publishing of this manuscript.

Conflicts of Interest: The authors declare no conflict of interest.

References

1. Sagar, N.A.; Pareek, S.; Sharma, S.; Yahia, E.M.; Lobo, M.G. Fruit and vegetable waste: Bioactive compounds, their extraction, and possible utilization. *Compr. Rev. Food Sci. Food Saf.* **2018**, *17*, 512–531. [CrossRef] [PubMed]
2. Ahmed, J.; Ramaswamy, H.S.; Hiremath, N. The effect of high pressure treatment on rheological characteristics and colour of mango pulp. *Int. J. Food Sci. Technol.* **2005**, *40*, 885–895. [CrossRef]
3. FAOSTAT. Food and Agriculture Data. 2020. Available online: http://www.fao.org/faostat/en/#data/QC (accessed on 5 October 2020).
4. Muchiri, D.R.; Mahungu, S.R.; Gituanja, S.N. Studies on mango (*Mangifera indica* L.) kernel fat of some Kenyan varieties in Meru. *J. Am. Oil. Chem. Soc.* **2012**, *89*, 1567–1575. [CrossRef]
5. Singh, D.; Siddiq, M.; Greiby, I.; Dolan, K.D. Total phenolics, antioxidant activity, and functional properties of 'Tommy Atkins' mango peel and kernel as affected by drying methods. *Food Chem.* **2013**, *141*, 2649–2655. [CrossRef]
6. Jahurul, M.H.A.; Zaidul, I.S.M.; Ghafoor, K.; Al-Juhaimi, F.Y.; Nyam, K.L.; Norulaini, N.A.; Sahena, F.; Omar, A.K.M. Mango (*Mangifera indica* L.) by-products and their valuable components: A review. *Food Chem.* **2015**, *183*, 173–180. [CrossRef]
7. Castro-vargas, H.I.; Vivas, D.B.; Barbosa, J.O.; Johanna, S.; Medina, M.; Aristizabal, F.; Parada-alfonso, F. Bioactive phenolic compounds from the agroindustrial waste of Colombian mango cultivars 'Sugar Mango' and 'Tommy Atkins'—An alternative for their use and valorization. *Antioxidants* **2019**, *8*, 41. [CrossRef]
8. Ranum, P.; Peña-Rosas, J.P.; Garcia-Casal, M.N. Global maize production, utilization, and consumption. *Ann. N. Y. Acad. Sci.* **2014**, *1312*, 105–112. [CrossRef]
9. Qamar, S.; Aslam, M.; Huyop, F.; Javed, M.A. Comparative study for the determination of nutritional composition in commercial and noncommercial maize flours. *Pak. J. Bot.* **2017**, *49*, 519–523.
10. Shiriki, D.; Igyor, M.A.; Gernah, D.I. Nutritional evaluation of complementary food formulations from maize, soybean and peanut fortified with moringa oleifera leaf powder. *Food Nutr. Sci.* **2015**, *6*, 494–500. [CrossRef]
11. Kruger, J.; Taylor, J.R.N.; Ferruzzi, M.G.; Debelo, H. What is food-to-food fortification? A working definition and framework for evaluation of efficiency and implementation of best practices. *Compr. Rev. Food Sci. Food Saf.* **2020**, *19*, 3618–3658. [CrossRef]
12. Pobee, R.A.; Johnson, P.N.T.; Akonor, P.T.; Buckman, S.E. Nutritional, pasting and sensory properties of a weaning food from rice (*Oryza sativa*), soybeans (*Glycine max*) and kent mango (*Mangifera indica*) flour blends. *Afr. J. Food Agric. Nutr. Dev.* **2017**, *17*, 11533–11551. [CrossRef]
13. Ashoush, I.S.; Gadallah, M.G.E. Utilization of mango peels and seed kernels powders as sources of phytochemicals in biscuit. *World J. Dairy Food Sci.* **2011**, *6*, 35–42.
14. Awolu, O.O. Influence of defatted mango kernel seed flour addition on the rheological characteristics and cookie making quality of wheat flour. *Food Sci. Nutr.* **2018**, *6*, 2363–2373. [CrossRef]
15. Blancas-benitez, F.J.; Avena-bustillos, R.D.J.; Montalvo-gonzález, E.; Sáyago-ayerdi, S.G.; Mchugh, T.H. Addition of dried 'Ataulfo' mango (*Mangifera indica* L.) by-products as a source of dietary fiber and polyphenols in starch molded mango snacks. *J. Food Sci. Technol.* **2015**, *52*, 7393–7400. [CrossRef]
16. AOAC. *Official Methods of Analysis of Association of Official Analytical Chemists (AOAC). Official Methods of Analysis of AOAC International*, 18th ed.; AOAC International: Arlington, VA, USA, 2010.
17. Ashoka, S.; Peake, B.M.; Bremner, G.; Hageman, K.J.; Reid, M.R. Comparison of digestion methods for ICP-MS determination of trace elements in fish tissues. *Anal. Chim. Acta* **2009**, *653*, 191–199. [CrossRef]
18. Gonzales, G.B.; Smagghe, G.; Raes, K.; Van camp, J. Combined alkaline hydrolysis and ultrasound-assisted extraction for the release of nonextractable phenolics from cauli flower (*Brassica oleracea var. botrytis*) waste. *J. Agric. Food Chem.* **2014**, *62*, 3371–3376. [CrossRef]
19. Huynh, N.T.; Smagghe, G.; Gonzales, G.B.; Van camp, J.; Raes, K. Enzyme-assisted extraction enhancing the phenolic release from cauliflower (*Brassica oleracea L. var. botrytis*) outer leaves. *J. Agric. Food Chem.* **2014**, *62*, 7468–7476. [CrossRef]
20. Singleton, V.L.; Orthofer, R.; Lamuela-Raventós, R.M. Analysis of total phenols and other oxidation substrates and antioxidants by means of folin-ciocalteu reagent. *Methods Enzymol.* **1999**, *299*, 152–178. [CrossRef]
21. Re, R.; Pellegrini, N.; Proteggente, A.; Pannala, A.; Yang, M.; Rice-Evans, C. Antioxidant activity applying an improved Abts radical cation decolorization assay. *Free Radic. Biol. Med.* **1999**, *26*, 1231–1237. [CrossRef]
22. Brand-Williams, W.; Cuvelier, M.E.; Berset, C. Use of a free radical method to evaluate antioxidant activity. *LWT–Food Sci. Technol.* **1995**, *28*, 25–30. [CrossRef]

23. Minekus, M.; Alminger, M.; Alvito, P.; Ballance, S.; Bohn, T.; Bourlieu, C.; Carrière, F.; Boutrou, R.; Corredig, M.; Dupont, D.; et al. Standardised static in vitro digestion method suitable for food-an international consensus. *Food Funct.* **2014**, *5*, 1113–1124. [CrossRef]
24. Mutua, J.K.; Imathiu, S.; Owino, W. Evaluation of the proximate composition, antioxidant potential, and antimicrobial activity of mango seed kernel extracts. *Food Sci. Nutr.* **2017**, *5*, 349–357. [CrossRef]
25. Bertha, C.T.; Alberto, S.B.J.; Tovar, J.; Sáyago-Ayerdi, S.G.; Zamora-Gasga, V.M. In vitro gastrointestinal digestion of mango by-product snacks: Potential absorption of polyphenols and antioxidant capacity. *Int. J. Food Sci. Technol.* **2019**, *54*, 3091–3098. [CrossRef]
26. Rai, S.; Kaur, A.; Singh, B. Quality characteristics of gluten free cookies prepared from different flour combinations. *J. Food Sci. Technol.* **2014**, *51*, 785–789. [CrossRef]
27. Celia, M.; Hauly, D.O. Inulin and oligofructosis: A review about functional properties, prebiotic effects and importance for food industry. *Semin. Ciênc. Exatas Tecnol.* **2002**, *23*, 105–117.
28. Kaur, H.; Gill, B.S.; Karwasra, B.L. In vitro digestibility, pasting, and structural properties of starches from different cereals. *Int. J. Food Prop.* **2018**, *21*, 70–85. [CrossRef]
29. Nguyen, N.M.P.; Le, T.T.; Vissenaekens, H.; Gonzales, G.B.; Van Camp, J.; Smagghe, G.; Raes, K. In vitro antioxidant activity and phenolic profiles of tropical fruit by-products. *Int. J. Food Sci. Technol.* **2019**, *54*, 1169–1178. [CrossRef]
30. Abdalla, A.E.M.; Darwish, S.M.; Ayad, E.H.E.; El-Hamahmy, R.M. Egyptian mango by-product 1. Compositional quality of mango seed kernel. *Food Chem.* **2007**, *103*, 1134–1140. [CrossRef]
31. Baye, K.; Guyot, J.P.; Mouquet-Rivier, C. The unresolved role of dietary fibers on mineral absorption. *Crit. Rev. Food Sci. Nutr.* **2017**, *57*, 949–957. [CrossRef]
32. Sanz-Penella, J.M.; Laparra, J.M.; Sanz, Y.; Haros, M. Bread supplemented with amaranth (*Amaranthus cruentus*): Effect of phytates on in vitro iron absorption. *Plant. Foods Hum. Nutr.* **2012**, *67*, 50–56. [CrossRef]
33. Fernández-García, E.; Carvajal-Lérida, I.; Pérez-Gálvez, A. In vitro bioaccessibility assessment as a prediction tool of nutritional efficiency. *Nutr. Res.* **2009**, *29*, 751–760. [CrossRef] [PubMed]
34. Gibson, R.S. Content and bioavailability of trace elements in vegetarian diets. *Am. J. Clin. Nutr.* **1994**, *59*, 1223S–1232S. [CrossRef] [PubMed]
35. Hemalatha, S.; Platel, K.; Srinivasan, K. Zinc and iron contents and their bioaccessibility in cereals and pulses consumed in India. *Food Chem.* **2007**, *102*, 1328–1336. [CrossRef]
36. Sandstrom, B. Micronutrient interactions: Effects on absorption and bioavailability. *Br. J. Nutr.* **2001**, *85*, S181–S185. [CrossRef]
37. Bourne, M.C. *Food Texture and Viscosity: Concept and Measurement*, 2nd ed.; Academic Press: New York, NY, USA, 2002.
38. Corradini, C.; Lantano, C.; Cavazza, A. Innovative analytical tools to characterize prebiotic carbohydrates of functional food interest. *Anal. Bioanal. Chem.* **2013**, *405*, 4591–4605. [CrossRef]
39. Arocha, M.; De, E.; Gómez, M.; Rosell, C.M. Effect of different fibers on batter and gluten-free layer cake properties. *LWT–Food Sci. Technol.* **2012**, *48*, 209–214. [CrossRef]

Article

Functionality and Palatability of Yogurt Produced Using Beetroot Pomace Flour Granulated with Lactic Acid Bacteria

Marina Jovanović [1,*], Snežana Zlatanović [1], Darko Micić [1], Dragan Bacić [2], Dragana Mitić-Ćulafić [3], Mihal Đuriš [4] and Stanislava Gorjanović [1]

1. Institute of General and Physical Chemistry, Studentski trg 12/V, 11158 Belgrade, Serbia; snezana.zlatanovic@gmail.com (S.Z.); micic83@gmail.com (D.M.); stasago@yahoo.co.uk (S.G.)
2. Faculty of Veterinary Medicine, University of Belgrade, Bulevar oslobođenja 18, 11000 Belgrade, Serbia; bacicd@vet.bg.ac.rs
3. Faculty of Biology, University of Belgrade, Studentski trg 16, 11158 Belgrade, Serbia; mdragana@bio.bg.ac.rs
4. Institute of Chemistry, Technology and Metallurgy—National Institute of The Republic of Serbia, University of Belgrade, Njegoševa 12, 11000 Belgrade, Serbia; mihal.djuris@ihtm.bg.ac.rs
* Correspondence: marina.rajic.jovanovic@gmail.com; Tel.: +38-163-744-3004

Citation: Jovanović, M.; Zlatanović, S.; Micić, D.; Bacić, D.; Mitić-Ćulafić, D.; Đuriš, M.; Gorjanović, S. Functionality and Palatability of Yogurt Produced Using Beetroot Pomace Flour Granulated with Lactic Acid Bacteria. *Foods* **2021**, *10*, 1696. https://doi.org/10.3390/foods10081696

Academic Editors: Simona Grasso, Konstantinos Papoutsis, Claudia Ruiz-Capillas and Ana Herrero Herranz

Received: 18 June 2021
Accepted: 19 July 2021
Published: 22 July 2021

Publisher's Note: MDPI stays neutral with regard to jurisdictional claims in published maps and institutional affiliations.

Copyright: © 2021 by the authors. Licensee MDPI, Basel, Switzerland. This article is an open access article distributed under the terms and conditions of the Creative Commons Attribution (CC BY) license (https://creativecommons.org/licenses/by/4.0/).

Abstract: Following the idea of sustainability in food production, a yogurt premix based on beetroot (*Beta vulgaris*) pomace flour (BPF) was developed. BPF was granulated with lactose solution containing lactic acid bacteria (LAB) by a fluidized bed. Particle size increased ~30%. A decrease in Carr Index from 21.5 to 14.98 and Hausner ratio from 1.27 to 1.18 confirmed improved flowability of granulated BPF, whereas a decrease in water activity implied better storability. Yogurts were produced weekly from neat starters and granulated BPF (3% w/w) that were stored for up to one month (4 °C). High viability of *Streptococcus thermophilus* was observed. Less pronounced syneresis, higher inhibition of colon cancer cell viability (13.0–24.5%), and anti-*Escherichia* activity were ascribed to BPF yogurts or their supernatants (i.e., extracted whey). Acceptable palatability for humans and dogs was demonstrated. A survey revealed positive consumers' attitudes toward the granulated BPF as a premix for yogurts amended to humans and dogs. For the first time, BPF granulated with LAB was used as a premix for a fermented beverage. An initial step in the conceptualization of a novel DIY (do it yourself) formula for obtaining a fresh yogurt fortified with natural dietary fiber and antioxidants has been accomplished.

Keywords: vegetable pomace; dairy beverage; fluidized bed; heat-sensitive compounds; functional food; palatability; by-products; *Canis familiaris*; DIY formula; sustainability

1. Introduction

The growing awareness of nutrition and the environment has directed the flow of current research toward the usability of agri-food by-products. The long-term focus on this approach is supported by the objectives of the Food and Agriculture Organization's (FAO) Sustainable Development Goals, Farm to Fork Strategy 2020, the FOOD Strategy 2030, and Serbian and European Bio-Economy Strategy [1–3]. The root vegetable *Beta vulgaris* L. (fam. Chenopodiaceae), commonly known as beetroot, is globally consumed as part of the normal diet and used to fabricate a natural food coloring agent known as E162 and ready-to-drink nonalcoholic beverages. Large quantities of beetroot pomace (BP) are available as a by-product [4,5].

The restoration of minimally processed BP back to the food chain can enhance the nutritional characteristics of processed food. The considerable potential of BP utilizations lays in the high dietary fiber (DF) content [5,6], a moderate caloric value, and the significant amount of antioxidants (AOs), such as nitrogenous pigments (betacyanins and betaxanthins) and polyphenolics (ferulic, protocatechuic, gallic and caffeic acid, catechin, and myricetin) [7–9]. To date, beetroot has been used as an ingredient in fermented juice, cereal bars, beetroot-enriched bread [10], chips [11], vegetable-based smoothies [12], and yogurts [13]. The majority of listed products contain lactic acid bacteria (LAB) [10,11,13]. However, none of them were obtained using BP with immobilized LAB. The immobilization can increase the viability of LAB by promoting the adherent properties of microorganisms and mimicking cell growth within natural structures. Known as good immobilization materials, DF, fruit, and vegetable powders were already used as immobilization support for yogurt starter cultures [14].

Fluidized bed granulation is often a method of choice for drying heat-sensitive compounds, including microbial cultures. Advantages, such as good temperature control, equal temperature distribution, high drying capacity, short drying time, low maintenance costs, and large-scale production, are noticeable [15–18]. To minimize the damage caused by drying and to achieve increased bacterial viability in storage time, fluidized bed immobilization of LAB is mainly performed in the presence of protective carbohydrates [15]. Lactose, for instance, interacts with the polar section of membrane lipids, providing protection from cell degradation during drying and storage [19,20].

Yogurt, a worldwide consumed beverage with beneficial effects on the immune system, gastrointestinal health, and a reduction of specific cancer risk [21,22], is a fermented dairy product that is acidified and coagulates owing to the activity of *Lactobacillus delbrueckii* subsp. *bulgaricus* and *Streptococcus thermophilus* [23]. It is highly recommended for immunocompromised populations and those suffering from lactose intolerance [21]. Various probiotics strains could also be added in the dairy products intended for humans and their companion animals [24]. LAB enables the proteolytic release of peptides with anticarcinogenic potential from yogurt proteins. Various food-derived peptides also exhibit cytotoxic activities against malignant cells, usually by disrupting cell membranes [22]. Fermentation improves the nutritional value, palatability, and preservative and medicinal properties [25]. Functional features can be further improved by adding fruits and vegetable pomace [13,26].

Modern consumers favor green consumerism, take special notice of their food choices, and raise the question of the eating habits of their companion animals, particularly dogs (*Canis familiaris* L.) [6]. Functional foods intended for dog nutrition have gained considerable popularity [25,27]. Moreover, pet parents prefer food that resembles their meals [6]. The concept of sharing food with pets, such as commercially available snack bars (Yaff BAR, Mudd & Wyeth, South Hero, VT, USA), has received increasing attention. Functional human or commercial pet food contains beet as a source of DF and AOs [6]. Therefore, a fermented dairy beverage with BP could have a positive tag appeal for pet owners and the wellness generation.

In this study, the main purpose was to granulate BP flour (BPF) with LAB and evaluate the applicability of the obtained granulate as a yogurt premix. Wet granulation of BPF in a fluidized bed was employed to improve the physical properties of BPF and to utilize it as LAB support. The physical characteristics of BPF before and after granulation were compared. For one month, neat and immobilized starters were stored and used weekly to produce plain and BPF yogurts. The functionality and palatability of fresh yogurts were investigated. A decrease in pH during fermentation and occurrence of syneresis was followed. To highlight functionality, plain and BPF yogurts were analyzed from the microbiological point of view. LAB viability and anti-*Escherichia* activity were estimated. Further, the cytotoxic effect of yogurts' supernatants (i.e., extracted whey) on the human colon cancer cell line (HCT116) was surveyed. An in vivo toxicity test was performed to confirm the novel product safety before palatability surveying by both humans and dog trials. Finally, to gain insight into the public attitudes toward the BPF premix for yogurts amended to humans and dogs, a survey was conducted. This study represents a step toward the conceptualization of an affordable DIY (do it yourself) formula for obtaining a fresh yogurt fortified with natural DF and AOs, suitable for both human and dog consumption.

2. Materials and Methods

2.1. Materials

Pasteurized cow milk (Dairy plant "Zapis Tare", Serbia) and red beetroot (Detroit variety) were purchased from the local markets (Belgrade, Serbia). Commercial raspberry yogurt for dogs ("Creamy Timmy ", Creamy Jimmy) was purchased from the local pet shop (Belgrade, Serbia).

Lyophilized starter culture YC-X11, containing *Lactobacillus delbrueckii* subsp. *bulgaricus* and *Streptococcus thermophilus* was obtained from Chr. Hansen (Hørsholm, Denmark). Bacterial strain *Escherichia coli* ATCC 8739 was obtained from the Department of Microbiology, University of Belgrade—Faculty of Biology, Serbia.

Human colon cancer cell line HCT116 (ATCC CCL-247) was obtained from the Oncology Institute of Vojvodina, Serbia.

M17 agar, De Man-Rogosa-Sharpe agar (MRS), Müller Hinton broth (MHB), and Mac Conkey Agar were obtained from HiMedia (Mumbai, India). Dulbecco's modified Eagle's medium (DMEM), penicillin–streptomycin mixtures, phosphate buffered saline (PBS), trypsin from the porcine pancreas, and 3-(4,5-dimethylthiazol-2-yl)-2,5-diphenyltetrazolium bromide (MTT) were purchased from Sigma-Aldrich (Steinheim, Germany).

2.2. Beetroot Pomace Flour Preparation and Characterization

The beetroot pomace (BP) collected immediately after squeezing thoroughly washed beets (Bosch GmbH, Renningen, Germany) was subjected to dehydration, in the laboratory dehydrator (Excalibur, model 3926 TB, Sacramento, CA, USA) for 480 min at ≤ 55 °C, and to subsequent grinding (>200 µm, determined by sieving). The proximate composition of BPF was determined. The moisture level was determined at 105 ± 5 °C in a thermostatically controlled dry oven until a constant weight was reached. The proximate composition of BPF, including protein (Nx5.30) (Official Method No 950.36), fat (Official Method No. 935.38), crude fiber (Official Method No. 962.09), and ash (Official Method No. 930.22), was determined by standard AOAC methods [28,29]. The carbohydrate content was calculated by subtracting the sum of moisture, ash, crude fiber, fat, and protein from 100%.

2.3. Batch Granulation of BPF by Fluidized Bed at Pilot-Scale Level

The process of wet granulation in a fluidized bed (pilot-scale device constructed and built within the scope of Innovation Project IP 16-10 2018 funded by the Ministry of Education, Science and Technological Development of the Republic of Serbia) was performed to obtain granulated BPF. BPF granulation implied immobilization of a yogurt starter culture (YC-X11) onto BPF. BPF (300 g) was sterilized in the preheated column (78 °C) of a fluidized bed dryer for 10 min. The temperature in the column was reduced to 48 °C. The BPF was granulated via top spraying of aqueous lactose solution (5% w/v; 150 mL), mixed with a bacterial starter culture (2 g). The pressure through a two-fluid nozzle was 1.8 bar. The mean fluid flow rate and air flow rate were 5 mL/min and 35 m^3/h, respectively. The granulated BPF was dried by fluidization at 40 °C for 15 min. To prevent disintegration of the formed granules due to collision and attrition, the air flow rate in the drying phase was reduced to 30 m^3/h. After the drying phase, the samples were cooled to room temperature, transferred into food-grade sterile polypropylene (PP) zip lock bags, and maintained at 4 °C for a one-month period.

2.4. Determination of Physical and Technological Properties of BPF before and after Granulation

Physical and technological properties of BPF before and after granulation, including bulk and tap density, flowability, particle size distribution, and water activity (aw), were compared.

2.4.1. The Flowability

Bulk density measurement was performed in a glass cylinder of 73.95 mL in volume (23 mm in diameter and 178 mm in height). The tapped density was obtained in the same cylinder by mechanically tapping until the flour surface reached the maximum packing condition (300–400 taps). The measurements of bulk and tapped densities were repeated three times, and the mean value was used to calculate the Carr Compressibility Index (C_I) and Hausner ratio (H) that defines the flowability.

$$C_I = 100 \times \frac{\rho_{tapped} - \rho_{bulk}}{\rho_{tapped}} \quad (1)$$

$$H = \frac{\rho_{tapped}}{\rho_{bulk}} \quad (2)$$

where ρ_{bulk} and ρ_{tapped} are bulk and tapped density of material in g/mL, respectively.

2.4.2. Particle Size Distribution

The particle size distribution was determined by analyzing a scanned sample image (between 1000–2000 particles) using 2D image analysis software ImageJ (Developed by University of Wisconsin-Madison, Madison, WI, USA, under Public Domain, BSD-2 license). Images were captured by a 2D scanner Hp Scanjet 300 and stored at the high resolution of 4800 dpi, as described [30]. Since the particles are irregular in shape, the size of the particles was expressed as the projected area diameter d_A.

$$d_A = \sqrt{\frac{4 \times A}{\pi}} \quad (3)$$

where A is the surface area of the projection of the particle; a sum of pixels areas in calibrated units (e.g., μm^2).

2.4.3. Water Activity

For one month, aw of non-granulated and granulated BPF was determined weekly by the aw meter Novasina LabSwift Bench-model Water Activity Meter (Neutec Group Inc., Farmingdale, NY, USA) at $25 \pm 2\ °C$. For each sample, three independent experiments in triplicates were performed.

2.5. Evaluation of Granulated BPF Toxicity

Acute toxicity was examined to confirm the safety of granulated BPF. The care and treatment of laboratory animals were performed according to the regulations and standards of the national (Serbian) Law on the Experimental Animal Treatment and European Directive 2010/63/EU (European Convention for the Protection of Vertebrate Animals used for Experimental and other Scientific Purposes). Four-week-old pathogen-free Han: NMRI mice (n = 5), with initial body weights of 18–22 g, were housed in standard cages in a room with a 12 h light-dark cycle at a temperature of $22 \pm 3\ °C$. Mice were fed with granulated BPF (50 mg) mixed with water once using a gastric probe. The observation period was 72 h.

2.6. Application of Granulated BPF as a Premix in Yogurt Production

2.6.1. Production of Yogurt at Laboratory Scale Level

Neat starter cultures and a BPF premix (i.e., granulated BPF) stored for 0, 7, 14, 21, and 28 days were used to prepare a control yogurt and BPF yogurt, respectively. Yogurts were prepared once a week from BPF premix in order to determine if there was a difference in fermentation rate. The cow milk (2.8% milk fat) was heat-treated at $85\ °C$ for 10 min, cooled down to $43\ °C$, and poured into glass containers (100 mL) and neat (0.02 g) starter culture or BPF premix (3% and 9% w/w) were added. After thorough mixing, the samples were subjected to fermentation at $43\ °C$ until a pH < 4.6 was reached. The pH values were determined using a pH meter with a gel-filled electrode (WTW™ SenTix™ 41 pH, California, CA, USA). Obtained yogurts were stirred and stabilized by cooling ($4\ °C$ for 24 h). Freshly prepared yogurts with 3% w/w BPF premix were used in all further testing except in the cytotoxicity test. Yogurt containing 9% BPF premix was further processed and used in the cytotoxicity assay. All samples were made in triplicate, and the samples preparation was repeated three times.

2.6.2. Syneresis of Yogurt Samples

The samples (2 × 25 g) from each batch of yogurt were weighed in centrifuge tubes, centrifuged at 3000× g for 10 min, and the whey was separated. Syneresis (%) was calculated as a weight of generated supernatant per weight of yogurt multiplied by 100 [31]. Experiments were performed in triplicate and repeated three times.

2.7. Determination of Functionality of Yogurt Produced Using Granulated BPF as Premix

2.7.1. LAB Viability

The viability of the LAB was evaluated upon fermentation using the pour plate technique and serial dilutions in phosphate buffer saline (1% PBS). *S. thermophilus* was enumerated using M17 agar (pH 7.2) under aerobic incubation at 37 °C for 48 h. After anaerobic incubation at 37 °C for 72 h, *L. bulgaricus* was counted on MRS agar (pH 6.2). The results were expressed as the log of the mean number of the colony-forming units (log CFU/mL).

2.7.2. In Situ Antibacterial Activity

Cow milk was inoculated with 6 log CFU/mL of *E. coli* ATCC 8739, fermentation with and without BPF premix was conducted, and the obtained yogurts were stabilized for 24 h. After the stabilization, yogurts' serial decimal dilutions were prepared, plated on Mac Conkey agar, and incubated for 24 h at 37 °C under aerobic conditions. The results were expressed as log CFU/mL.

2.7.3. Cytotoxicity

The cytotoxic effect was estimated using an MTT assay [26]. For the purpose of cytotoxicity testing, yogurt supernatants (i.e., extracted whey) were prepared as previously described [26]. In brief, yogurt samples were centrifuged at 20,000× g for 60 min at 4 °C and filtered using a 0.45 µm syringe filter. The pH was adjusted to 6.8. The filtrates obtained from the control yogurt (without BPF) and yogurt with 9% BPF premix were diluted three times in DMEM. HCT116 cells were treated with diluted supernatants and incubated for 24 h. After an additional incubation (3 h) of the cells with MTT (3-(4,5-dimethylthiazol-2-yl)-2,5-diphenyltetrazolium bromide) solution and dissolution of the formed formazan crystals in DMSO, the viability of the cells was determined by measuring the absorbance at 570 nm using a microplate reader (Multiskan FC, Thermo Scientific, Shanghai, China). Three independent experiments in sextuplicate were performed.

2.8. Yogurt Palatability Assessment

2.8.1. Trial with Humans—JAR Test

A consumer test was conducted with 40 untrained panelists (22 females and 18 males, mean age years 40.7 ± 16.4). Fresh yogurt prepared with a 3% BPF premix was evaluated using 5-point just-about-right (JAR) scales (1 = too little, 3 = JAR, and 5 = too much) for the intensity of 'color' (too light–JAR–too dark), 'sourness' (not sour enough–JAR–too sour), 'beetroot taste' (too weak–JAR–too strong), and also, the 5-point hedonic scale (1 = dislike extremely, 3 = neither like nor dislike and 5 = like extremely) for 'color acceptance', 'odor acceptance',' taste acceptance', 'acceptance of sourness', and 'texture acceptance'. Consumer acceptance data were subjected to mean drop analysis, as described by Tomić et al. [32]. In addition, three background questions were asked: (1) What is your frequency of consumption (less than once a week or more than once a week) of yogurt? (2) What is your frequency of consumption of beetroot-based products? and (3) Would you share a BPF yogurt with a dog?

2.8.2. Trial with Dogs—The Two-Bowl Test

Eleven healthy, vaccinated, and dewormed adult dogs of varying ages (4 males, 7 females; mean age years 4.28 ± 3.08 years) were included in the present study. Dog breeds and neutered status are presented in Supplementary Table S1. The body condition of all dogs was considered to be ideal (ribs can be palpated through slight fat cover) or moderately overweight (difficult to palpate ribs due to moderate fat cover). Dogs were housed individually in controlled home environments under the supervision of trained staff.

The trial was performed using the two-bowl test [33] administered in the morning for two consecutive days. Yogurt fermented with 3% BPF and commercial yogurt with raspberry were pairwise compared. The ingredients of the commercial dog yogurt are presented in Supplementary Table S2. The amount of provided meal each morning, including tested and regular food, did not exceed the dogs' recommended total daily intake. Two tested, pre-weighed diets (50 g each) were simultaneously offered to the dogs in two identical bowls. The position of the bowls was changed daily to prevent side bias (left or right bowl preference). The food the dog chose first (first food visited) and the amount of eaten food were noted. Prior to data collection, a two-day adaptation period was provided to familiarize the dogs with the new food and to examine whether dogs display a neophobic behavior and a tendency to side bias. For these purposes, imitating the conditions of the two-bowl test, a common diet that the dogs were accustomed to, and the yogurt fermented with 3% BPF were served. Dogs that exhibited position bias were excluded. Water was offered ad libitum.

The intake ratio of the tested diets was calculated as follows:

$$\text{Intake ratio of BPF yogurt} = \frac{\text{intake (g) of BPF yogurt}}{\text{intake (g) of BPF yogurt} + \text{intake (g) of C yogurt}} \quad (4)$$

$$\text{Intake ratio of C yogurt} = \frac{\text{intake (g) of C yogurt}}{\text{intake (g) of BPF yogurt} + \text{intake (g) of C yogurt}} \quad (5)$$

Fresh urine and fecal samples were analyzed before and after the two-bowl tests. The urine collected from each animal by natural voiding was visually inspected for color and described as: light-yellow, yellow, dark-yellow, and pink or red (colored urine sample due to beetroot intake). Fecal samples score was determined as described previously [34].

2.9. Surveying of Consumers' Attitude towards Novel Product and Accompanied Concept

To gain insight into the owners' attitudes toward yogurt intended for dog nutrition and current feeding practices, a paper questionnaire and an online questionnaire were administrated (Google form). The link to the open survey was shared through a public dog owners' group on a social media website (Facebook), and the paper questionnaire has been placed at a veterinary clinic (Line Alba, Belgrade, Serbia). Owners completed the paper questionnaire at the clinic at the time of the appointment. One response per participant was allowed. The data collection ran for one month. The questionnaire contained 32 closed questions (all mandatory), about the owner (personal and household data), the dog's signalment (age, sex, body weight, and health status) and the issues associated with animal's welfare, owner's preferences regarding pet food, and the finances allocated for dog-maintenance. The survey participation was anonymous and on a voluntary basis. The survey was created as described by Morelli et al. [35] and is available in translated form as Supplementary material.

2.10. Statistical Analysis

The following programs for statistical data analysis were combined and used: software GraphPad Prism 6.0 (GraphPad Software Inc., San Diego, CA, USA) and Excel 2016 (Microsoft). The data obtained from pH, syneresis, bacterial counts, in situ antibacterial activity, and MTT assay were analyzed by analysis of variance (one-way ANOVA, Dunnett's multiple comparisons test, and Tukey's honestly significant difference test (HSD)). The level of statistical significance was defined as $p < 0.05$. The data collected from the nutritional questionnaire were transferred into a spreadsheet (Excel, Microsoft) and submitted to descriptive analysis.

3. Results and Discussion

3.1. Production and Proximate Composition of BPF

The proximate composition of BPF obtained by dehydration and grinding to particle size >200 μm was in line with the previously published results [5]. Carbohydrate content was 34.58 ± 0.23%, whereas the contents of protein and fat were 13.85 ± 0.34% and 0.92 ± 0.12%, respectively. Relatively rich ash portion (8.73 ± 0.21%) can be attributed to high mineral content. A low moisture level (5.85 ± 0.14%) was achieved. A significant amount of crude fiber (36.1 ± 0.47%) confirmed that BPF represents the outstanding source of DF. Owing to the important role in digestibility, laxation, and stool quality, beet pulp has often been used as the primary DF source in dogs' diets [6].

3.2. Granulation of BPF with LAB Using the Fluidized Bed Technique

The BPF produced was granulated successfully with LAB by using fluidized bed, chosen as it consumes less energy for the immobilization of beneficial microorganisms than freeze-drying [15]. In contrast to air drying, fluidized bed is characterized by optimal heat and equal temperature distribution that allows efficient drying of sensitive compounds and microorganisms [15]. The fluidized bed technique was applied to the commercial production of dried baker's yeast [18], and probiotic encapsulates (e.g., Probiocap™) [36]. However, immobilization of LAB on a minimally processed pomace via fluidized bed has not been conducted to date. To our knowledge, fluidized bed was used until now only as one of the processing steps to obtain carrot pomace powder [37].

3.3. Physical and Technological Properties of BPF before and after Granulation

3.3.1. Flowability

The bulk density before and after wet fluidized bed granulation of BPF with 5% w/v of lactose solution containing LAB was 0.44 g/mL and 0.41 g/mL, respectively, whereas tapped density was 0.55 g/mL and 0.48 g/mL. Based on Carr Index and Hausner Ratio (C_I = 21.54, H = 1.27), the flowability of BPF was defined as possible [38]. Significantly improved flowability was ascribed to granulated BPF (C_I = 14.98 and H = 1.18). According to [38], flowability expected for powder characterized by C_I = 14.98 and H = 1.18 belongs to Good/Free flow. Classification of powders according to Carr Index and H ratio enable better insight into the improvement of the flowability of initial flour achieved by granulation. Flowability of granulated BPF secures easier manipulation during transport, packaging without dusting, congestion, or spilling.

3.3.2. Particle Size Distribution

The comparison of fractional particle size distribution with respect to the projected particle diameter d_A for BPF before and after granulation is shown in Figure 1. The particle distribution of non-granulated BPF was in the interval 94 < d_A < 916 μm; whereas after the granulation the range was 163 < d_A < 1172 μm. The particle size increased by approximately 30%. The particle size enlargement via wet granulation improved the flowability of BPF.

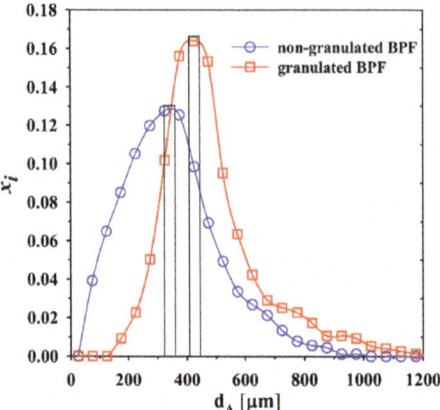

Figure 1. Fractional particle size distribution of BPF before and after wet fluidized bed granulation with 5% w/v of lactose solution containing LAB at a pilot-scale level.

3.3.3. Storability of Non-Granulated and Granulated BPF

Water activity followed weekly, during a month of storage, confirmed superior storability of granulated BPF (Table 1).

Table 1. Water activity of non-granulated and granulated BPF during a month of storage at 4 °C.

	Water Activity of (aw)	
Day	Non-Granulated BPF (10 g)	Granulated BPF (10 g)
0	0.311 ± 0.004 [D,a]	0.208 ± 0.001 [E,b]
7	0.317 ± 0.001 [CD,a]	0.239 ± 0.002 [D,b]
14	0.322 ± 0.001 [C,a]	0.259 ± 0.001 [C,b]
21	0.333 ± 0.001 [B,a]	0.279 ± 0.001 [B,b]
28	0.343 ± 0.001 [A,a]	0.290 ± 0.002 [A,b]

BPF samples were stored for up to 28 days (4 °C). Values are presented as mean ± standard error ($n = 3$); different uppercase superscript in the same column indicates a significant difference in means during time, and different lowercase superscript within the same row indicates a significant difference in means at the same time, according to HSD test ($p < 0.05$).

3.4. Evaluation of Granulated BPF Toxicity

Han: NMRI mice (18–22 g) fed with 50 mg of granulated BPF did not display any signs of toxicity within 72 h of observation. Therefore, BPF could be labeled as safe for consumption.

3.5. Application of Granulated BPF as Premix for Yogurt Production

Influence of BPF Premix on Yogurt pH and Syneresis

The pH, acidity, and syneresis are important indicators of yogurt quality. Acidity, more precisely the characteristic acidic taste of yogurt, is formed as a result of the presence of lactic acid, diacetyl, and acetaldehyde produced during the fermentation process [21]. Fermentation time required for the preparation of yogurts with neat starter cultures and the BPF premix moderately differed (Figure 2). Time required to reach pH 4.6 with samples stored up to 3 weeks were between 2 h 40 and 3 h 20 min for neat starters and between 3 h 50 and 4 h 30 min for the BPF premix. However, neat and immobilized cultures stored for 4 weeks had slightly slower fermentation. Contrary to our results, beetroot enriched yogurt had lower pH values in comparison to plain yogurt [13]. Yadav et al. [13] reported that the pH of the yogurt samples decreasing upon beetroot powder addition was related to the acidic nature of beetroot. However, the amount of beetroot powder added to achieve such results was two folds higher (6%).

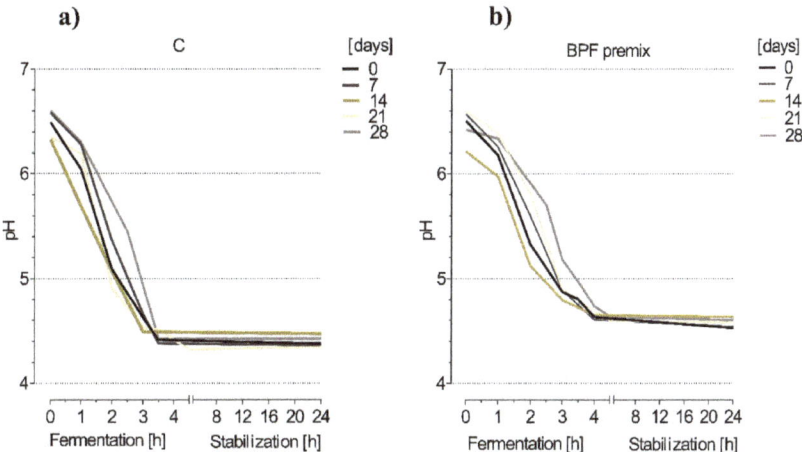

Figure 2. Decrease in pH values during yogurt production (**a**) without (control (C)) and (**b**) with 3% BPF premix. Yogurt samples were prepared weekly with neat starters and the BPF premix stored for up to 28 days (4 °C).

Higher syneresis rates indicate fermented dairy products' inferior quality and are attributed to milk preparation, the coagulation process, and the ingredients used [26]. Yogurts produced with the BPF premix had less pronounced syneresis than the plain one (Figure 3). In particular, yogurts produced at the beginning of the starter's storage (upon 0 and 7 days) exhibited such a property. The lower percent of syneresis was already ascribed to the yogurts supplemented with immobilized cells rather than to those receiving free ones [14]. Fruit and vegetable segments, as well as vegetable powder, can reduce syneresis owing to their hygroscopic nature [14]. Accordingly, the high water-holding capacity of beetroot pomace could explain lower syneresis rates.

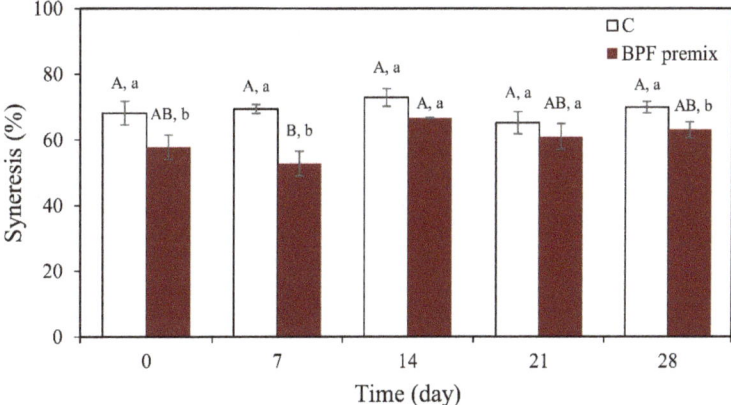

Figure 3. Amount of whey segregated (%) from yogurt prepared by applying neat (control (C)) starter cultures and the 3% BPF premix. Values are presented as mean ± standard error (n = 3); [AB] a significant difference in means during time within the same sample, [ab] a significant difference of means between samples at the same time according to the HSD test ($p < 0.05$).

3.6. Functionality of Yogurt Prepared Using Granulated BPF as a Premix

3.6.1. Bacterial Viability

The change in the microbial counts was monitored in fresh yogurts prepared weekly for one month with the stored BPF premix and neat starters. In accordance with previous findings [14,21], the viability of *S. thermophilus* was higher than that of *L. bulgaricus* (Table 2). Different oxygen tolerance of strains caused the survival rate distinction. The ability to neutralize reactive oxygen species (ROS) depends on enzymatic equipment. The low capacity of *L. bulgaricus* to eliminate H_2O_2, or other ROSs could be related to poor equipment with antioxidative enzymes, such as catalase, peroxidase, and superoxide dismutase. Paradoxically, under the microaerobic conditions, LAB, including *L. bulgaricus*, produced H_2O_2, which resulted in growth arrest. LAB is often exposed to oxygen during fermentation, product handling, and storage [39]. Importantly, neat and immobilized starters were stored in oxygen permeable bags. Thus, the oxygen exposure might have negatively affected the growth of *L. bulgaricus*, with a less significant impact on the viable count of *S. thermophilus* [21].

Table 2. Viable counts (log CFU/mL) of *S. thermophilus* and *L. bulgaricus* in yogurts prepared by applying starter cultures (control) and granulated BPF as a premix.

LAB S *	*L. bulgaricus* 7.54 ± 0.05		*S. thermophilus* 10.18 ± 0.15	
Day	Control	BPF	Control	BPF
0	7.07 ± 0.11 [A,a]	7.00 ± 0.29 [A,a]	9.21 ± 0.11 [A,a]	9.10 ± 0.01 [AB,a]
7	6.82 ± 0.10 [AB,a]	6.11 ± 0.30 [AB,a]	9.15 ± 0.13 [AB,a]	9.19 ± 0.08 [A,a]
14	6.37 ± 0.29 [AB,a]	6.26 ± 0.44 [AB,a]	8.99 ± 0.12 [AB,a]	9.13 ± 0.12 [AB,a]
21	5.85 ± 0.18 [BC,a]	5.77 ± 0.18 [B,a]	8.70 ± 0.22 [B,a]	8.68 ± 0.16 [BC,a]
28	4.83 ± 0.24 [C,a]	5.67 ± 0.23 [B,a]	8.87 ± 0.02 [AB,a]	8.51 ± 0.17 [C,a]

* S-initial counts (log CFU/mL) of strains in solution before fluidized bed drying; values are presented as mean ± standard error (n = 3), different uppercase superscript in the same column indicates a significant difference of means during that time; different lowercase superscript within the same row indicates a significant difference in means at the same time, according to the HSD test (p < 0.05).

In line with [14], the high viability of *S. thermophilus* was maintained upon immobilization and after the storage (CFU > log 8.5 after 4 weeks of storage). Moreover, *L. bulgaricus* counts were higher, by 18.34%, when a 4-week stored BPF premix (log 5.67) was used instead of neat starters (log 4.83). The result could be considered to be a promising initial step to overcoming the challenge of sustaining the high viability of *L. bulgaricus*, especially for a longer period of storage. However, the link between the length of LAB storage and their reduced survival rates was evident. Similar to our findings, *L. plantarum* cells fluidized with glucose, trehalose, sucrose, and maltodextrin as protectants and lost their initial viability after one month of storage [15].

These results confirmed that BPF could be used as matrices for the immobilization of yogurt starter cultures. Presumably, BPF enabled the natural entrapment of LAB, providing physical adsorption by electrostatic forces and covalent binding between the cell wall and the carrier [14]. However, higher viability of LAB might be achieved by an additional coating of BPF granulate. Alginate, chitosan, or lipid-based coating already suggested for that purpose [36,40,41] remained to be used in the following phase of the investigation.

3.6.2. In Situ Anti-Escherichia Activity

E. coli is often presented as a foodborne gram-negative bacteria that can contaminate milk-based products [42]. Here, the viability of *E. coli* in yogurts fermented with neat starters and the BPF premix was investigated, and anti-*Escherichia* properties were detected in all samples (Table 3). The yogurts fermented with neat starters and the BPF premix stored for up to 14 days had the most evident response. Reduction in *E. coli* numbers may be related to the large initial LAB counts required to produce adverse effects on the pathogen [43]. LAB starter culture could exert an antagonistic influence against bacterial pathogens [42]. The LAB metabolic activity responsible for pH dropping and liberating potent antibacterial peptides from milk proteins could be a limiting factor for the growth of *E. coli* [22,42]. Supernatants of plain yogurt and yogurt supplemented with pineapple peel powder also inhibited the growth of *E. coli* [22].

Table 3. In situ anti-*Escherichia* activity of yogurt samples fermented with neat starters and granulated BPF as a premix in comparison to growth control (GC).

Days of BPF Storage	Viable Counts of *E. coli* (log CFU/mL)	
	Yogurt with Neat Starters	Yogurt with BPF
0	5.96 ± 0.02 [AB,a,*]	5.76 ± 0.12 [A,a,*]
7	5.57 ± 0.14 [B,a,*]	5.26 ± 0.21 [A,a,*]
14	5.48 ± 0.07 [B,a,*]	5.81 ± 0.16 [A,a,*]
21	6.74 ± 0.25 [A,a]	6.67 ± 0.38 [A,a]
28	6.74 ± 0.17 [A,a]	6.39 ± 0.47 [A,a]
GC	6.85 ± 0.14	

Values are presented as mean ± standard error ($n = 3$); [AB] a significant difference in means between samples in the same column; [ab] a significant difference in means between samples in the same row according to HSD test ($p < 0.05$); *significant difference between samples and GC (growth control) according to the Dunnett's test ($p < 0.05$).

BP usage as a functional ingredient with antibacterial activity was suggested based on the finding that a high concentration of BP extract (100 mg/mL) inhibited *E. coli* viability [43]. A slight increase in the antimicrobial activity of BPF yogurt observed here can be related to the low share of BPF (3%). It might be supposed that a higher portion of BPF would further increase the antimicrobial properties.

3.6.3. Cytotoxic Properties

The cytotoxic potential of the supernatants of yogurts prepared with neat starters and the BPF premix were tested on the human colon cancer cell line HCT116. For all tested samples, a statistically significant reduction of cell viability was observed (13.02–24.47%) (Figure 4). However, cells were more susceptible to supernatants obtained from BPF yogurts. The cytotoxic effect of the yogurt supernatants on HCT116 cells was established earlier [26]. Supernatants of yogurts fermented with apple pomace [26] and pineapple peel powder [22] exhibited more pronounced inhibition of colon cancer cell viability (HCT116 and HT29, respectively) than plain yogurt supernatants. Several bioactive peptides liberated from milk proteins exhibit cytotoxic activity against malignant cells by disrupting cell membranes, modulating the cell cycle, and promoting apoptosis [22]. Betalains were also found responsible for potent anticancer activities [7,11]. A few in vivo studies confirmed that LAB exerts anticarcinogenic effects and that diet-induced microfloral alteration may prevent the development of colon cancer [44].

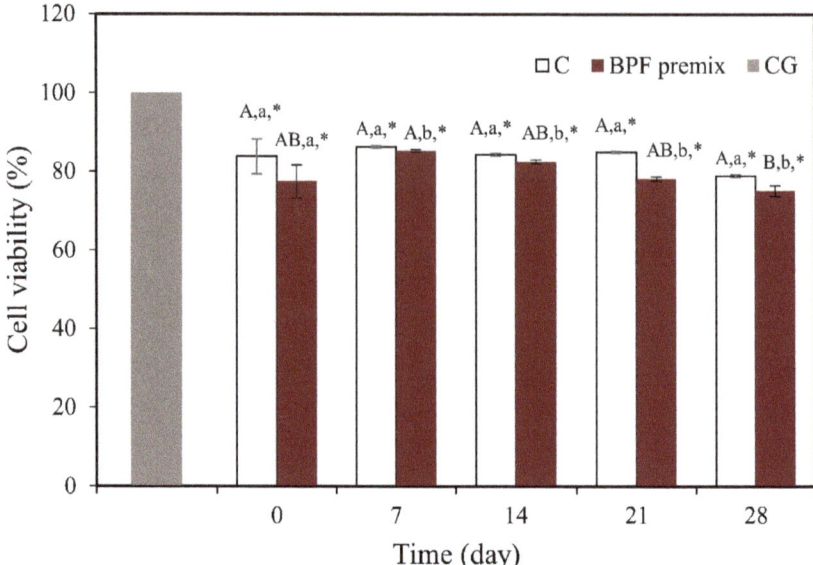

Figure 4. Inhibition rates of HCT116 cells treated for 24 h with supernatants from yogurts prepared by applying neat starter cultures (control (C)) and BPF premix. CG; cell growth control. Values are presented as mean ± standard error (n = 3); ^{AB} a significant difference in means during time within the same sample; ^{ab} a significant difference in means between samples at the same time according to HSD test ($p < 0.05$); * a significant difference in means between all samples and CG according to the Dunnett's test ($p < 0.05$).

3.7. Palatability Study

3.7.1. Trial with Human Panelists

Data reported in the literature related to the sensory analysis of beet dairy beverages are quite diverse. Smoothies with beets had high acceptability in terms of color, but the intense earthy flavor was not described as pleasant [12]. However, yogurt enriched with beetroot powder gained higher acceptance for color, texture, flavor, and overall acceptability [13]. In accordance with these findings, average hedonic scores for "color acceptance", "odor acceptance", "taste acceptance", "acceptance of sourness", and "texture acceptance" of BPF yogurt were 4.68, 4.05, 3.43, 3.30, 3.50, respectively, i.e., more or less within the range of "like a little" category. Thus, the majority of tested respondents liked the BPF yogurt.

The results of the mean drop analysis are shown in Figure 5. Statistically significant mean drops ($p < 0.05$) in sufficiently large consumer groups (≥20%) were noted only for beetroot taste (42.5%). The "beetroot taste" could be softened by using pomace flour obtained from a different selection of beetroot varieties and by adding other kinds of vegetable or fruit pomace flour. Further, adding probiotic strains could modify the taste and contribute to masking the taste of beets.

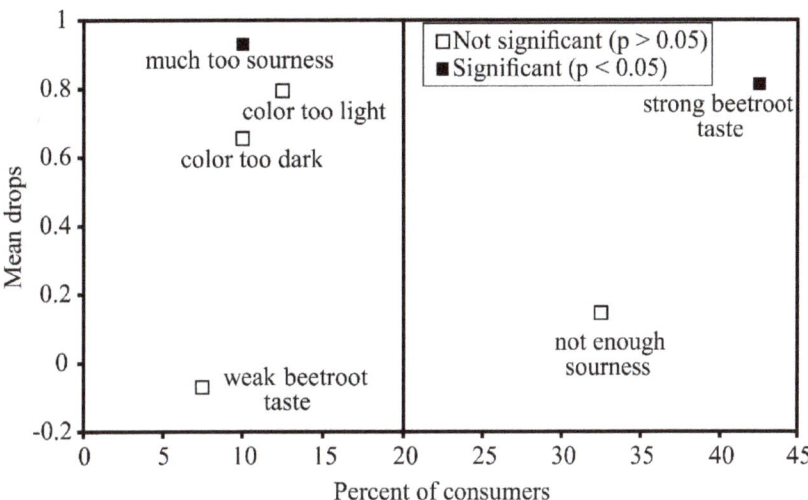

Figure 5. Mean drop analysis of the BPF yogurt sample ($N = 40$).

3.7.2. Feeding Trial with Dogs

The yogurt prepared with BPF premixes was chosen first more often than commercial yogurt with raspberry (13:5). "First bite" is commonly assumed to be related to the aromatic characteristics of the food, and thus it is a critical feature in palatability assessment [33]. Eight out of ten dogs showed no resistance to eating; there were no leftovers from tested foods. On the second day of testing, two dogs tasted neither BPF yogurt nor the control product. Unlike other canines' diet that is based on dry food, their meals mostly include raw or boiled meat. Total food intake was 0.45 for both products. Other important criteria for pet foods' evaluation are stool consistency and quality. The consumption of BPF yogurt did not affect fecal consistency. Stool quality was rated with values 3 or 4 according to the Bristol stool scale. Further, none of the urine samples appeared colored (pink or red) due to BPF yogurt intake.

3.8. Consumers Attitude Survey

As a potential link between the basal canine diet and their preferences for functional beverages was detected, a questionnaire related to the owners' attitudes towards functional treats, particularly yogurt intended for dog nutrition and current dogs' feeding practices, was administrated. A total of one hundred dog owners participated in this survey. The owner's data and canine signalment are presented in Supplementary Table S3. Importantly, most dogs (90%) had an ideal body condition according to their owner's estimation. Furthermore, overweight condition and obesity are the most underrated common disorders. According to the data reported in the literature [45], the body conditions of the dogs are probably underestimated when they are based on the subjective assessments of the owners.

A significant percent of the survey respondents are aware that dogs could consume yogurt (67%) and that fermented dairy products may have beneficial effects (49%) on canine health. However, only 17% of dog owners are familiar with the fact that yogurts designed for dog nutrition exist on the market. Moreover, only 20% of dog owners were informed that dogs are allowed to consume beets. Therefore, it is important that information considering the health benefits of functional, fermented dairy products for dog feed becomes widely available to owners. Above all, it is crucial that owners become aware of the actual body condition of their companion animals. Moreover, the biggest financial burden when maintaining a dog for most people is related to pet healthcare costs (43%). A healthy dog diet can prevent the development of various diseases. Food containing fibers could increase satiety, decrease blood cholesterol concentrations, and promote gut commensal bacteria growth [6]. Our previous study demonstrated that the addition of apple pomace into high fat and sugar, as well as standard diet, decreased feed efficiency ratio significantly and improved glucose tolerance [46]. Thus, yogurts with BPF premix should be considered as adequate functional treats and included in an appropriate feeding plan. Furthermore, the owners are largely responsible for the obesity of their companion animals. Moreover, humans and canines share most of the causative factors and the mental/physical epidemiology associated with overweight problems [47]. Therefore, the concept of sharing a fiber-enriched meal between dogs and owners is imposed. According to our results, most of the respondents (80%) who participated in the palatability study/trial with a human panelist stated that they would like to share BPF yogurt with a dog.

4. Conclusions

An innovative way of beetroot pomace utilization in human and animal nutrition was demonstrated. As a source of DF and LAB carrier, BPF was successfully employed to obtain a fermented beverage with improved functional properties. Wet fluidized bed was estimated as a feasible technique to achieve better physical properties and storability of BPF and to employ it as adequate support of LAB starter culture. The applicability of granulated BPF conceptualized as a novel DIY formula for obtaining a fresh, nutritious yogurt suitable for human and dog consumption was confirmed. To our knowledge, this is the first study that focuses on the application of pomace granulated with LAB as a premix for fermented beverages. The core concepts of BPF premix were the functionality and palatability of yogurts amended to both humans and dogs. Enrichment of BPF yogurt with 1% DF w/w improved cytotoxic activity against the human colon cancer cell line and anti-*Escherichia* activity, leading to the conclusion that its introduction into the diet might contribute to the well-being of consumers and their animal companions. Acceptable sensory properties for both humans and dogs, and positive consumers' attitude toward granulated BPF as a premix, provided useful guidance and initial evaluation of its market prospective. The concept of sharing food with pets was also positively received by consumers. Additional improvement of yogurt functionality and palatability, as well as LAB viability by the further coating of granulated BPF with various protectants, remains to be investigated in the near future.

5. Patents

Gorjanović, S., Zlatanović, S., Jovanović, M., Đurišić, M., Micić, D., Šoštarić, T., Lopičić Z. Granulates of fruit and vegetable flours with prebiotics and lactic acid bacteria P-2021/0695, 2021. Patent Application.

Supplementary Materials: The following are available online at https://www.mdpi.com/2304-8158/10/8/1696/s1, Table S1: The dog's signalment, Table S2: Ingredients of the commercial yogurt, Table S3: Nutritional survey. Survey respondents' demographics (A), dogs signalment (N = 100) (B), financial requirements (C), and dog feeding practices (D), Material S Survey.

Author Contributions: Conceptualization, M.J. and S.G.; methodology, M.J., S.Z., M.Đ. and D.B.; software, D.M.; validation, S.Z., D.B. and D.M.-Ć.; formal analysis, M.J., M.Đ. D.B. and S.Z.; investigation, M.J. and S.G.; resources, S.G.; data curation, D.M.; writing—original draft preparation, M.J. and M.Đ.; writing—review and editing, S.G.; visualization, M.J. and D.M.; supervision, S.G. and D.M.-Ć.; project administration, S.G.; funding acquisition, S.G. All authors have read and agreed to the published version of the manuscript.

Funding: This research was funded by the Ministry of Education, Science, and Technological Development of the Republic of Serbia, contract number 200051. The APC was funded by Institute of General and Physical Chemistry, Belgrade, Serbia.

Institutional Review Board Statement: The care and treatment of laboratory animals were performed according to the regulations and standards of the national (Serbian) Law on the Experimental Animal Treatment and European Directive 2010/63/EU (European Convention for the Protection of Vertebrate Animals used for Experimental and other Scientific Purposes).

Informed Consent Statement: Informed consent was obtained from all subjects involved in the study.

Data Availability Statement: The data presented in this study are available on request from the corresponding author.

Acknowledgments: The authors are grateful to Slavica Ristić, Faculty of Medicine, University of Belgrade, Belgrade, Serbia for the performed toxicity evaluation and Dragana Četojević-Simin, Oncology Institute of Vojvodina, Sremska Kamenica, Serbia, for providing HCT116 (ATCC CCL-247) cells.

Conflicts of Interest: The authors declare no conflict of interest. The funders had no role in the design of the study, in the collection, analyses, or the interpretation of data, in the writing of the manuscript, nor in the decision to publish the results.

References

1. OECD/FAO. *OECD-FAO Guidance for Responsible Agricultural Supply Chains*; OECD Publishing: Paris, France, 2016.
2. Schebesta, H.; Candel, J.J. Game-changing potential of the EU's Farm to Fork Strategy. *Nat. Food* **2020**, *1*, 586–588. [CrossRef]
3. Kok, K.P.; Den Boer, A.C.; Cesuroglu, T.; Van Der Meij, M.G.; de Wildt-Liesveld, R.; Regeer, B.J.; Broerse, J.E. Transforming Research and Innovation for Sustainable Food Systems—A Coupled-Systems Perspective. *Sustainability* **2019**, *11*, 7176. [CrossRef]
4. Clifford, T.; Howatson, G.; West, D.J.; Stevenson, E.J. The potential benefits of red beetroot supplementation in health and disease. *Nutrients* **2015**, *7*, 2801–2822. [CrossRef]
5. Kohajdová, Z.; Karovičová, J.; Kuchtová, V.; Lauková, M. Utilisation of beetroot powder for bakery applications. *Chem. Pap.* **2018**, *72*, 1507–1515. [CrossRef]
6. De Godoy, M.R.; Kerr, K.R.; Fahey, G.C., Jr. Alternative dietary fiber sources in companion animal nutrition. *Nutrients* **2013**, *5*, 3099–3117. [CrossRef]
7. Vulić, J.; Čanadanović-Brunet, J.; Ćetković, G.; Tumbas, V.; Djilas, S.; Četojević-Simin, D.; Čanadanović, V. Antioxidant and cell growth activities of beet root pomace extracts. *J. Funct. Foods* **2012**, *4*, 670–678. [CrossRef]
8. Tumbas Šaponjac, V.; Čanadanović-Brunet, J.; Ćetković, G.; Jakišić, M.; Djilas, S.; Vulić, J.; Stajčić, S. Encapsulation of beetroot pomace extract: RSM optimization, storage and gastrointestinal stability. *Molecules* **2016**, *21*, 584. [CrossRef]
9. Wadhwa, M.; Bakshi, M.P.S.; Makkar, H.P.S. Wastes to worth: Value added products from fruit and vegetable wastes. *CAB Int.* **2015**, *43*, 1–25. [CrossRef]
10. Baião, D.D.S.; da Silva, D.V.; Paschoalin, V.M. Beetroot, a Remarkable Vegetable: Its Nitrate and Phytochemical Contents Can be Adjusted in Novel Formulations to Benefit Health and Support Cardiovascular Disease Therapies. *Antioxidants* **2020**, *9*, 960. [CrossRef]
11. Barbu, V.; Cotârleț, M.; Bolea, C.A.; Cantaragiu, A.; Andronoiu, D.G.; Bahrim, G.E.; Enachi, E. Three Types of Beetroot Products Enriched with Lactic Acid Bacteria. *Foods* **2020**, *9*, 786. [CrossRef]
12. Cano-Lamadrid, M.; Tkacz, K.; Turkiewicz, I.P.; Clemente-Villalba, J.; Sánchez-Rodríguez, L.; Lipan, L.; García-García, E.; Carbonell-Barrachina, Á.A.; Wojdyło, A. How a Spanish Group of Millennial Generation Perceives the Commercial Novel Smoothies? *Foods* **2020**, *9*, 1213. [CrossRef]
13. Yadav, M.; Masih, D.; Sonkar, C. Development and quality evaluation of beetroot powder incorporated yogurt. *Int. J. Eng. Sci.* **2016**, *4*, 582–586.
14. Phuapaiboon, P.; Leenanon, B.; Levin, R.E. Effect of Lactococcus lactis immobilized within pineapple and yam bean segments, and jerusalem artichoke powder on its viability and quality of yogurt. *Food Bioproc. Technol.* **2013**, *6*, 2751–2762. [CrossRef]
15. Strasser, S.; Neureiter, M.; Geppl, M.; Braun, R.; Danner, H. Influence of lyophilization, fluidized bed drying, addition of protectants, and storage on the viability of lactic acid bacteria. *J. Appl. Microbiol.* **2009**, *107*, 167–177. [CrossRef] [PubMed]
16. Mujaffar, S.; Ramsumair, S. Fluidized bed drying of pumpkin (*Cucurbita* sp.) seeds. *Foods* **2019**, *8*, 147. [CrossRef]
17. Cil, B.; Topuz, A. Fluidized bed drying of corn, bean and chickpea. *J. Food Process Eng.* **2010**, *33*, 1079–1096. [CrossRef]
18. Ghandi, A.; Adhikari, B.; Powell, I.B. Powders containing microorganisms and enzymes. In *Handbook of Food Powders*, 1st ed.; Bhandari, B., Bansal, N., Zhang, M., Schuck, P., Eds.; Woodhead Publishing: Philadelphia, PA, USA, 2013; pp. 593–624.
19. Rama, G.R.; Kuhn, D.; Beux, S.; Maciel, M.J.; de Souza, C.F.V. Potential applications of dairy whey for the production of lactic acid bacteria cultures. *Int. Dairy J.* **2019**, *98*, 25–37. [CrossRef]
20. Nag, A.; Das, S. Effect of trehalose and lactose as cryoprotectant during freeze-drying, in vitro gastro-intestinal transit and survival of microencapsulated freeze-dried Lactobacillus casei 431 cells. *Int. J. Dairy Technol.* **2013**, *66*, 162–169. [CrossRef]
21. Kaur Sidhu, M.; Lyu, F.; Sharkie, T.P.; Ajlouni, S.; Ranadheera, C.S. Probiotic Yogurt Fortified with Chickpea Flour: Physico-Chemical Properties and Probiotic Survival during Storage and Simulated Gastrointestinal Transit. *Foods* **2020**, *9*, 1144. [CrossRef]
22. Sah, B.N.P.; Vasiljevic, T.; McKechnie, S.; Donkor, O.N. Antibacterial and antiproliferative peptides in synbiotic yogurt—Release and stability during refrigerated storage. *J. Dairy Sci.* **2016**, *99*, 4233–4242. [CrossRef]
23. Brodziak, A.; Król, J.; Barłowska, J.; Teter, A.; Florek, M. Changes in the physicochemical parameters of yoghurts with added whey protein in relation to the starter bacteria strains and storage time. *Animals* **2020**, *10*, 1350. [CrossRef]

24. Park, H.E.; Kim, Y.J.; Kim, M.; Kim, H.; Do, K.H.; Kim, J.K.; Ham, J.S.; Lee, W.K. Effects of Queso Blanco cheese containing Bifidobacterium longum KACC 91563 on fecal microbiota, metabolite and serum cytokine in healthy beagle dogs. *Anaerobe* **2020**, *64*, 102234. [CrossRef]
25. Park, D.H.; Kothari, D.; Niu, K.M.; Han, S.G.; Yoon, J.E.; Lee, H.G.; Kim, S.K. Effect of Fermented Medicinal Plants as Dietary Additives on Food Preference and Fecal Microbial Quality in Dogs. *Animals* **2019**, *9*, 690. [CrossRef] [PubMed]
26. Jovanović, M.; Petrović, M.; Miočinović, J.; Zlatanović, S.; Laličić Petronijević, J.; Mitić-Ćulafić, D.; Gorjanović, S. Bioactivity and Sensory Properties of Probiotic Yogurt Fortified with Apple Pomace Flour. *Foods* **2020**, *9*, 763. [CrossRef]
27. Corsato Alvarenga, I.; Aldrich, C.G. The effect of increasing levels of dehulled faba beans (*Vicia faba* L.) on extrusion and product parameters for dry expanded dog food. *Foods* **2019**, *8*, 26. [CrossRef] [PubMed]
28. AOAC. *Official Methods of Analysis of AOAC International*, 17th ed.; Sections 925.10, 992.23, 920.85, 923.03; Association of Official Analytical Chemists: Gaithersburg, MD, USA, 2000.
29. AOAC. *Official Methods of Analysis of AOAC International*, 18th ed.; Section 962.09; Association of Official Analytical Chemists: Gaithersburg, MD, USA, 2006.
30. Đuriš, M.; Arsenijević, Z.; Jaćimovski, D.; Radoičić, T.K. Optimal pixel resolution for sand particles size and shape analysis. *Powder Technol.* **2016**, *302*, 177–186. [CrossRef]
31. Wang, X.; Kristo, E.; LaPointe, G. Adding apple pomace as a functional ingredient in stirred-type yogurt and yogurt drinks. *Food Hydrocoll.* **2020**, *100*, 105343. [CrossRef]
32. Tomic, N.; Djekic, I.; Hofland, G.; Smigic, N.; Udovicki, B.; Rajkovic, A. Comparison of Supercritical CO_2-Drying, Freeze-Drying and Frying on Sensory Properties of Beetroot. *Foods* **2020**, *9*, 1201. [CrossRef] [PubMed]
33. Aldrich, G.C.; Koppel, K. Pet food palatability evaluation: A review of standard assay techniques and interpretation of results with a primary focus on limitations. *Animals* **2015**, *5*, 43–55. [CrossRef]
34. Kim, D.H.; Jeong, D.; Kang, I.B.; Lim, H.W.; Cho, Y.; Seo, K.H. Modulation of the intestinal microbiota of dogs by kefir as a functional dairy product. *J. Dairy Sci.* **2019**, *102*, 3903–3911. [CrossRef]
35. Morelli, G.; Bastianello, S.; Catellani, P.; Ricci, R. Raw meat-based diets for dogs: Survey of owners' motivations, attitudes and practices. *BMC Vet. Res.* **2019**, *15*, 74. [CrossRef] [PubMed]
36. Manojlović, V.; Nedović, V.A.; Kailasapathy, K.; Zuidam, N.J. Encapsulation of probiotics for use in food products. In *Encapsulation Technologies for Active Food Ingredients and Food Processing*, 1st ed.; Nedović, V.A., Zuidam, N.J., Eds.; Springer: New York, NY, USA, 2010; pp. 269–302.
37. Nasir, G.; Chand, K.; Azad, Z.A.A.; Nazir, S. Optimization of Finger Millet and Carrot Pomace based fiber enriched biscuits using response surface methodology. *J. Food Sci. Technol.* **2020**, *57*, 4613–4626. [CrossRef]
38. Carr, R.L. Evaluating flow properties of solids. *Chem. Eng.* **1965**, *18*, 163–168.
39. Rochat, T.; Gratadoux, J.J.; Gruss, A.; Corthier, G.; Maguin, E.; Langella, P.; van de Guchte, M. Production of a heterologous nonheme catalase by Lactobacillus casei: An efficient tool for removal of H_2O_2 and protection of Lactobacillus bulgaricus from oxidative stress in milk. *Appl. Environ. Microbiol.* **2006**, *72*, 5143–5149. [CrossRef] [PubMed]
40. González-Forte, L.; Bruno, E.; Martino, M. Application of coating on dog biscuits for extended survival of probiotic bacteria. *Anim. Feed Sci. Technol.* **2014**, *195*, 76–84. [CrossRef]
41. Albadran, H.A.; Chatzifragkou, A.; Khutoryanskiy, V.V.; Charalampopoulos, D. Stability of probiotic Lactobacillus plantarum in dry microcapsules under accelerated storage conditions. *Food Res. Int.* **2015**, *74*, 208–216. [CrossRef] [PubMed]
42. Rios, E.A.; Ramos-Pereira, J.; Santos, J.A.; López-Díaz, T.M.; Otero, A.; Rodríguez-Calleja, J.M. Behaviour of Non-O157 STEC and Atypical EPEC during the Manufacturing and Ripening of Raw Milk Cheese. *Foods* **2020**, *9*, 1215. [CrossRef]
43. Vulić, J.J.; Ćebović, T.N.; Čanadanović, V.M.; Ćetković, G.S.; Djilas, S.M.; Čanadanović-Brunet, J.M.; Velićanski, A.S.; Cvetković, D.D.; Tumbas, V.T. Antiradical, antimicrobial and cytotoxic activities of commercial beetroot pomace. *Food Funct.* **2013**, *4*, 713–721. [CrossRef]
44. Meydani, S.N.; Ha, W.K. Immunologic effects of yogurt. *Am. J. Clin. Nutr.* **2000**, *71*, 861–872. [CrossRef]
45. Morelli, G.; Marchesini, G.; Contiero, B.; Fusi, E.; Diez, M.; Ricci, R. A survey of dog owners' attitudes toward treats. *J. Appl. Anim. Welf. Sci.* **2020**, *23*, 1–9. [CrossRef]
46. Gorjanović, S.; Micić, D.; Pastor, F.; Tosti, T.; Kalušević, A.; Ristić, S.; Zlatanović, S. Evaluation of apple pomace flour obtained industrially by dehydration as a source of biomolecules with antioxidant, antidiabetic and antiobesity effects. *Antioxidants* **2020**, *9*, 413. [CrossRef] [PubMed]
47. Orsolya Julianna, T.; Kata, V.; Vanda Katalin, J.; Péter, P. Factors Affecting Canine Obesity Seem to Be Independent of the Economic Status of the Country—A Survey on Hungarian Companion Dogs. *Animals* **2020**, *10*, 1267. [CrossRef] [PubMed]

Article

The Use of Winery by-Products to Enhance the Functional Aspects of the Fresh Ovine "Primosale" Cheese

Raimondo Gaglio [1,*], Pietro Barbaccia [1], Marcella Barbera [1], Ignazio Restivo [2], Alessandro Attanzio [2], Giuseppe Maniaci [1], Antonino Di Grigoli [1], Nicola Francesca [1], Luisa Tesoriere [2], Adriana Bonanno [1], Giancarlo Moschetti [1] and Luca Settanni [1]

[1] Dipartimento Scienze Agrarie, Alimentari e Forestali, Ed. 5, Università degli Studi di Palermo, Viale delle Scienze, 90128 Palermo, Italy; pietro.barbaccia@unipa.it (P.B.); marcella.barbera@unipa.it (M.B.); giuseppe.maniaci@unipa.it (G.M.); antonino.digrigoli@unipa.it (A.D.G.); nicola.francesca@unipa.it (N.F.); adriana.bonanno@unipa.it (A.B.); giancarlo.moschetti@unipa.it (G.M.); luca.settanni@unipa.it (L.S.)

[2] Dipartimento di Scienze e Tecnologie Biologiche, Chimiche e Farmaceutiche, Università degli Studi di Palermo, Via Archirafi 32, 90123 Palermo, Italy; ignazio.restivo@unipa.it (I.R.); alessandro.attanzio@unipa.it (A.A.); luisa.tesoriere@unipa.it (L.T.)

* Correspondence: raimondo.gaglio@unipa.it

Abstract: Fresh ovine "primosale" cheese was processed with the addition of grape pomace powder (GPP). Cheese making was performed using pasteurized ewes' milk and four selected *Lactococcus lactis* strains (Mise36, Mise94, Mise169 and Mise190) inoculated individually. For each strain the control cheese (CCP) was not added with GPP, while the experimental cheese (ECP) was enriched with 1% (w/w) GPP. GPP did not influence the starter development that reached levels of 10^9 CFU/g in all final cheeses. The comparison of the bacterial isolates by randomly amplified polymorphic DNA (RAPD)-PCR showed the dominance of the added strains over indigenous milk bacteria resistant to pasteurization. GPP addition reduced fat content and determined an increase of protein and of secondary lipid oxidation. Sensory tests indicated that cheeses CCP94 and ECP94, produced with the strain Mise94, reached the best appreciation scores. Following in vitro simulated human digestion, bioaccessible fraction of ECP94 showed antioxidant capacity, evaluated as radical scavenging activity and inhibition of membrane lipid oxidation, significantly higher than that from CCP94, with promising increase in functional properties. Thus, the main hypothesis was accepted since the functional aspects of the final cheeses improved, confirming that GPP is relevant for sustainable nutrition by using winemaking by-products.

Keywords: functional ovine cheese; grape pomace powder; *Lactococcus lactis*; physicochemical properties; polyphenols; volatile organic compounds; antioxidant properties

1. Introduction

The production of wine generates a large amount of by-products, known as grape pomace, that are composed by a mix of grape skins and seeds [1]. Grape pomace constitutes a relevant environmental issue related to the production of wine [2] even though it represents a consistent source of functional compounds such as polyphenols and dietary fiber [3]. These compounds can exert a positive impact on human health, especially for the prevention of diseases associated with oxidative stress such as cancer, stroke and coronary heart disease [4]. Others main constituents of grape pomace are colorants, minerals and organic acids [5,6]. Moreover, considering that grape pomace is classified as a Generally Recognized As Safe (GRAS) matrix by the U.S. Food and Drug Administration [7], this by-product possess a high potential to be used as alternative to synthetic antioxidants in food processing such as butylated hydroxytoluene and butylated hydroxyanisole, coinciding with consumers demand for healthy and functional foods with no chemical additives [8]. For this reason, academic and industrial research is focusing on the use of grape pomace powder (GPP) as a food additive or novel ingredient in different food productions [9].

Dairy products contain a low concentration of phenolic compounds, antioxidants and fibers [10]. To this purpose, the fortification of cheeses with non-dairy ingredients represents an improved strategy to enhance the functional and bioactive properties of the final products [11]. So far, the fortification with GPP was proposed for various dairy products obtained with bovine milk such as fermented milk beverages [12], yogurt [13] and processed semi-hard and hard cheeses [14].

Sicily is a region of southern Italy characterized by an intense breed of sheep that play an important role in the protection of the local cultural heritage related to the traditions of shepherds, uses and habits of the mountain populations and typical cuisine. In this region, sheep milk is almost totally processed into traditional cheese types, but several cheese producers pushed research institutes to develop dairy functional foods in order to enlarge the number of dairy products and, especially, to provide cheese with a positive image among consumers aware of the effects of bioactive compounds on the human body.

The objective of the present work was to produce a novel fresh ewes' milk pressed cheese with the addition of GPP and selected *Lactococcus lactis* strains resistant to the main grape polyphenols [15]. The final cheeses were subjected to the evaluation of the microbiological, physicochemical, sensory and functional aspects.

2. Materials and Methods

2.1. Grape Pomace Powder Production

Red wine grape pomace of Nero d'Avola cultivar composed by a mix of grape seeds and skins was provided by a winemaking company of Trapani (Southern Italy) at the end of the vintage 2019. Grape pomace was collected after 150 d of post-fermentation maceration. Following the methodology described by Marchiani et al. [14], grape pomace were dried at 54 °C for 48 h in a semi-industrial oven (Compact Combi, Electrolux, Pordenone, Italy). The dry grape pomace was reduced to a particle size of 250 µm through a Retsch centrifugal Mill ZM1 (Haan, Germany).

2.2. Strains and Development of Natural Milk Starter Cultures

The strains *Lactococcus lactis* Mise36, Mise94 Mise169 and Mise190, belonging to the culture collection of the Department of Agricultural, Food and Forest Sciences (University of Palermo, Italy), were used in this study. These strains were previously tested for their resistance to GPP and for their main dairy traits [15]. The cultures were individually overnight grown at 30 °C in M17 broth (Biotec, Grosseto, Italy), centrifuged at $10,000\times g$ for 5 min to separate the cells from supernatant, washed and re-suspended in Ringer's solution (Sigma-Aldrich, Milan, Italy). All strains were individually inoculated into 1 L of whole fat UHT milk (Conad, Mantova, Italy) at the final concentration of about 10^6 CFU/mL obtaining four distinct natural milk starter cultures (NMSC). After incubation for 24 h at 30 °C, the NMSCs were separately used for cheese making.

2.3. Experimental Cheese Production and Sample Collection

Cheese productions were carried out under controlled conditions at a dairy pilot plant (Biopek, Gibellina, Italy) using ewes' milk from the indigenous Sicilian sheep breed "Valle del Belice" during February 2020. The experimental plan included eight different cheese productions as reported in Figure 1. For each strain, 40 L of pasteurized ewes' milk was divided into two plastic vats (20 L each) representing two different trials. Both vats were inoculated with 200 mL of the corresponding NMSC (Mise36, Mise94, Mise169 and Mise190) to reach a final cell density of 10^7 CFU/mL. One vat represented the control cheese production (CCP) while the second vat represented the experimental cheese production (ECP) that, after curd extraction, was added with 1% (w/w) GPP.

Both CCP and ECP were performed applying "primosale" pressed cheese technology as reported in Figure 2.

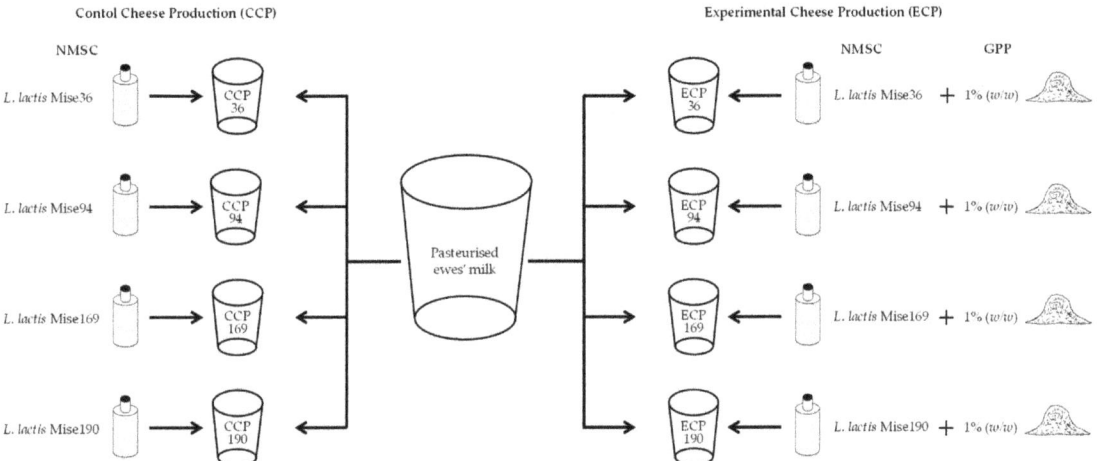

Figure 1. Experimental design of cheese productions. Abbreviations: NMSC, natural milk starter culture; GPP, grape pomace powder; *L. Lactococcus*.

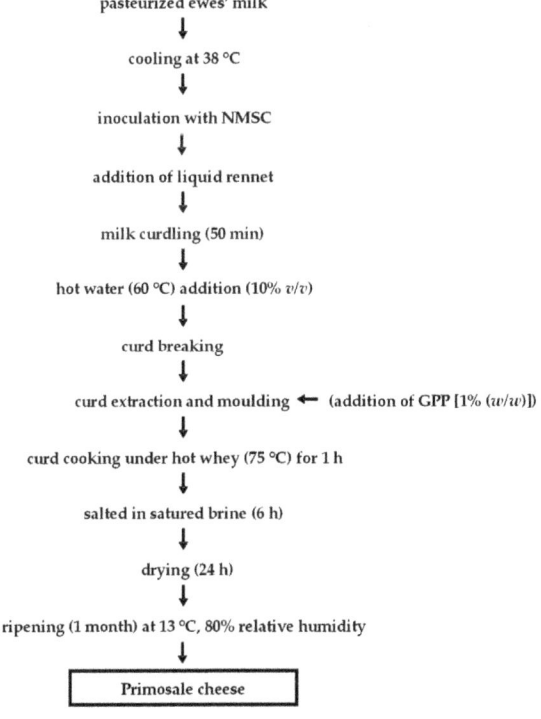

Figure 2. Flow diagrams of Primosale cheese production. Abbreviations: NMSC, natural milk starter culture; GPP, grape pomace powder.

Cheese productions were carried out in duplicate in two consecutive weeks. The measurement of pH during cheese making (from milk to curd) was carried out with a portable pH-meter pH 70 + DHS (XS Instruments, Carpi, Italy). In order to follow the curd acidification, one sample of curd was collected from each production and kept at ambient temperature for 7 days. Each curd sample was subjected to the monitoring of pH at 2-h intervals for the first 8 h and, then, after 1, 2, 3 and 7 days from milk curdling.

The following matrices were sampled during cheese production: bulk milk, pasteurized bulk milk, inoculated milk after addition of NMSC, curd, GPP and cheese after 1 month of ripening occurred at 13 °C and 85% relative humidity.

2.4. Microbiological Analyses

Cell suspensions of milk samples were subjected to decimal serial dilutions in Ringer's solution (1:10), while GPP, curd and cheese samples were first homogenized in Ringer's solution by a stomacher (Bag-Mixer 400; Interscience, Saint Nom, France) for 2 min at the maximum speed (blending power 4) and then serially diluted. Cell suspensions of GPP were subjected to plate count for the main microbial groups belonging to the pro-technological, spoilage and pathogenic populations following the approach of Cruciata et al. [16].

Cell suspensions of raw milk and pasteurized milk were analyzed for total mesophilic microorganisms (TMM), mesophilic rods and cocci as reported by Barbaccia et al. [15].

Milk inoculated with each NMSC, curd and cheese samples were analyzed only for the levels of TMM and *L. lactis* on skim milk agar [17] incubated aerobically at 30 °C for 72 h and M17 agar incubated anaerobically at 30 °C for 48 h, respectively. All media and supplements were purchased from Biotec. Plate counts were performed in duplicate.

2.5. Phenotypic Grouping, Genotypic Differentiation and Identification of Thermoduric LAB

After growth, all colonies with different morphologies [in order to collect the total lactic acid bacteria (LAB) biodiversity] from the highest dilutions of pasteurized milk sample suspensions were isolated from M17 agar plates and purified by sub-culturing. After microscopic inspection, the pure cultures were tested by KOH assay to determine Gram type and for the presence of catalase by suspension of colonies into H_2O_2 5% (v/v). All presumptive LAB cultures (Gram-positive and catalase-negative) were grouped on their morphological/physiological/biochemical traits as described by Gaglio et al. [18]. Cell lysis for DNA extraction was performed by using DNA-SORB-B kit (Sacace Biotechnologies Srl, Como, Italy) following the protocol provided by the manufacturer. The differentiation at strain level was performed by randomly amplified polymorphic DNA (RAPD)-PCR analysis as reported by Gaglio et al. [19]. RAPD profiles were analyzed through Gelcompare II software version 6.5 (Applied-Maths, Sin Marten Latem, Belgium). Genotypic identification was performed by sequencing the 16S rRNA gene following the procedures applied by Gaglio et al. [20].

2.6. Persistence of the Added Strains

The dominance of the strains individually inoculated as starter cultures (*L. lactis* Mise36, Mise94, Mise169 and Mise190) over LAB resistant to pasteurization was confirmed, after colony isolation, by microscopic inspection and RAPD-PCR profile comparison between LAB collected during cheeses making and those of *L. lactis* Mise36, Mise94, Mise169 and Mise190 pure cultures.

2.7. Physicochemical Analysis of Cheeses

Color of external and internal surfaces of the cheeses of the cheeses was assessed by a Minolta Chroma Meter CR300 (Minolta, Osaka, Japan) using the illuminant C; measurements of lightness (L*, from 0 = black, to 100 = white), redness (a*, from red = +a, to green = −a) and yellowness (b*, from yellow = +b, to blue = −b) were performed according to the CIE L* a* b* system [21].

The maximum resistance to compression (compressive stress, N/mm^2) of samples (2 cm × 2 cm × 2 cm) kept at room temperature (22 °C) was measured, as index of cheese hardness, with an Instron 5564 tester (Instron, Trezzano sul Naviglio, Milan, Italy).

The freeze-dried cheese samples were analyzed for the content of dry matter (DM), fat, protein (N × 6.38) and ash as reported by Bonanno et al. [22].

The products of secondary lipid oxidation were determined as thiobarbituric acid-reactive substances (TBARS), expressed as µg malonylaldehyde (MDA)/kg DM, as reported by Bonanno et al. [22]. Each physicochemical determination was assessed in duplicate.

2.8. Volatile Organic Compounds Emitted from Cheeses

Three grams of dried grape pomace and 5 g of chopped cheese samples, were put into 25 mL glass vials sealed with silicon septum. Extraction of volatile compounds were performed through the headspace solid phase microextraction SPME (DVB/CAR/PDMS, 50 mm, Supelco) fiber. The samples were exposed to the fiber under continuous stirring at 60 °C for 15 min. After adsorption, the SPME fiber was thermally desorbed for 1 min through a splitless GC injector at 250 °C. The chromatographic analyses was performed by a gas chromatograph (Agilent 6890) equipped with a mass selective detector (Agilent 5975 c) and a DB-624 capillary column (Agilent Technologies, 60 m, 0.25 mm, 1.40 µm). Chromatographic conditions were as follows: helium carrier gas at 1 mL/min and an oven temperature program with a 5 min isotherm at 40 °C followed by a linear temperature increase of 5 °C min up to 200 °C, where it was held for 2 min. The MS scan conditions applied were: scan acquisition mode; 230 °C Interface temperature; acquisition mass range from 40 to 400. For each sample three replicates were analyzed. The identification of significant volatile compounds were performed through a comparison of the MS spectra with NIST05 library. The relative proportions of the identified constituents were expressed as percentages obtained by GC-MS peak area normalization with total area of significant peaks.

2.9. Sensory Evaluation

All cheeses were evaluated by sensory analysis in order to define and detect differences between CCP and ECP. The cheese samples were cubed (approximately 1 cm each side) and then coded and presented on white paperboard plates in a random order. The judges also had available an entire transverse slice of each cheese for evaluating appearance attributes. A total of twelve descriptive attributes were judged by a panel of 11 assessors members (six men and five woman, from 21 to 65 years old). The judges had several years of experience in sensory evaluation of dairy products; however, they were specifically trained for cheese attribute evaluation following the ISO 8589 [23] indications. Each attribute was chosen among those reported by Niro et al. [24] and evaluated by Costa et al. [25]. The intensity of each attribute was quantified using a line scale from 0 to 7 (cm) as reported by Faccia et al. [26].

2.10. Simulated Gastrointestinal Digestion

Simulated in vitro human digestion procedure, including the oral, gastric and small intestine phases, was performed three times according to Attanzio et al. [27].

Oral Phase. Samples of 15.0 g of cheese were homogenized using a Waring blender (Waring, New Hartford, CT, USA) in 40 mL of a buffered pH 6.8 solution simulating saliva. Artificial saliva, prepared following official pharmacopoeia, contained: NaCl (0.126 g), KCl (0.964 g) KSCN (0.189 g), KH_2PO_4 (0.655 g), urea (0.200 g), $Na_2SO_4 \cdot 10H_2O$ (0.763 g), NH_4Cl (0.178 g), $CaCl_2 \cdot 2H_2O$ (0.228 g) and $NaHCO_3$ (0.631 g) in 1 L of distilled water. The final pH of the preparations (post-oral digest, PO) ranged between 4.0 and 4.5. An aliquot of 5 mL was stored at −80 °C until analysis.

Gastric and Small Intestinal Phase. The sample from the oral phase was acidified at pH 2.0 with HCl, and 8 mg/mL porcine pepsin (3200–4500 units/mg) was added. The sample was transferred in an amber bottle, sealed, and incubated in a shaking (100 rpm) water bath (type M 428-BD, Instruments s.r.l., Bernareggio, Mi, Italy) at 37 °C, for 2 h. Then the reaction mixture was placed on ice, and a 5 mL aliquot was stored at −80 °C (post-gastric digest, PG). The pH of the remaining sample was immediately brought to 7.5 with 0.5 N $NaHCO_3$, and 2.4 mg/mL porcine bile extract and 0.4 mg/mL of pancreatin from hog pancreas (amylase activity >100 units/mg) were added to initiate the small intestinal phase of digestion. The amber bottle was sealed and incubated in the shaking water bath for 2 h at 37 °C. At the end of the incubation, 5 mL of the reaction mixture (post-intestinal digest, PI) were stored at −80°C until analysis.

Preparation of the Bioaccessible Fraction. The PI digest was centrifuged at 167,000× g, for 35 min at 4 °C in a Beckman Optima TLX ultracentrifuge, equipped with an MLA-55 rotor (Beckman Instruments, Inc., Palo Alto, CA, USA), to separate the aqueous fraction (bioaccessible fraction, BF) from particulate material.

Before analysis, samples from each digestion step were centrifuged at 1500× g for 10 min at 4 °C and supernatants were brought at pH 2.0 to stabilize polyphenols.

2.11. Total Antioxidant Activity

The total antioxidant activity (TAA) of samples was measured using the ABTS radical cation decolorization assay [28]. ABTS•+ was prepared by reacting ABTS with $K_2S_2O_4$ [29]. Samples were analyzed in duplicate, at three different dilutions, within the linearity range of the assay. The vitamin E hydro-soluble analog, Trolox, was used as reference antioxidant and results were expressed as micromoles of Trolox equivalents per gram of cheese weight.

2.12. Membrane Lipid Peroxidation Assay

Pig's brain was homogenized in 10 mM phosphate buffer saline, pH 7.4 (PBS) and submitted to centrifugation at 9000× g for 20 min at 4 °C. Post-mitochondrial supernatant was then centrifuged at 105,000× g for 60 min at 4 °C in a Beckman Optima TLX ultracentrifuge. Microsomal pellet was resuspended in PBS and proteins were determined by the Bio Rad colorimetric method [30]. Microsomes, at 2 mg protein/mL concentration, were pre-incubated for 5 min at 37 °C either in the absence (control) or in the presence of variable amounts of the bioaccessible fraction of cheeses. Lipid oxidation was induced by 20 mM 2,2′-azobis (2-amidino-propane) dihydrochloride (AAPH, Sigma) for 60 min at 37 °C following Attanzio et al. [31]. Oxidized lipid formation was monitored after reaction with thiobarbituric acid (TBA), as TBARS [31]. Prior to sample processing, a calibration analytical curve was prepared at concentrations of 1, 5, 10 and 25 nmol, using tetraethoxypropane (TEP) as the standard. The absorbance was measured using a DU 640 Beckmanspectrophotometer (Beckman, Milan, Italy) at the wavelength of 532 nm. The results were expressed as nmol TBARS/mg protein.

2.13. Statistical Analysis

Microbiological data and antioxidant capacity were subjected to One-Way Variance Analysis (ANOVA) using XLStat software version 7.5.2 for Excel (Addinsoft, NY, USA) and the differences between mean were determined by Tukey's test at $p < 0.05$.

The generalized linear model (GLM) procedure in SAS 9 (Version 9.2, SAS Institute Inc., Campus Drive Cary, NC, USA) was used to analyze physicochemical data of cheeses; the model included the effects of cheese trial (1, 2), treatment (TR) with GPP (control, experimental), starter culture (NMSC: Mise36, Mise94 Mise169, Mise190) and the interaction TR*NMSC. When the effect of NMSC and TR*NMSC resulted significant ($p \leq 0.05$), means comparisons were performed by the Tukey–Kramer multiple test.

Data on sensory evaluations were tested by a 2-factor analysis of variance (ANOVA), using XLStat software version 2020.3.1 for excel, with judges (i = 1 ... 11) and cheeses (j = 1 ... 8) as fixed factors. Least square means (LSM) were compared using T test ($p < 0.05$).

3. Results and Discussion

3.1. Acidification Kinetics

The value of pH of pasteurized bulk milk was 6.88, while NMSCs reached values ranging between 4.24–4.28. After the addition of the NWSC, that represent 0.1% cows' milk in ewe's milk, bulk milk pH dropped, on average, to 6.71.

The average values of the early acidification process for both control and experimental curds are reported in Figure 3.

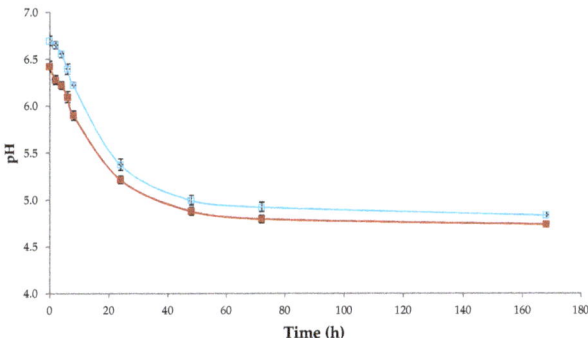

Figure 3. pH decrease during 7-days of curd acidification. Empty symbols: control curds. Full symbols: experimental curds. Results indicate mean values ± SD of the four determinations (carried out in duplicate for two independent productions) of all trials.

According to Tukey's test, statistical significant differences ($p < 0.0001$) were found between control and experimental trials for all measuring time. In particular, the experimental trials showed values 0.3–0.4 points lower than control trials. These differences are mainly imputable to the presence of organic acids such as tartaric acid, malic acid and citric acid in GPP [14].

3.2. Microbiological Analyses

The microbiological counts of GPP did not reveal the presence of any of the microbial groups object of investigation. Mainente et al. [32] assessed that the absence of microorganisms in the GPP is due to the oven-drying treatment performed on grape pomace. The ewes' milk before pasteurization was characterized by a concentration of TMM of 6.32 Log CFU/mL that is higher than the limit of <500,000 CFU/mL reported by the Commission Regulation (EC) No 853/2004 for raw ewes' milk. High levels of TMM in raw ewe's milk before processing into cheese are often detected [33,34]; LAB cocci were found at the same level (10^6 CFU/mL) of TMM, while LAB rods were one Log unit lower. Similar results were previously reported by Guarcello et al. [35] in raw ewes' milk used for PDO Pecorino Siciliano cheese production. These results indicated that milking month exerts a limited influence on the microbial load of bulk milk. After pasteurization, TMM and coccus LAB were found at 10^3 CFU/mL while LAB rods were not detected. These results confirmed what previously reported by Gaglio et al. [36] and Rynne et al. [37] that thermoduric indigenous milk LAB are able to survive the pasteurization process. No statistical significant differences ($p > 0.05$) were found for the levels of microorganisms object of investigation during all steps of cheese making. After inoculation with NMSC, all milks showed approximately 7.0 Log cycles of TMM and almost the same levels of mesophilic coccus LAB, confirming that *L. lactis* inoculums occurred at 10^7 CFU/mL. After coagulation, all control and experimental curds reached values of TMM and LAB cocci of about 10^8 CFU/g showing an increase of microbial counts as a consequence of whey draining [34].

These data also confirmed the dominance of lactococci among the microbial community of the curds, reaching values of about 9 Log CFU/g in all control and experimental primosale cheeses. These results highlighted that the addition of 1% (w/w) of GPP did not influence the fermentation process, carried out by the four strains of *L. lactis* (Mise36, Mise94 Mise169 and Mise190) used individually.

3.3. Identification of LAB Resistant to Pasteurization Process

After enumeration, presumptive LAB (Gram-positive and catalase negative) were isolated, purified and analyzed by RAPD-PCR in order to recognize the different strains that overcome the thermal treatment. RAPD analysis showed the presence of 4 different strains (Figure 4a) from a total of 25 presumptive LAB isolates that formed four distinct phenotypical groups 1 for rods and 3 for cocci (Figure 4b).

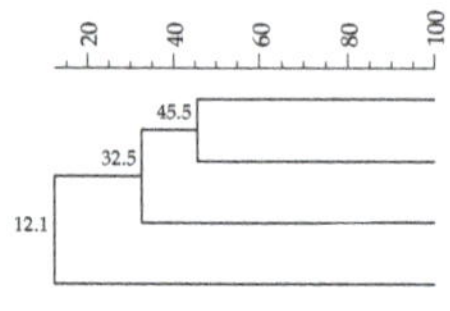

Strain	Phenotypic group	Species	Acc. No.
BK2	1	E. faecalis	MW353141
BK19	2	E. faecium	MW353142
BK7	3	E. casseliflavus	MW353143
BK14	4	Ls. fermentum	MW353144

(a)

Characters	Clusters			
	1	2	3	4
Morphology	R	C	C	C
Cell disposition	sc	sc	sc	sc
Growth:				
15 °C	-	+	+	+
45 °C	+	+	+	+
pH 9.2	nd	+	+	+
6.5% NaCl	nd	+	+	+
Resistance to 60 °C	-	-	-	-
Hydrolysis of:				
arginine	+	+	+	-
aesculin	-	+	+	+
Acid production from:				
arabinose	+	+	+	+
ribose	+	+	+	+
xylose	+	+	+	+
fructose	+	+	+	+
galactose	+	+	+	+
lactose	+	+	+	+
sucrose	+	+	-	+
glycerol	+	+	+	+
CO_2 from glucose	+	-	-	-

(b)

Figure 4. Differentiation of lactic acid bacteria (LAB) isolates from pasteurized ewes' milk. (**a**) Dendrogram obtained with combined randomly amplified polymorphic DNA (RAPD)-PCR patterns of the LAB strains identified; (**b**) phenotypic grouping of the LAB isolates based of morphological, physiological and biochemical traits. Abbreviations: E., *Enterococcus*; Ls., *Limosilactobacillus*; R, rod; C, coccus; s.c., short chain; n.d., not determined.

The sequencing of 16S rRNA gene indicated that the LAB community resistant to pasteurization process was represented by the species *Enterococcus casseliflavus*, *Enterococcus faecalis*, *Enterococcus faecium* and *Lactobacillus fermentum* (recently reclassified as *Limosilactobacillus fermentum*) [38]. All these species are part of the non-starter LAB community implicated in the ripening process of cheeses [39] and represent an essential part of the microbiota of raw ewes' milk cheeses. In particular, enterococci are often isolated from raw ewe's milk [40], while *Ls. fermentum* are during cheese ripening [41].

3.4. Monitoring of the Added Strains

The development of the added strains was monitored at several cheese production steps collecting 347 isolates which were identified and typed using a polyphasic approach combining microscopic inspection and RAPD-PCR analysis. This approach is commonly applied to monitor the added starter cultures in dairy products [20,42]. Microscopic inspection confirmed that all isolates were cocci with cells organized in short chains, typical of lactococci [15]. The direct comparison of the polymorphic profiles clearly showed the dominance of *L. lactis* Mise36, Mise94, Mise169 and Mise190 both in control and experimental cheeses, excluding any negative influence of GPP.

3.5. Physicochemical Analysis of Cheeses

The physical properties and the chemical composition of the cheeses (Table 1) were affected only by the treatment, because no significant variations caused by the starter cultures emerged. Due to the reddish color of GPP, both external and internal surfaces of experimental cheeses showed a low lightness and yellowness and a high redness. However, the indices of internal color recorded in control cheeses were comparable to those measured for primosale cheese after 21 d from production [22], as well as ripened Pecorino cheese [43].

The chemical components of cheese ranged into the levels observed in other investigations for the same cheese typology [22,44]. GPP are poor in lipid components; thus, GPP inclusion in cheese decreased fat level and, as a consequence, protein content increased. The levels of fat in control and experimental cheeses were in the range 44.52–46.31% DM and 39.71–41.83% DM, respectively, while protein content ranged between 43.55% and 46.62% DM in control cheeses and between 47.50% and 50.19% DM in experimental cheeses. A similar behavior was also observed by Marchiani et al. [14] and Frühbauerová et al. [45] in GPP added cow's cheeses. Moreover, the lower fat content of GPP added cheeses explains their higher hardness, evaluated as resistance to compression, than that registered in control cheeses.

TBARS values registered for experimental cheeses were higher than those recorded for control cheeses. These results depended on the major oxidation sensitivity of polyunsaturated fatty acids that characterize the lipid profile of GPP [46]. Thus, the antioxidant activity of the phenolic compounds of GPP [14,46] seems not to have preserved cheese from lipid oxidation.

3.6. Volatile Organic Compounds Composition of Cheeses

VOC profiles generated by GPP and cheese samples are reported in Table 2.

The compounds identified belonged to alkanes, aldehydes, monoterpenes, esters, acids, ketones, alcohols and diols. GPP VOC profile was characterized by 20 main compounds and the most abundant belonged to the classes of alcohols, diols and esters groups. The cheeses processed with and without GPP addition emitted 16 and 14 VOCs, respectively with acids, ketones, alcohols and aldehydes being the most represented groups. The higher differences imputable to GPP addition regarded octanoic acid-ethyl ester and 2-phenylethanol. The main acids identified in cheese samples were acetic, hexanoic, butyric and 2-hydroxy4-methyl-pentanoic acids, generally recognized in ewe's milk cheeses [41,47,48]. Acetic acid may be produced by carbohydrate catabolism by LAB, 2-hydroxy-4-methyl pentanoic acid is formed enzymatically from the corresponding amino acid (L-leucine) [48]. Hexanoic and butyric acids derive mainly from the action of the lamb rennet used for curdling, responsible for the high amounts of short-chain free fatty acids [49]. Free fatty acids (FFA) are responsible for cheese flavor both directly and indirectly. FFA are precursors of odor-active compounds such as methyl ketones, aldehydes, esters and lactones [50]. Even though FAA were present in all cheese samples, esters were poorly detected. No ester compound was identified in control cheeses probably because the ripening period is particularly short [48,51,52]. Two aldehydes (hexanal and heptanal) were detected in cheese with and without GPP addition. In general, cheese VOC profile derives from the hydrolysis or metabolism of carbohydrates, proteins and fats due to the activity of LAB [53]. 2,3 butanediol and 3 hydroxy 2 butanone were found in all cheeses independently on the presence of GPP. Both compounds are generated from the metabolism of carbohydrates (lactose and citrate) by LAB [54]. As a matter of fact, GPP addition contributed to the presence of octanoic acid, ethyl ester and 2-phenylethanol.

3.7. Sensory Test

All cheeses were subjected to the sensory analysis and the results are reported in Table 3.

Table 1. Physicochemical traits of primosale cheeses at 1-month of ripening.

		Treatment (TR)	Starter Culture (NMSC)				SEM	Significance $p <$		
			MISE36	MISE94	MISE169	MISE190		TR	NMSC	TR*NMSC
External color	lightness (L*)	Control Experimental Total[a]	62.24 31.40	62.17 26.93 44.55	61.15 33.39 47.27	60.96 33.21 47.08	64.70 32.07 48.38	2.34	<0.0001	0.4294	0.3086
	redness (a*)	Control Experimental Total	−5.82 1.77	−5.82 2.08 −1.87	−5.99 1.71 −2.14	−5.79 1.43 −2.18	−5.68 1.86 −1.91	0.26	<0.0001	0.5537	0.6203
	yellowness (b*)	Control Experimental Total	13.64 1.82	13.57 1.73 7.65	14.04 1.89 7.97	12.54 1.67 7.11	14.40 1.97 8.19	0.47	<0.0001	0.1419	0.3863
Internal color	lightness (L*)	Control Experimental Total	70.39 43.22	71.31 42.70 57.00	70.47 43.72 57.10	69.16 41.35 55.25	70.61 45.13 57.87	1.46	<0.0001	0.3541	0.7342
	redness (a*)	Control Experimental Total	−4.90 3.45	−4.95 3.12 −0.91	−5.20 3.37 −0.91	−4.62 4.13 −0.25	−4.82 3.16 −0.83	0.36	<0.0001	0.2181	0.6629
	yellowness (b*)	Control Experimental Total	12.60 3.67	13.04 4.09 8.57	13.82 3.61 8.71	11.53 3.27 7.40	12.01 3.69 7.85	0.68	<0.0001	0.2068	0.4600
Hardness, N/mm^2		Control Experimental Total	1.03 1.23	1.01 1.31 1.16	0.98 1.15 1.06	1.10 1.26 1.18	1.02 1.19 1.11	0.048	<0.0001	0.0906	0.4099
Chemical composition	Dry matter (DM), %	Control Experimental Total	63.23 63.65	63.19 64.05 63.62	62.75 63.59 63.17	63.46 63.49 63.47	63.51 63.49 63.50	0.48	0.2497	0.8107	0.6814
	Ash, % DM	Control Experimental Total	6.37 6.47	6.32 6.62 6.47	6.22 6.43 6.33	6.51 6.43 6.47	6.45 6.40 6.42	0.21	0.5420	0.8912	0.7650
	Protein, % DM	Control Experimental Total	45.45 48.72	45.43 49.56 47.49	43.55 47.64 45.59	46.62 50.19 48.40	46.61 47.50 46.85	1.44	0.0147	0.3339	0.7330
	Fat, % DM	Control Experimental Total	45.69 40.63	46.31 39.71 43.01	46.06 41.64 43.85	44.52 39.35 41.93	45.87 41.83 43.85	1.03	0.0002	0.2815	0.6349
TBARS, µg MDA/kg DM		Control Experimental Total	31.49 38.41	33.30 39.46 36.38	31.47 37.45 34.46	31.25 32.98 32.11	29.94 43.75 36.85	3.8123	0.0173	0.5980	0.4708

[a] Total means of starter cultures. Abbreviations: SEM, standard error of mean; TBARS, thiobarbituric acid-reactive substances; MDA, malonylaldehyde.

Table 2. Volatile organic compounds emitted from GPP and primosale cheeses at 1-month of ripening.

Chemical Compounds [a]	Samples								
	GPP	CPC36	EPC36	CPC94	EPC94	CPC169	EPC169	CPC190	EPC190
Acids									
Acetic acid	n.d.	7.4	15.6	13.6	16.8	12.6	12.6	11.1	13.9
Butanoic acid	n.d.	4.1	10.6	12.5	7.3	7.3	7.3	6.0	9.9
4-Hydroxybutanoic acid	4.1	n.d.	n.d.	n.d.	n.d.	n.d.	n.d.	n.d.	n.d.
Hexanoic acid	1.6	1.9	4.6	5.7	3.4	5.8	4.2	3.6	4.8
Pentanoinc acid-2-hydroxy-4-methyl	n.d.	1.7	3.4	1.4	1.9	1.4	1.2	1.2	1.2
Nonanoic acid	3.4	n.d.	n.d.	n.d.	n.d.	n.d.	n.d.	n.d.	n.d.
ketones									
2-Pentanone	n.d.	2.2	0.8	1.5	0.9	2.7	0.8	2.7	0.8
3-Hydroxy-2-butanone	n.d.	23.8	41.8	31.1	49.6	7.6	22.3	14.0	34.6
2-Heptanone	n.d.	0.2	0.4	0.8	0.4	0.5	0.6	1.3	0.3
p-Phenylacetophenone	4.2	n.d.	n.d.	n.d.	n.d.	n.d.	n.d.	n.d.	n.d.
Alcohol									
Isoamyl alcohol	4.9	7.8	2.7	6.7	8.2	27.2	37.7	10.4	15.0
2-Butanol	n.d.	0.7	2.6	4.8	2.1	2.5	2.6	1.7	4.1
2-Phenylethanol	11.3	n.d.	1.4	n.d.	1.0	n.d.	1.5	n.d.	0.9
Hydrocarbons									
Hexane 2-methyl	n.d.	3.0	0.7	1.0	0.6	2.2	0.7	2.6	0.8
Heptane 2,4-dimethyl	3.2	2.0	2.7	3.0	3.2	2.6	1.4	1.6	3.2
Nonane	2.2	n.d.	n.d.	n.d.	n.d.	n.d.	n.d.	n.d.	n.d.
Nonane 2,5-methyl	2.3	n.d.	n.d.	n.d.	n.d.	n.d.	n.d.	n.d.	n.d.
Decane	1.8	n.d.	n.d.	n.d.	n.d.	n.d.	n.d.	n.d.	n.d.
Dodecane	2.3	n.d.	n.d.	n.d.	n.d.	n.d.	n.d.	n.d.	n.d.
Hexadecane	1.7	n.d.	n.d.	n.d.	n.d.	n.d.	n.d.	n.d.	n.d.
Aldeyde									
Hexanal	3.2	22.0	2.8	5.6	0.2	18.1	1.0	15.8	2.4
Heptanal	n.d.	21.8	4.4	6.6	0.5	6.6	1.5	25.0	2.6
Nonanal	1.7	n.d.	n.d.	n.d.	n.d.	n.d.	n.d.	n.d.	n.d.
Monoterpene									
D-Limonene	6.3	n.d.	n.d.	n.d.	n.d.	n.d.	n.d.	n.d.	n.d.
α-Pinene	2.1	n.d.	n.d.	n.d.	n.d.	n.d.	n.d.	n.d.	n.d.
Carene	1.5	n.d.	n.d.	n.d.	n.d.	n.d.	n.d.	n.d.	n.d.
Esters									
Octanoinc acid, ethyl ester	9.6	n.d.	1.3	n.d.	1.1	n.d.	1.2	n.d.	0.7
Butanedioic acid, diethyl ester	2.2	n.d.	n.d.	n.d.	n.d.	n.d.	n.d.	n.d.	n.d.
Decanoic acid, ethyl ester	9.7	n.d.	n.d.	n.d.	n.d.	n.d.	n.d.	n.d.	n.d.
Diol									
2,3-Butanediol	20.6	1.4	4.1	5.6	2.9	2.9	3.4	2.8	4.9

[a] Data are means percentage of three replicate expressed as (peak area of each compound/total area of significant peaks) × 100. Abbreviations: GPP, grape pomace powder; CPC36, CPC94, CPC169 and CPC190, control primosale cheese with *L. lactis* MISE36, MISE94, MISE169 and MISE190, respectively; EPC36, EPC94, EPC169 and EPC190 experimental primosale cheese with 1% of GPP and *L. lactis* MISE36, MISE94, MISE169 and MISE190, respectively; n.d., not detectable.

As reported by Torri et al. [55] the addition of GPP exerts a strong effect on the sensory parameters of dairy products. In this study, except for salt attribute, which was reported not significantly different for judges and cheeses, all other sensory attributes were scored different for cheeses, and not significantly different for judges. In detail, the addition of GPP increased odor and aroma intensity, acid perception, fiber, friability, adhesiveness and humidity, but influenced negatively sweet and hardness. Similar results were observed by Costa et al. [25] and Lucera et al. [56], who tested white and red wine grape pomace to enrich bovine primosale cheese and spreadable cheese, respectively. The overall assessment clearly indicated the cheeses from the trials inoculated with the strain Mise94 (with and without GPP) as those more appreciated.

Table 3. Evaluation of the sensory attributes of primosale cheeses at 1-month of ripening.

Attributes	Trial								SEM	p-Value	
	CPC36	EPC36	CPC94	EPC94	CPC169	EPC169	CPC190	EPC190		Judges	Cheeses
Intensity of odor	5.32 [cd]	5.94 [bc]	6.22 [b]	6.78 [a]	5.06 [d]	5.76 [bcd]	5.26 [cd]	5.91 [bc]	0.07	0.053	<0.0001
Intensity of aroma	5.41 [cde]	5.95 [b]	5.99 [b]	6.55 [a]	4.95 [e]	5.46 [cd]	5.16 [de]	5.82 [bc]	0.06	0.099	<0.0001
Sweet	5.02 [b]	4.47 [c]	5.57 [a]	5.05 [b]	5.19 [b]	4.71 [c]	5.15 [b]	4.62 [c]	0.04	0.627	<0.0001
Salt	3.53 [a]	3.48 [a]	3.43 [a]	3.46 [a]	3.47 [a]	3.45 [a]	3.44 [a]	3.46 [a]	0.03	0.999	0.999
Acid	2.42 [b]	3.28 [a]	2.44 [b]	3.22 [a]	2.37 [b]	3.20 [a]	2.48 [b]	3.32 [a]	0.05	0.733	<0.0001
Astringent	0.00 [b]	1.66 [a]	0.00 [b]	1.64 [a]	0.00 [b]	1.60 [a]	0.00 [b]	1.59 [a]	0.08	0.999	<0.0001
Friability	1.53 [b]	2.42 [a]	1.50 [b]	2.46 [a]	1.42 [b]	2.31 [a]	1.56 [b]	2.33 [a]	0.05	0.860	<0.0001
Fiber	1.39 [b]	2.56 [a]	1.30 [b]	2.48 [a]	1.36 [b]	2.56 [a]	1.42 [b]	2.58 [a]	0.06	0.952	<0.0001
Adhesiveness	2.41 [b]	3.49 [a]	2.45 [b]	3.54 [a]	2.38 [b]	3.42 [a]	2.46 [b]	3.44 [a]	0.05	0.998	<0.0001
Hardness	4.18 [a]	2.45 [b]	4.05 [a]	2.56 [b]	4.02 [a]	2.46 [b]	4.06 [a]	2.38 [b]	0.08	0.985	<0.0001
Humidity	2.53 [b]	3.62 [a]	2.40 [b]	3.51 [a]	2.28 [b]	3.56 [a]	2.60 [b]	3.65 [a]	0.06	0.971	<0.0001
Overall assessment	4.36 [c]	4.56 [c]	5.74 [b]	6.07 [a]	4.33 [c]	4.36 [c]	4.46 [c]	4.62 [c]	0.06	0.999	<0.0001

Results indicate mean value. Data within a line followed by the same letter are not significantly different according to Tukey's test. Abbreviations: CPC36, CPC94, CPC169 and CPC190, control primosale cheese with *L. lactis* MISE36, MISE94, MISE169 and MISE190, respectively; EPC36, EPC94, EPC169 and EPC190 experimental primosale cheese with 1% of GPP and *L. lactis* MISE36, MISE94, MISE169 and MISE190, respectively.

3.8. Antioxidant Properties

Bioactive peptides and phenolic compounds released during digestion of the cheese are considered to be the components primarily responsible for its antioxidative properties [57,58]. GPP is a very rich source of polyphenol compounds with potential health-promoting effects due to the compound's ability to counteract with body oxidative stress [59]. Contribution of GPP components to the reducing potential of the cheese was estimated while evaluating the antioxidant capacity of GPP-fortified primosale EPC94 compared to the unenriched cheese CPC94, which were found to be the most appreciated by the judges. To simulate the degradation of the matrix in a gastrointestinal environment, samples of both the cheeses were submitted to in vitro digestion and the Total Antioxidant Activity (TAA) in the different digestion phases was measured by the ABTS$^{+\bullet}$ decolorization assay. As shown in Figure 5, although post-oral fractions of both the cheeses showed reducing activity not significantly different, simulated PG fraction of EPC94 had a TAA value (0.342 ± 0.028 µmol TE/g) higher ($p < 0.001$) than that of the CPC94 cheese (0.234 ± 0.019 µmol TE/g). It is plausible that digestion of casein micelles by gastric pepsin has solubilized the incorporated polyphenols in GPP-enriched cheese, resulting in an increase of the reducing activity of the fraction [60]. After intestinal digestion, antioxidant activity of the fractions from both cheeses was about 50% higher than that measured in the relevant gastric digesta, possibly because of release of antioxidant fat-soluble vitamins or amino acids from the dairy matrix [61,62]. Finally, reducing compounds appeared entirely portioned in the BF, i.e., the soluble fraction of the intestinal digesta available for the absorption, and antioxidant capacity of BF from GPP-enriched cheese (0.590 ± 0.033 µmol TE/g) were 60% higher than that of the unenriched cheese (0.371 ± 0.029 µmol TE/g) (Figure 5).

The antioxidant potential of the bioaccessible fractions of the cheeses was also assessed utilizing an in vitro model of membrane lipid oxidation. Lipid oxidation was induced in bovine brain microsomes (2 mg protein/mL) by AAPH-derived peroxyl radicals (20 mM) and oxidized lipids were spectrophotometrically measured as TBARS. In these conditions, after 60 min incubation at 37 °C, an amount of 0.91 ± 0.07 nmoles TBARS /mg protein was detected (control, Figure 6). When BF from 0.1 g or 0.2 g of CPC94 cheese was added to the microsomal preparation before AAPH, slightly lower amounts of TBARS were measured (0.82 ± 0.06 and 0.75 ± 0.05 nmoles/mg protein, respectively). Interestingly, in the presence of BF from 0.1 g or 0.2 g GPP-fortified primosale EPC94, much more marked dose-dependent inhibition of the lipid oxidation was evident (0.55 ± 0.04 and 0.29 ± 0.01 nmoles TBARS/mg protein, respectively) (Figure 6).

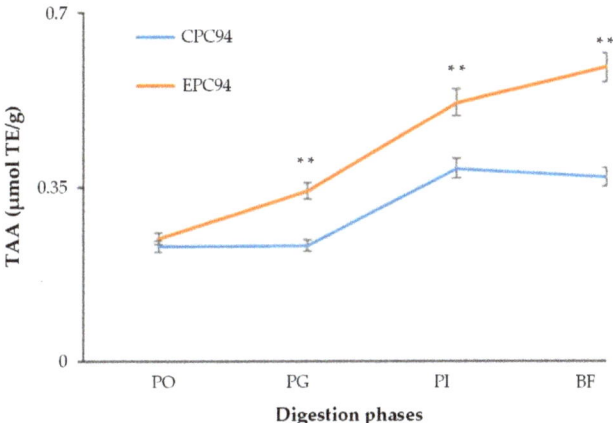

Figure 5. Antioxidant activity of cheeses during in vitro simulated human digestion. In vitro digestion conditions and measurement of total antioxidant activity (TAA) of the different digestion phases. Within the same digestion phase, values are significantly different with ** $p < 0.001$. Abbreviations: PO, post-oral digest; PG, post-gastric digest; PI, post-intestinal digest; BF, bioaccessible fraction; CPC190, control primosale cheese with *L. lactis* MISE94; EPC190 experimental primosale cheese with 1% of GPP and *L. lactis* MISE94.

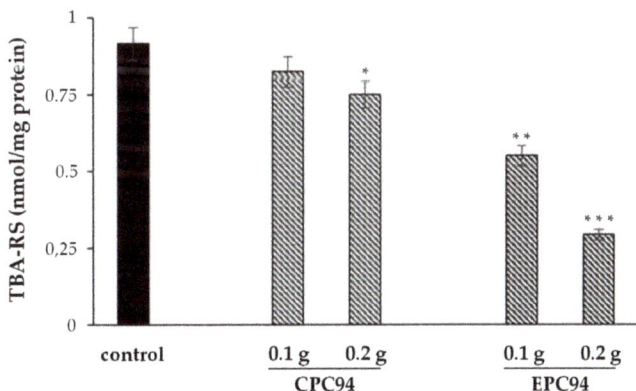

Figure 6. Thiobarbituric acid-reactive substances (TBARS) formation after AAPH-induced oxidation of microsomal membrane either in the absence (control) or in the presence of bioaccessible fraction obtained following simulated human digestion of cheeses. Microsomes, at 2 mg of protein per mL of reaction mixture, were incubated at 37 °C in the presence of 20 mM AAPH for 60 min. TBARS was spectrophotometrically measured as reported in the methods. Value are the mean ± SD of three determinations performed in duplicate. In comparison to the control, values are significantly different with * $p < 0.05$, ** $p < 0.001$; *** $p < 0.0001$ (Student's *t*-test). Abbreviations: CPC190, control primosale cheese with *L. lactis* MISE94; EPC190 experimental primosale cheese with 1% of GPP and *L. lactis* MISE94.

Collectively our results demonstrate that polyphenols from GPP added into primosale cheese significantly boost antioxidant properties of the product, conferring to it potential capacity to control oxidative stress. As reported by Ianni and Martino [63] winemaking by-products enhanced the antioxidant capacity of several beverages, but to the best of our knowledge, no previous work evaluated this aspects in GPP-enriched cheeses.

4. Conclusions

The investigation on GPP-enriched ovine primosale cheese revealed that winemaking by-products did not alter the microbiological parameters during the ripening carried out with *L. lactis*. The chemical composition of the final cheeses clearly showed that the enrichment with GPP decreased the fat content and increased the protein content as well as the values of secondary lipid oxidation. GPP addition impacted cheese VOC profiles with 2-phenylethanol and octanoinc acid, ethyl ester. The sensory analysis evidenced the highest overall acceptability for the cheeses produced with the strain MISE94 (with and without GPP). From the functional point of view, GPP addition increased antioxidant activity of the cheese after that the dairy matrix was degraded by simulated digestive process. Further studies will be carried out for a more accurate validation of this manufacturing method for ovine cheeses considering also the addition of selected LAB in multi-strain combination.

Author Contributions: Conceptualization, R.G., N.F., G.M. (Giancarlo Moschetti) and L.S.; methodology, R.G., M.B., L.T. and A.B.; software, R.G. and A.D.G.; validation, R.G., A.A., L.T. and A.B.; formal analysis, A.A., A.D.G., L.T., A.B. and L.S.; investigation, R.G., P.B., M.B., I.R. and G.M. (Giuseppe Maniaci); resources, N.F.; data curation, R.G., A.A. and A.D.G.; writing—original draft preparation, R.G., M.B., L.T. and A.B.; writing—review and editing, R.G. and L.S.; project administration, N.F.; funding acquisition, N.F. All authors have read and agreed to the published version of the manuscript.

Funding: This work was financially supported by the project for industrial research "Integrated approach to product development innovations in the leading sectors of the Sicilian agri-food sector" Prog. F/050267/03/X32—COR 109494—CUP: B78I17000260008 of the Ministry of the Economic Development, General Management for Business Incentives.

Data Availability Statement: All data included in this study are available upon request by contacting the corresponding author.

Acknowledgments: The authors are grateful to the seasonal fixed-term agricultural staff of the Department of Agricultural, Food and Forest Science—University of Palermo who provided assistance with laboratory analytical activities to this research under the technical and scientific responsibility of the structured personnel of the department.

Conflicts of Interest: The authors declare that they have no conflict of interest.

References

1. Beres, C.; Costa, G.N.; Cabezudo, I.; da Silva-James, N.K.; Teles, A.S.; Cruz, A.P.; Mellinger-Silva, C.; Tonon, R.V.; Cabral, L.M.C.; Freitas, S.P. Towards integral utilization of grape pomace from winemaking process: A review. *Waste Manag.* **2017**, *68*, 581–594. [CrossRef] [PubMed]
2. Ruggieri, L.; Cadena, E.; Martínez-Blanco, J.; Gasol, C.M.; Rieradevall, J.; Gabarrell, X.; Gea, T.; Sort, X.; Sánchez, A. Recovery of organic wastes in the Spanish wine industry. Technical, economic and environmental analyses of the composting process. *J. Clean. Prod.* **2009**, *17*, 830–838. [CrossRef]
3. Fontana, A.R.; Antoniolli, A.; Bottini, R. Grape pomace as a sustainable source of bioactive compounds: Extraction, characterization, and biotechnological applications of phenolics. *J. Agric. Food Chem.* **2013**, *61*, 8987–9003. [CrossRef] [PubMed]
4. Krishnaswamy, K.; Orsat, V.; Gariépy, Y.; Thangavel, K. Optimization of Microwave-Assisted Extraction of Phenolic Antioxidants from Grape Seeds (*Vitis vinifera*). *Food Bioproc. Technol.* **2013**, *6*, 441–455. [CrossRef]
5. Kokkinomagoulos, E.; Kandylis, P. Sustainable Exploitation of By-Products of Vitivinicultural Origin in Winemaking. *Proceedings* **2020**, *67*, 5.
6. Antonić, B.; Jančíková, S.; Dordević, D.; Tremlová, B. Grape Pomace Valorization: A Systematic Review and Meta-Analysis. *Foods* **2020**, *9*, 1627. [CrossRef]
7. Food and Drug Administration (FDA). Agency Response Letter GRAS Notice No. GRN 000125. CFSAN/Office of Food Additive Safety. Available online: https://www.polyphenolics.com/wp-content/uploads/2015/08/Gras-2003.pdf (accessed on 11 January 2021).
8. Mikovà, K. The regulation of antioxidants in foods. In *Handbook of Food Preservation*; Rahman, S., Ed.; Taylor & Francis Group: Milton Park, UK, 2007; pp. 83–267.
9. García-Lomillo, J.; González-SanJosé, M.L. Applications of wine pomace in the food industry: Approaches and functions. *Compr. Rev. Food Sci. Food Saf.* **2017**, *16*, 3–22. [CrossRef]
10. O'connell, J.E.; Fox, P.F. Significance and applications of phenolic compounds in the production and quality of milk and dairy products: A review. *Int. Dairy J.* **2001**, *11*, 103–120. [CrossRef]

11. Shan, B.; Cai, Y.Z.; Brooks, J.D.; Corke, H. Potential application of spice and herb extracts as natural preservatives in cheese. *J. Med. Food* **2011**, *14*, 284–290. [CrossRef]
12. Freire, F.C.; Adorno, M.A.T.; Sakamoto, I.K.; Antoniassi, R.; Chaves, A.C.S.D.; Dos Santos, K.M.O.; Sivieri, K. Impact of multi-functional fermented goat milk beverage on gut microbiota in a dynamic colon model. *Food Res. Int.* **2017**, *99*, 315–327. [CrossRef]
13. Marchiani, R.; Bertolino, M.; Belviso, S.; Giordano, M.; Ghirardello, D.; Torri, L.; Piochi, M.; Zeppa, G. Yogurt enrichment with grape pomace: Effect of grape cultivar on physicochemical, microbiological and sensory properties. *J. Food Qual.* **2016**, *39*, 77–89. [CrossRef]
14. Marchiani, R.; Bertolino, M.; Ghirardello, D.; McSweeney, P.L.; Zeppa, G. Physicochemical and nutritional qualities of grape pomace powder-fortified semi-hard cheeses. *J. Food Sci. Technol.* **2016**, *53*, 1585–1596. [CrossRef] [PubMed]
15. Barbaccia, P.; Francesca, N.; Gerlando, R.D.; Busetta, G.; Moschetti, G.; Gaglio, R.; Settanni, L. Biodiversity and dairy traits of indigenous milk lactic acid bacteria grown in presence of the main grape polyphenols. *FEMS Microbiol. Lett.* **2020**, *367*, fnaa066. [CrossRef] [PubMed]
16. Cruciata, M.; Gaglio, R.; Scatassa, M.L.; Sala, G.; Cardamone, C.; Palmeri, M.; Moschetti, G.; La Mantia, T.; Settanni, L. Formation and characterization of early bacterial biofilms on different wood typologies applied in dairy production. *Appl. Environ. Microbiol.* **2018**, *84*, e02107-17. [CrossRef]
17. ISO. *Enumeration of Colony-Forming Units of Micro-Organisms—Colony-Count Technique at 30 Degrees C. Milk and Milk Products*; International Standardization Organization (ISO): Geneva, Switzerland, 1992; Volume ISO 6610.
18. Gaglio, R.; Francesca, N.; Di Gerlando, R.; Cruciata, M.; Guarcello, R.; Portolano, B.; Portolano, G.; Settanni, L. Identification, typing and investigation of the dairy characteristics of lactic acid bacteria isolated from "Vastedda della valle del Belice" cheeses. *Dairy Sci. Technol.* **2014**, *94*, 157–180. [CrossRef]
19. Gaglio, R.; Francesca, N.; Di Gerlando, R.; Mahony, J.; De Martino, S.; Stucchi, C.; Moschetti, G.; Settanni, L. Enteric bacteria of food ice and their survival in alcoholic beverages and soft drinks. *Food Microbiol.* **2017**, *67*, 17–22. [CrossRef] [PubMed]
20. Gaglio, R.; Cruciata, M.; Di Gerlando, R.; Scatassa, M.L.; Mancuso, I.; Sardina, M.T.; Moschetti, G.; Portolano, B.; Settanni, L. Microbial activation of wooden vats used for traditional cheese production and evolution of the neo-formed biofilms. *Appl. Environ. Microbiol.* **2016**, *82*, 585–595. [CrossRef]
21. CIE. *Colorimetry*; Commission Internationale de l'Eclairage (CIE): Vienna, Austria, 1986; Volume CIE 15.2.
22. Bonanno, A.; Di Grigoli, A.; Vitale, F.; Di Miceli, G.; Todaro, M.; Alabiso, M.; Gargano, M.L.; Venturella, G.; Anike, F.N.; Isikhuemhenal, O.S. Effects of feeding diets supplemented with medicinal mushrooms myceliated grains on some production, health and oxidation traits of dairy ewes. *Int. J. Med. Mushrooms* **2019**, *21*, 89–103. [CrossRef] [PubMed]
23. ISO. *Sensory Analysis e General Guidance for the Design of Test. Rooms*; International Standardization Organization (ISO): Geneva, Switzerland, 2007; Volume ISO 8589.
24. Niro, S.; Fratianni, A.; Tremonte, P.; Sorrentino, E.; Tipaldi, L.; Panfili, G.; Coppola, R. Innovative Caciocavallo cheeses made from a mixture of cow milk with ewe or goat milk. *J. Dairy Sci.* **2014**, *97*, 1296–1304. [CrossRef] [PubMed]
25. Costa, C.; Lucera, A.; Marinelli, V.; Del Nobile, M.A.; Conte, A. Influence of different by-products addition on sensory and physicochemical aspects of Primosale cheese. *J. Food Sci. Technol.* **2018**, *55*, 4174–4183. [CrossRef] [PubMed]
26. Faccia, M.; Angiolillo, L.; Mastromatteo, M.; Conte, A.; Del Nobile, M.A. The effect of incorporating calcium lactate in the saline solution on improving the shelf life of fiordilatte cheese. *Int. J. Dairy Technol.* **2013**, *66*, 373–381.
27. Attanzio, A.; Diana, P.; Barraja, P.; Carbone, A.; Spanò, V.; Parrino, B.; Cascioferro, S.M.; Allegra, M.; Cirrincione, G.; Tesoriere, L.; et al. Quality, functional and sensory evaluation of pasta fortified with extracts from *Opuntia ficus-indica* cladodes. *J. Sci. Food Agric.* **2019**, *99*, 4242–4247. [CrossRef]
28. Miller, N.J.; Rice-Evans, C.A. Spectrophotometric determination of antioxidant activity. *Redox Rep.* **1996**, *2*, 161–171. [CrossRef]
29. Pellegrini, N.; Ke, R.; Yang, M.; Rice-Evans, C. Screening of dietary carotenoids and carotenoid-rich fruit extracts for antioxidant activities applying 2,2-azinobis(3-ethylenebenzothiazoline- 6-sulfonic acid) radical cation decolorization assay. *Meth. Enzymol.* **1999**, *299*, 379–389.
30. Bradford, M.M. A rapid and sensitive method for quantitation of microgram quantities of protein utilizing the principle of protein-dye binding. *Anal. Biochem.* **1976**, *72*, 248–254. [CrossRef]
31. Attanzio, A.; Tesoriere, L.; Allegra, M.; Livrea, M.A. Monofloral honeys by Sicilian black honeybee (*Apis mellifera* ssp. *sicula*) have high reducing power and antioxidant capacity. *Heliyon* **2016**, *2*, e00193. [PubMed]
32. Mainente, F.; Menin, A.; Alberton, A.; Zoccatelli, G.; Rizzi, C. Evaluation of the sensory and physical properties of meat and fish derivatives containing grape pomace powders. *Int. J. Food Sci. Technol.* **2019**, *54*, 952–958. [CrossRef]
33. Gaglio, R.; Todaro, M.; Scatassa, M.L.; Franciosi, E.; Corona, O.; Mancuso, I.; Di Gerlando, R.; Cardamone, C.; Settanni, L. Transformation of raw ewes' milk applying "Grana" type pressed cheese technology: Development of extra-hard "Gran Ovino" cheese. *Int. J. Food Microbiol.* **2019**, *307*, 108277. [CrossRef] [PubMed]
34. Settanni, L.; Gaglio, R.; Guarcello, R.; Francesca, N.; Carpino, S.; Sannino, C.; Todaro, M. Selected lactic acid bacteria as a hurdle to the microbial spoilage of cheese: Application on a traditional raw ewes' milk cheese. *Int. Dairy J.* **2013**, *32*, 126–132. [CrossRef]
35. Guarcello, R.; Carpino, S.; Gaglio, R.; Pino, A.; Rapisarda, T.; Caggia, C.; Marino, G.; Randazzo, C.L.; Settanni, L.; Todaro, M. A large factory-scale application of selected autochthonous lactic acid bacteria for PDO Pecorino Siciliano cheese production. *Food Microbiol.* **2016**, *59*, 66–75. [CrossRef] [PubMed]

36. Gaglio, R.; Gentile, C.; Bonanno, A.; Vintaloro, L.; Perrone, A.; Mazza, F.; Barbaccia, P.; Settanni, L.; Di Grigoli, A. Effect of saffron addition on the microbiological, physicochemical, antioxidant and sensory characteristics of yoghurt. *Int. J. Dairy Technol.* **2019**, *72*, 208–217. [CrossRef]
37. Rynne, N.M.; Beresford, T.P.; Kelly, A.L.; Guinee, T.P. Effect of milk pasteurisation temperature on age-related changes in lactose metabolism, pH and the growth of non-starter lactic acid bacteria in half-fat Cheddar cheese. *Food Chem.* **2007**, *100*, 375–382. [CrossRef]
38. Zheng, J.; Wittouck, S.; Salvetti, E.; Franz, C.M.; Harris, H.M.; Mattarelli, P.; O'Toole, P.W.; Pot, B.; Vandamme, P.; Walter, J.; et al. A taxonomic note on the genus *Lactobacillus*: Description of 23 novel genera, emended description of the genus *Lactobacillus* Beijerinck 1901, and union of *Lactobacillaceae* and *Leuconostocaceae*. *Int. J. Syst. Evol. Microbiol.* **2020**, *7*, 2782–2858. [CrossRef] [PubMed]
39. Settanni, L.; Moschetti, G. Non-starter lactic acid bacteria used to improve cheese quality and provide health benefits. *Food Microbiol.* **2010**, *27*, 691–697. [CrossRef] [PubMed]
40. Kopčáková, A.; Dubíková, K.; Šuľák, M.; Javorský, P.; Kmeť, V.; Lauková, A.; Pristaš, P. Restriction-modification systems and phage resistance of enterococci from ewe milk. *LWT-Food Sci. Technol.* **2018**, *93*, 131–134. [CrossRef]
41. Gaglio, R.; Cruciata, M.; Scatassa, M.L.; Tolone, M.; Mancuso, I.; Cardamone, C.; Corona, O.; Todaro, M.; Settanni, L. Influence of the early bacterial biofilms developed on vats made with seven wood types on PDO Vastedda della valle del Belìce cheese characteristics. *Int. J. Food Microbiol.* **2019**, *291*, 91–103. [CrossRef] [PubMed]
42. Fusco, V.; Quero, G.M.; Poltronieri, P.; Morea, M.; Baruzzi, F. Autochthonous and probiotic lactic acid bacteria employed for production of "advanced traditional cheeses". *Foods* **2019**, *8*, 412. [CrossRef]
43. Todaro, M.; Francesca, N.; Reale, S.; Moschetti, G.; Vitale, F.; Settanni, L. Effect of different salting technologies on the chemical and microbiological characteristics of PDO Pecorino Siciliano cheese. *Eur. Food Res. Technol.* **2011**, *233*, 931–940. [CrossRef]
44. Bonanno, A.; Di Grigoli, A.; Mazza, F.; De Pasquale, C.; Giosuè, C.; Vitale, F.; Alabiso, M. Effects of ewes grazing sulla or ryegrass pasture for different daily durations on forage intake, milk production and fatty acid composition of cheese. *Animal* **2016**, *10*, 2074–2082. [CrossRef]
45. Frühbauerová, M.; Červenka, L.; Hájek, T.; Salek, R.N.; Velichová, H.; Buňka, F. Antioxidant properties of processed cheese spread after freeze-dried and oven-dried grape skin powder addition. *Potravin. S. J. Food Sci.* **2020**, *14*, 230–238. [CrossRef]
46. Ribeiro, L.F.; Ribani, R.H.; Francisco, T.M.G.; Soares, A.A.; Pontarolo, R.; Haminiuk, C.W.I. Profile of bioactive compounds from grape pomace (*Vitis vinifera* and *Vitis labrusca*) by spectrophotometric, chromatographic and spectral analyses. *J. Chromatogr. B* **2015**, *1007*, 72–80. [CrossRef]
47. Kırmacı, H.A.; Hayaloğlu, A.A.; Özer, H.B.; Atasoy, A.F.; Levent, O. Effects of Wild-Type Starter Culture (Artisanal Strains) on Volatile Profile of Urfa Cheese Made from Ewe Milk. *Int. J. Food Prop.* **2015**, *18*, 1915–1929. [CrossRef]
48. Todaro, M.; Palmeri, M.; Cardamone, C.; Settanni, L.; Mancuso, I.; Mazza, F.; Scatassa, M.L.; Corona, O. Impact of packaging on the microbiological, physicochemical and sensory characteristics of a "pasta filata" cheese. *Food Packag. Shelf Life* **2018**, *17*, 85–90. [CrossRef]
49. Virto, M.; Chavarri, F.; Bustamante, M.A.; Barron, L.J.R.; Aramburu, M.; Vicente, M.; Pérez-Elortondo, F.J.; Albisu, M.; de Renobales, M. Lamb rennet paste in ovine cheese manufacture. Lipolysis and flavour. *Int. Dairy J.* **2003**, *13*, 391–399. [CrossRef]
50. Thierry, A.; Collins, Y.F.; Mukdsi, M.C.A.; McSweeney, P.L.H.; Wilkinson, M.G.; Spinnler, H.E. Lipolysis and metabolism of fatty acids in cheese. In *Cheese: Chemistry, Physics and Microbiology*; McSweeney, P.L.H., Fox, P.F., Cotter, P.D., Everett, D.W., Eds.; Academic Press: Cambridge, MA, USA, 2017; pp. 423–444.
51. Battelli, G.; Scano, P.; Albano, C.; Cagliani, L.R.; Brasca, M.; Consonni, R. Modifications of the volatile and nonvolatile metabolome of goat cheese due to adjunct of non-starter lactic acid bacteria. *LWT-Food Sci. Technol.* **2019**, *116*, 108576. [CrossRef]
52. Fernandez-Garcia, E.; Gaya, P.; Medina, M.; Nunez, M. Evolution of the volatile components of raw ewe's milk Castellano cheese: Seasonal variation. *Int. Dairy J.* **2004**, *14*, 39–46. [CrossRef]
53. Guarrasi, V.; Sannino, C.; Moschetti, G.; Bonanno, A.; Di Grigoli, A.; Settanni, L. The individual contribution of starter and non-starter lactic acid bacteria to the volatile organic compound composition of Caciocavallo Palermitano cheese. *Int. J. Food Microbiol.* **2017**, *259*, 35–42. [CrossRef]
54. Fox, P.F.; Guinee, T.P.; Cogan, T.M.; McSweeney, P.L.H. Biochemistry of cheese ripening. In *Fundamentals of Cheese Science*; Fox, P.F., Guinee, T.P., Cogan, T.M., McSweeney, P.L.H., Eds.; Springer: Berlin, Germany, 2020; pp. 391–442.
55. Torri, L.; Piochi, M.; Marchiani, R.; Zeppa, G.; Dinnella, C.; Monteleone, E. A sensory-and consumer-based approach to optimize cheese enrichment with grape skin powders. *J. Dairy Sci.* **2016**, *99*, 194–204. [CrossRef]
56. Lucera, A.; Costa, C.; Marinelli, V.; Saccotelli, M.A.; Del Nobile, M.A.; Conte, A. Fruit and vegetable by-products to fortify spreadable cheese. *Antioxidants* **2018**, *7*, 61. [CrossRef]
57. Hilario, M.C.; Puga, C.D.; Ocana, A.N.; Romo, F.P.G. Antioxidant activity, bioactive polyphenols in Mexican goats' milk cheeses on summer grazing. *J. Dairy Res.* **2010**, *77*, 20–26. [CrossRef] [PubMed]
58. Helal, A.; Tagliazucchi, D.; Verzelloni, E.; Conte, A. Gastro-pancreatic release of phenolic compounds incorporated in a polyphenols-enriched cheese-curd. *Food Sci. Technol.* **2015**, *60*, 957–963. [CrossRef]
59. Mattos, G.N.; Tonon, R.V.; Furtadob, A.A.; Cabralb, L.M.C. Grape by-product extracts against microbial proliferation and lipid oxidation: A review. *J. Sci. Food Agric.* **2017**, *97*, 1055–1064. [CrossRef] [PubMed]

60. Tagliazucchi, D.; Helal, A.; Verzelloni, E.; Conte, A. The type and concentration of milk increased the in vitro bioaccessibility of coffee chlorogenic acids. *J. Agric. Food Chem.* **2012**, *60*, 11056–11064. [CrossRef] [PubMed]
61. Lucas, A.; Agabriel, C.; Martin, B.; Ferlay, A.; Verdier-Metz, I.; Coulon, J.B.; Rock, E. Relationships between the conditions of cow's milk production and the contents in components of nutritional interest in raw milk farmhouse cheese. *Lait* **2006**, *86*, 177–202. [CrossRef]
62. Gupta, A.; Mann, B.; Kumar, R.; Sangwan, R.B. Antioxidant activity of Cheddar cheeses at different stages of ripening. *Int. J. Dairy Technol.* **2009**, *62*, 339–347. [CrossRef]
63. Ianni, A.; Martino, G. Dietary grape pomace supplementation in dairy cows: Effect on nutritional quality of milk and its derived dairy products. *Foods* **2020**, *9*, 168. [CrossRef]

Article

Revalorization of Cava Lees to Improve the Safety of Fermented Sausages

Salvador Hernández-Macias [1,2,3], Núria Ferrer-Bustins [4], Oriol Comas-Basté [1,2,3], Anna Jofré [4], Mariluz Latorre-Moratalla [1,2,3], Sara Bover-Cid [4] and María del Carmen Vidal-Carou [1,2,3,*]

1. Departament de Nutrició, Ciències de l'Alimentació i Gastronomia, Facultat de Farmàcia i Ciències de l'Alimentació, Campus de l'Alimentació de Torribera, Universitat de Barcelona (UB), Av. Prat de la Riba 171, 08921 Santa Coloma de Gramenet, Spain; salva.hernandez@ub.edu (S.H.-M.); oriolcomas@ub.edu (O.C.-B.); mariluzlatorre@ub.edu (M.L.-M.)
2. Institut de Recerca en Nutrició i Seguretat Alimentària (INSA·UB), Universitat de Barcelona (UB), Av. Prat de la Riba 171, 08921 Santa Coloma de Gramenet, Spain
3. Xarxa d'Innovació Alimentària (XIA), C/Baldiri Reixac 7, 08028 Barcelona, Spain
4. Food Safety and Functionality Programme, Institute of Agrifood Research and Technology (IRTA), Finca Camps i Armet s/n, 17121 Monells, Spain; nuria.ferrer@irta.cat (N.F.-B.); anna.jofre@irta.cat (A.J.); sara.bovercid@irta.cat (S.B.-C.)
* Correspondence: mcvidal@ub.edu; Tel.: +34-934-031-984

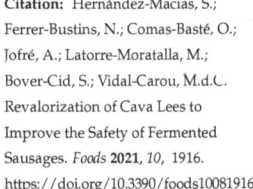

Citation: Hernández-Macias, S.; Ferrer-Bustins, N.; Comas-Basté, O.; Jofré, A.; Latorre-Moratalla, M.; Bover-Cid, S.; Vidal-Carou, M.d.C. Revalorization of Cava Lees to Improve the Safety of Fermented Sausages. *Foods* **2021**, *10*, 1916. https://doi.org/10.3390/foods10081916

Academic Editors: Simona Grasso, Konstantinos Papoutsis, Claudia Ruiz-Capillas and Ana Herrero Herranz

Received: 27 July 2021
Accepted: 17 August 2021
Published: 18 August 2021

Publisher's Note: MDPI stays neutral with regard to jurisdictional claims in published maps and institutional affiliations.

Copyright: © 2021 by the authors. Licensee MDPI, Basel, Switzerland. This article is an open access article distributed under the terms and conditions of the Creative Commons Attribution (CC BY) license (https:// creativecommons.org/licenses/by/ 4.0/).

Abstract: The revalorization of food processing by-products not only reduces the environmental impact of their disposal, but also generates added economic value. Cava lees consist of inactive cells of *Saccharomyces cerevisiae*, and though regarded as a valueless winery by-product, they are rich in fiber and phenolic compounds. In this study, a challenge test was performed to assess the effect of cava lees and a phenolic extract (LPE) derived therefrom on the behaviour of technological microbiota (lactic acid bacteria used as a starter culture) and the foodborne pathogens *Salmonella* spp. and *Listeria monocytogenes* during the fermentation and ripening of pork sausages. Ten batches of fermented sausages were prepared with and without cava lees or the LPE, and with or without different strains of *Latilactobacillus sakei* (CTC494 or BAP110). The addition of cava lees reduced the pH values of the meat batter throughout the fermentation and ripening process. No growth-promoting effect on spontaneous lactic acid bacteria (LAB) or the starter culture was observed. In contrast, the presence of cava lees prevented the growth of the tested pathogens (*Salmonella* and *L. monocytogenes*), as did the starter culture, resulting in significantly lower counts compared to the control batch. In addition, the combination of cava lees with *L. sakei* CTC494 had a bactericidal effect on *Salmonella*. LPE supplementation did not affect the pH values or LAB counts but reduced the mean counts of *Salmonella*, which were 0.71 $\log_{10}$ lower than the control values at the end of the ripening. The LPE did not exert any additional effect to that of the starters applied alone. The revalorization of cava lees as a natural ingredient to improve the microbiological safety of fermented sausages is a feasible strategy that would promote a circular economy and benefit the environment.

Keywords: cava lees; phenolic extract; food by-product; lactic acid bacteria; fermented sausages; *Salmonella* spp.; *Listeria monocytogenes*; revalorization

1. Introduction

The food industry generates large amounts of by-products, whose disposal is costly from both an economic and environmental point of view [1]. Nowadays, there is growing interest in the revalorization of by-products rich in components such as polyphenols, proteins, fiber or lipids, which may have technological, nutritional and food safety applications [2].

The use of fiber-rich by-products as natural ingredients is being widely evaluated as an innovative reformulation strategy of fermented foods to achieve positive nutritional effects, such as the reduction of fat and/or the increase of dietary fiber content [3–8]. From a technological perspective, plant-derived by-products have been used in fermented food manufacture to promote the growth of lactic acid bacteria (LAB) and thus accelerate the fermentation process, with promising preliminary results [8–11]. Another widely explored revalorization strategy has focused on upcycling phenolic compounds from plant by-products as natural antioxidants or antimicrobial compounds for the formulation of different food products [12–19]. It has been demonstrated, although mostly in vitro assays, that phenolic extracts from by-products, such as grape (seeds, skins and stems), olive and apple pomace, and shitake stems, have a protective effect against foodborne pathogens [18,20–26]. In fact, it has been verified that some phenolic extracts help reduce the growth of some of the most common foodborne pathogenic bacteria (i.e., *Salmonella* spp., *Escherichia coli*, *Staphylococcus aureus* and *Listeria monocytogenes*) [27–30]. However, the minimum inhibitory concentration against a specific pathogen can vary depending on factors such as the type of polyphenol or the bacterial strain [27].

Cava lees are a by-product of the second fermentation of cava (Spanish sparkling wine), with an estimated annual production of ca. 300 tons [31]. Cava lees consist of inactive and plasmolyzed cells of *Saccharomyces cerevisiae* and are naturally rich in fiber (β-glucans and mannan-oligosaccharides). Moreover, sustained contact with the wine during the aging process enriches cava lees with significant amounts of phenolic compounds and organic acids. Despite their interesting composition, and the fact that cava lees account for a high percentage of total winery by-products (ca. 25%), their revalorization in food applications has not been explored to date [32]. Our research group recently demonstrated that under in vitro laboratory conditions cava lees have a growth-promoting effect on specific strains of LAB species commonly used as probiotics and/or starter cultures [33].

Considering the richness of cava lees in different types of fiber, polyphenols and organic acids, they have potential for revalorization in food applications. A possible function is to improve the microbiological safety of fermented sausage, as cava lees can enhance the implantation and growth of fermentative LAB and have an antimicrobial effect against pathogenic bacteria. In this framework, the aim of the present study was to assess the effect of cava lees and a derived phenolic extract on technological microbiota (i.e., LAB) and the foodborne pathogens *Salmonella* spp. and *Listeria monocytogenes* during the fermentation and ripening of pork sausages using a challenge test. To the best of our knowledge, this is the first time that the use of cava lees and its phenolic extract has been studied with an application in food safety through a challenge test in a more complex food matrix.

2. Materials and Methods

2.1. Bacterial Strains

Listeria monocytogenes strains CTC1034 (serotype 4b), 12MOB045LM (genoserotype II) and Scott A (serotype 4b) and *Salmonella enterica* strains CTC1003 (serotype London), CTC1756 (serotype Derby) and CCUG34136 (serotype Enteritidis, Type strain) were used for the challenge test. Cultures were prepared by growing each strain independently in brain heart infusion (BHI, Beckton Dickinson, Sparks, MD, USA) at 37 °C for 7 h and subsequently sub-cultured again at the same temperature for 18 h to reach the stationary phase. The bioprotective *Latilactobacillus sakei* (formerly *Lactobacillus sakei*) CTC494, a meat isolate producing sakacin k [34], and *L. sakei* BAP110 grown at 30 °C for 19 h in MRS broth were used as starter cultures. All cultures were preserved frozen at −80 °C in the growth medium supplemented with 20% glycerol as the cryoprotectant until used.

2.2. Preparation of Cava Lees

Cava lees were provided by the winery Freixenet S.A. (Sant Sadurní d'Anoia, Spain). The characterization of the composition of cava lees is detailed by Aguilera-Curiel [35] and they are mainly composed of polysaccharides (72.3% in wet weight) and a lesser percentage of proteins (8.5% in wet weight). Wet lees were centrifuged at 18,000× g for

10 min to remove the remaining cava. The lees were subsequently frozen in an ultra-low temperature freezer (−80 °C), freeze-dried (Cryodos-50, Telstar, Terrassa, Spain) and ground. Lyophilized lees (pH = 3.2 ± 0.02) were preserved in sealed tubes protected from the light and humidity.

2.3. Phenolic Extract Preparation

The cava lees phenolic extract (LPE) was obtained according to the method described by Silva et al. [24] with some modifications. Thus, 1 g of powder lees was added to 10 mL of a mixture of ethanol/water/acetic acid (80/20/0.05) and sonicated for 30 min. The supernatant was isolated by centrifugation (2500× g for 10 min at 4 °C), transferred into a flask, and the pellet was re-extracted. The collected supernatants were evaporated under vacuum on a rotatory evaporator at 37 °C. The dry residue was weighted and stored at 4 °C until used in the sausage elaboration. The total phenolic content of the extract, expressed as mg of gallic acid equivalents (GAE)/g, was determined using the method described by Vallverdú-Queralt et al. [36].

2.4. Elaboration of Inoculated Fermented Pork Sausages

Meat batter was prepared on a pilot scale under biosafety conditions by mixing minced lean pork meat and fat (8:2) ground through a 6 mm plate and inoculating it with a mixture of the three *L. monocytogenes* and three *Salmonella* strains (same amount for each strain) at a level of ca. 6 $\log_{10}$ CFU/g. Subsequently, the ground meat was mixed with (in g/kg) sodium chloride (25), dextrose (7), black pepper (3), sodium ascorbate (0.5), sodium nitrite (0.15) and potassium nitrate (0.15). In the corresponding batches, 5% (*w/w*) of cava lees or 0.3% (*w/w*) of LPE was also added (corresponding to the content of phenolic compounds expected in 5% of lees). According to a previous study, 5% of cava lees was the most effective concentration for enhancing the in vitro bacterial growth [33]. In addition, this percentage of lees is also similar to those of others plant-based by-products used in some other studies [4–6,10]. The total amount of water of the sausage recipe was 2.6 mL/kg, including the volume used as a vehicle to add the pathogen mixture to the ground meat and the starter culture if required. In total, 10 batches were prepared for the two experiments (Table 1).

For each batch, 80 g portions of the prepared meat batter were stuffed into Tublin10 (Tub-Ex, Tass, Denmark) permeable plastic bags using vacuum packaging and were submitted to a process of fermentation and drying consisting of 2 days at 23 °C and subsequently 19 days at 15 °C.

Table 1. Batches of fermented sausages formulated with or without cava lees or the lees phenolic extract (LPE) and/or different strains of *L. sakei* as the starter culture.

Experiment	Batch	Ingredient	Starter Culture
1	C1		
	L1	Cava lees	
	C1 + CTC494		*L. sakei* CTC494 [1]
	L1 + CTC494	Cava lees	*L. sakei* CTC494 [1]
2	C2		
	E2	LPE	
	C2 + CTC494		*L. sakei* CTC494
	E2 + CTC494	LPE	*L. sakei* CTC494
	C2 + BAP110		*L. sakei* BAP110
	E2 + BAP110	LPE	*L. sakei* BAP110

[1] Producer of the bacteriocin sakacin K [34].

2.5. Microbiological and Physicochemical Analysis

For the microbiological analysis, ca. 15 g of sausage was diluted 10-fold in saline solution (0.85% NaCl and 0.1% Bacto Peptone (Beckton Dickinson, Franklin Lakes, NJ,

USA), homogenized in a Blender Smasher® bag (bioMérieux, Marcy-l'Etoile, France) for 1 min and again 10-fold serially diluted in saline solution. *L. monocytogenes* was enumerated on the chromogenic agar CHROMagar Listeria (CHROMagar, Paris, France) after incubation at 37 °C for 48 h. *Salmonella* was enumerated on the chromogenic agar CHROMagar Salmonella Plus (CHROMagar, Paris, France) after incubation at 37 °C for 24 h. LAB were enumerated in MRS (de Man, Rogosa and Sharpe; Merck, Darmstadt, Germany) agar plates incubated at 30 °C for 72 h under anaerobiosis using sealed jars with an AnaeroGen sachet (Oxoid Ltd., Altrincham, UK).

The pH was measured with a puncture electrode 5232 (Crison Instruments S.A., Alella, Spain) and a portable pHmeter PH25 (Crison Instruments S.A., Alella, Spain) and a_w with an Aqualab 3TE device (Decagon Devices, Inc. Pullman, WA, USA) at 25 °C. Analysis was performed in duplicate at selected sampling times throughout the fermentation and ripening process.

2.6. Isolation and Monitoring of Starter Culture Strains

To monitor the implantation of the starter cultures, eight colonies per batch were isolated from MRS plates at day 0, 8 or 9 and 21 and submitted to Repetitive Extragenic Palindromic(REP)-PCR and Enterobacteria Repetitive Intergenic Consensus (ERIC)-PCR with primers FW-REP1R-I (5′-IIIICGICGICATCIGGC-3′) and RV-REP2-I1 (5′-ICGICTTATCI GGCCTAC-3′), and FW-ERIC R1 (5′-ATGTAAGCTCCTGGGGATTCAC-3′) and RV-ERIC 2 (5′-AAGTAAGTGACTGGGGTGAGCG-3′), respectively, under the conditions described in Rubio et al. [37].

2.7. Statistical Analysis

Analysis of variance (ANOVA) and the post-hoc Tukey HSD test at a $p < 0.05$ significance level was done using JMP software (SAS Institute Inc, Cary, NC, USA). To determine statistical differences in bacterial counts, pH and a_w of each batch during the manufacturing or storage period, one-way ANOVA was performed, using "Time" as a fixed factor.

3. Results and Discussion

3.1. Effect of Cava Lees Applied in Fermented Pork Sausages (Experiment 1)

Firstly, a challenge test with *Salmonella* spp. and *L. monocytogenes* was carried out in fermented sausages spontaneously fermented or inoculated with the starter culture *L. sakei* CTC494, both with and without the addition of 5% of lyophilized cava lees.

3.1.1. Characterization of Physicochemical Parameters

Sausages supplemented with 5% (w/w) of cava lees initially had lower pH values than the unsupplemented batches due to the acidity of this winery by-product (pH 3.2 ± 0.02) (Table 2). During fermentation, the pH of sausages inoculated with the starter culture dropped significantly to values <5.3 ($p < 0.05$), while spontaneously fermented sausages underwent slower and slighter acidification (L1 and C1), due to the initial low levels of LAB (Figure 1). The subsequent increase in pH values in all batches throughout the ripening process could be explained by the formation of alkaline compounds during proteolysis [38]. In all cases, the presence of lees was associated with lower pH values. The difference in pH units in spontaneously fermented sausages with and without lees (L1 and C1, respectively) ranged from 0.46 at time zero to 0.85 at the end of the ripening. Studies on the use of citrus by-products in fermented sausages also report lower pH values due to their intrinsic acidity (e.g., orange fiber by-product pH = 3.28) [8,10,39]. A synergic effect was observed when cava lees were combined with *L. sakei* CTC494 (L1 + CTC494), resulting in a final pH value 1.23 units lower than in the control (C1, without cava lees or starter culture).

Values of a_w gradually decreased over the 21 days of ripening (Table 2) due to the sausage drying process, with no significant differences among batches ($p > 0.05$), neither by the inoculation of the starter culture nor by the addition of cava lees.

Table 2. Values of pH and a_w (mean ± standard deviation) during the fermentation and ripening of pork sausages. Batches included sausages formulated without (C1) or with (L1) cava lees, spontaneously fermented or with the addition of a starter culture (*L. sakei* CTC494).

		Batch			
	Day	C1	L1	C1 + CTC494	L1 + CTC494
pH	0	5.67 ± 0.01 [a]	5.20 ± 0.14 [b]	5.66 ± 0.02 [a]	5.22 ± 0.09 [b]
	2	5.74 ± 0.01 [a]	5.43 ± 0.04 [b]	5.07 ± 0.01 [c]	4.89 ± 0.01 [d]
	4	5.62 ± 0.02 [a]	5.39 ± 0.01 [b]	4.93 ± 0.01 [c]	4.76 ± 0.01 [d]
	8	5.44 ± 0.02 [a]	5.35 ± 0.01 [b]	5.01 ± 0.02 [c]	4.80 ± 0.01 [d]
	14	5.36 ± 0.05 [a]	5.18 ± 0.03 [ab]	5.40 ± 0.15 [a]	4.91 ± 0.00 [b]
	21	6.26 ± 0.38 [a]	5.41 ± 0.05 [b]	5.28 ± 0.05 [b]	5.03 ± 0.05 [b]
a_w	0	0.973 ± 0.001 [a]	0.972 ± 0.000 [a]	0.973 ± 0.000 [a]	0.972 ± 0.001 [a]
	2	0.972 ± 0.000 [a]	0.971 ± 0.000 [a]	0.972 ± 0.000 [a]	0.972 ± 0.000 [a]
	4	0.974 ± 0.000 [a]	0.972 ± 0.000 [a]	0.969 ± 0.000 [a]	0.970 ± 0.000 [a]
	8	0.969 ± 0.000 [a]	0.969 ± 0.000 [a]	0.967 ± 0.000 [b]	0.964 ± 0.000 [c]
	14	0.963 ± 0.000 [a]	0.964 ± 0.000 [b]	0.965 ± 0.000 [ab]	0.960 ± 0.000 [c]
	21	0.960 ± 0.000 [b]	0.966 ± 0.000 [a]	0.962 ± 0.000 [ab]	0.961 ± 0.000 [b]

Values are mean ± standard deviation of triplicates. For each sampling day, significant differences between batches are indicated by different superscript letters ($p < 0.05$).

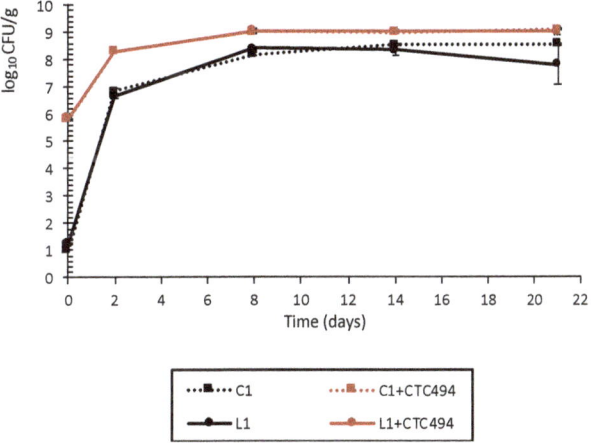

Figure 1. Growth of LAB in spontaneously fermented pork sausages with (L1) or without (C1) the addition of 5% of cava lees or fermented with the starter culture *L. sakei* CTC494, with (L1 + CTC494) or without (C1 + CTC494) cava lees.

3.1.2. Behavior of LAB during Fermentation and Ripening

Figure 1 shows the growth of LAB in the different batches of sausages during fermentation and ripening. The batches formulated with a starter culture (C1 + CTC494 and L1 + CTC494) exhibited the highest LAB counts throughout the process, ranging from the initial inoculated level of 5.9 $\log_{10}$ CFU/g to more than 9 $\log_{10}$ CFU/g from day 8, and remaining stable thereafter. The implantation of the *L. sakei* CTC494 starter culture was confirmed by RAPD-PCR, with 100% (eight out of eight) of the isolated colonies showing the same RAPD profile as the starter culture strain at the end of the ripening process. In contrast, LAB levels in sausages produced without a starter culture (C1 and L1) were initially ca. 1.2 $\log_{10}$ CFU/g and reached 8.3 $\log_{10}$ CFU/g at day 8. During the subsequent ripening process, the levels remained slightly lower than in sausages with a starter culture.

With the current study design and matrix composition, the addition of cava lees did not promote LAB growth compared to the control batches throughout the manufacturing process, whether using spontaneous fermentation or *L. sakei* CTC494. These results are not in accordance with those previously obtained in vitro, also using *L. sakei* CTC494. In that study, the supplementation of the culture medium with the same amount of cava lees (5%) resulted in a significantly higher concentration of cells in different LAB strains compared to the control (without lees); in the case of *L. sakei* CTC494, the maximum population density was 0.8 $\log_{10}$ units higher [33]. The lower amount of readily fermentable substrate in the fermented sausage formulation (i.e., 0.7% dextrose) compared to the in vitro culture media (i.e., MRS broth with 2% dextrose [33]) did not favor the use of cava lees fiber by LAB to promote their growth. A significant growth-promoting effect of other fiber-rich by-products on specific LAB strains has been demonstrated in laboratory media [40,41], whereas the addition of various by-products (from lemon, orange, tiger nut, peach or apple) in fermented sausages that also contained easily fermentable carbohydrates (e.g., glucose, dextrose, sucrose, lactose) did not improve LAB growth [3,5,8,42]. On the other hand, in the study of Yalınkılıç et al. [10], higher LAB counts were obtained in fermented sausages with 4% orange fiber compared to the control (without added fiber), although the difference in mean counts was low (0.24 $\log_{10}$ units).

3.1.3. Impact of Cava Lees on Pathogenic Bacteria

Figure 2 shows the behavior of the pathogenic bacteria during the fermentation and ripening of the four batches of pork sausages. The sakacin k-producing strain *L. sakei* CTC494 was selected as a bioprotective culture able to inhibit the growth of *L. monocytogenes* [34]. Although sakacin has no specific inhibitory effect on Gram-negative bacteria such as *Salmonella*, the presence of the starter culture accelerated acidification and resulted in a lower pH, which is known to enhance the inactivation of *Salmonella* [43]. The presence of 5% of cava lees also had an anti-pathogenic effect, reducing the load of *Salmonella* and *L. monocytogenes* in both types of fermented sausages (Figure 2).

Regarding the antimicrobial (growth inhibition) effect against *Salmonella*, significantly lower counts were recorded in sausages formulated with lees (L1) at all sampling times, being up to 2.7 $\log_{10}$ and 0.6 $\log_{10}$ lower than in C1 ($p < 0.05$) at day 8 and 21, respectively. It is important to highlight that the effect of cava lees on the *Salmonella* levels was similar to that exerted by the starter culture. Moreover, combining cava lees with *L. sakei* CTC494 (L1 + CTC494) enhanced the antimicrobial effect, resulting in a reduction of 3 $\log_{10}$ in *Salmonella* during the fermentation and ripening, which was due to both bacteriostatic and strong bactericidal effects. At the end of the process (day 21), *Salmonella* counts were 4.3 $\log_{10}$ lower than in control sausages (C1, $p < 0.05$).

The growth inhibitory effect of cava lees against *L. monocytogenes* was similar to that of bacteriocin-producing *L. sakei* CTC494. Compared to the control, *L. monocytogenes* counts were 2.3 $\log_{10}$ and 2.9 $\log_{10}$ lower in fermented sausages formulated with cava lees applied alone or together with *L. sakei* CTC494, respectively ($p < 0.05$).

To date, few studies have focused on the revalorization of by-products with antimicrobial effects against food-borne bacteria, especially in fermented products. A recent study revealed that a celery by-product powder produced a significant decrease in total *Enterobacteriaceae* counts in cooked sausages [44]. An inhibitory effect against pathogenic and opportunistic bacteria of an apple by-product added to fermented milk permeate beverages has also been recently reported [45]. Conversely, in a study on fermented sausages supplemented with a lemon by-product, higher levels of *Listeria innocua* (used as a surrogate of *L. monocytogenes*) were recorded in comparison with the unsupplemented sausages [8]. In contrast, far more studies have assessed the antimicrobial effect of food by-product extracts rich in bioactive compounds such as polyphenols [14,18,19,24,46].

In the current study, besides the growth-inhibitory effect of the starter culture *L. sakei* CTC494, the lower pH values achieved at the beginning of fermentation (ca. 5.2 at day 0) in sausages supplemented with cava lees (L1 and L1 + CTC494) could be another major factor responsible for the lower pathogen counts in these batches. In fact, in hurdle technology for food preservation, the pH is considered a crucial hurdle in the control of pathogenic bacteria in fermented sausages, especially in combination with a lower a_w [47]. Compared to *Salmonella*, *L. monocytogenes* is more tolerant of the harsh environment usually found at the last stages of ripening, characterized by a low pH and a_w [43,48,49], which could explain its lower reduction in the batches formulated with cava lees. Finally, besides the effect of pH, components of cava lees such as polyphenols and/or organic acids could also play a role in the antimicrobial activity of this by-product. It is worth highlighting that polyphenols tend to be more active against Gram-positive than Gram-negative bacteria, which can be attributed to the different bacterial cell wall structures [19,24,27,29,50].

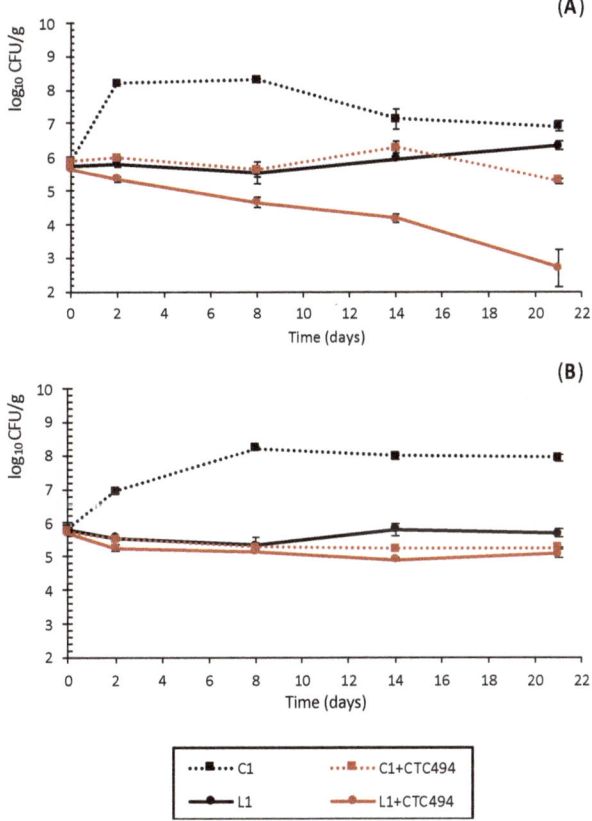

Figure 2. Counts of *Salmonella* (**A**) and *Listeria monocytogenes* (**B**) strains in pork sausages spontaneously fermented with (L1) and without (C1) the addition of 5% of cava lees or fermented with the starter culture *L. sakei* CTC494 with (L1+ CTC494) or without (C1+ CTC494) lees.

3.2. Effect of the LPE in Fermented Pork Sausages (Experiment 2)

In order to investigate whether the antimicrobial activity of cava lees could be also attributed to its phenolic fraction, a second challenge test with *Salmonella* and *L. monocytogenes* was carried out in fermented pork sausages formulated with LPE instead of cava lees. Additionally, the effect of a bacteriocinogenic (*L. sakei* CTC494) and a non-bacteriocinogenic (*L. sakei* BAP110) starter culture was evaluated.

The total phenolic content of the LPE was 152.2 ± 3.5 mg GAE/g. According to the literature, the total phenolic content of cava or wine lees differs widely, even among studies using the same extraction and determination methodology, with mean values ranging from 26 to 254 mg GAE/g [35,51,52]. Jara-Palacios et al. describe that the phenolic content in wine lees depends on the grape variety and other factors related with the vinification process [52]. The main phenolic compounds found in cava lees are caftaric acid, catechin and epicatechin, which are also the most abundant phenols in sparkling wines [35]. It has been reported that yeast cell walls possess a high capacity to adsorb phenolic compounds from wine [53–56], the contents often being greater than in other types of winery by-products (e.g., grape seeds, stems and skin) [57].

3.2.1. Characterization of Physicochemical Parameters and LAB Counts

Table 3 shows the pH and a_w values of the different batches of fermented sausages. Overall, supplementation with the LPE did not affect the acidity of the product at day 0. Batches inoculated with starter cultures underwent a faster acidification. The lowest pH values were reached at days 5 and 14 of ripening in sausages prepared with *L. sakei* CTC494 and *L. sakei* BAP110, respectively. A reducing effect on the pH was also observed when the starter culture was applied with the LPE. Acidification in non-inoculated fermented sausages (C2 and E2) was slower and weaker, the lowest pH value being recorded at the end of the ripening process (ca. 5.3), as usually occurs in spontaneously fermented sausages [43,58]. The rise in pH is due to proteolysis phenomena that take place at the end of the ripening period. pH values up to 6.5 have been reported in Spanish fermented sausages [59]. However, this process is highly variable and not very controllable, especially in spontaneous fermented sausages. In fact, in the current study, a great variability could be observed at this point (day 21) in batch C1 (6.26 ± 0.38) than in C2 (5.28 ± 0.17). The use of different meat raw materials in the preparation of experiments 1 and 2 could also explain these differences, even though the batches were prepared under the same conditions. No differences in a_w were observed between batches during the process, as the same drying conditions were applied in each case.

Table 3. Values of pH and a_w (mean ± standard deviation) during the fermentation and ripening of pork sausages. Batches included sausages elaborated without (C2) or with (E2) a cava lees phenolic extract, spontaneously fermented or with the addition of a starter culture (*L. sakei* CTC494 or *L. sakei* BAP110).

	Batch						
	Day	C2	E2	C2 + CTC494	E2 + CTC494	C2 + BAP110	E2 + BAP110
pH	0	5.88 ± 0.02 [a]	5.86 ± 0.01 [a]	5.88 ± 0.01 [a]	5.86 ± 0.02 [a]	5.85 ± 0.01 [a]	5.77 ± 0.05 [b]
	2	5.84 ± 0.01 [a]	5.81 ± 0.02 [a]	5.10 ± 0.03 [c]	5.10 ± 0.03 [c]	5.42 ± 0.10 [b]	5.38 ± 0.11 [b]
	5	5.71 ± 0.07 [a]	5.73 ± 0.04 [a]	4.96 ± 0.03 [c]	4.95 ± 0.01 [c]	5.24 ± 0.03 [b]	5.33 ± 0.04 [b]
	9	5.56 ± 0.07 [a]	5.56 ± 0.02 [a]	4.98 ± 0.01 [b]	4.96 ± 0.01 [b]	4.99 ± 0.02 [b]	4.98 ± 0.02 [b]
	14	5.38 ± 0.04 [b]	5.46 ± 0.03 [a]	5.01 ± 0.01 [c]	5.00 ± 0.01 [cd]	4.96 ± 0.02 [cd]	4.95 ± 0.00 [d]
	21	5.28 ± 0.17 [a]	5.38 ± 0.09 [a]	5.01 ± 0.01 [b]	5.01 ± 0.04 [b]	5.02 ± 0.02 [b]	4.99 ± 0.02 [b]
a_w	0	0.972 ± 0.001 [a]	0.971 ± 0.001 [a]	0.972 ± 0.001 [a]	0.971 ± 0.001 [a]	0.972 ± 0.001 [a]	0.971 ± 0.001 [a]
	2	0.970 ± 0.001 [a]	0.970 ± 0.001 [a]	0.972 ± 0.001 [a]	0.971 ± 0.001 [a]	0.971 ± 0.001 [a]	0.971 ± 0.001 [a]
	5	0.971 ± 0.002 [a]	0.969 ± 0.002 [a]	0.970 ± 0.001 [a]	0.970 ± 0.001 [a]	0.969 ± 0.001 [a]	0.969 ± 0.001 [a]
	9	0.972 ± 0.001 [a]	0.971 ± 0.001 [a]	0.971 ± 0.002 [a]	0.972 ± 0.001 [a]	0.969 ± 0.001 [a]	0.970 ± 0.001 [a]
	14	0.966 ± 0.001 [bc]	0.970 ± 0.002 [ab]	0.968 ± 0.001 [abc]	0.971 ± 0.002 [a]	0.964 ± 0.002 [c]	0.969 ± 0.001 [ab]
	21	0.964 ± 0.001 [a]	0.963 ± 0.001 [a]	0.963 ± 0.001 [a]	0.963 ± 0.002 [a]	0.962 ± 0.001 [a]	0.963 ± 0.001 [a]

Values are mean ± standard deviation of triplicates. For each sampling day, significant differences between batches are indicated by different superscript letters ($p < 0.05$).

As shown in Figure 3, batches with a fermentation process driven by all the strains of *L. sakei* starter culture (C2 + CTC494, E2 + CTC494, C2 + BAP110 and E2 + BAP110) achieved values of up to 9 $\log_{10}$ CFU/g of LAB at the first 2 days of ripening. These counts remained more or less stable until the end of the ripening process. The implantation of starter cultures was monitored by RAPD-PCR analysis, which showed that 100% (eight out of eight) of the isolates from the MRS plates at the end of the ripening had the same RAPD profile as the corresponding starter culture strain, thus confirming their competitiveness and dominance over the endogenous LAB. In batches without a starter culture, the initial levels of LAB were <1 $\log_{10}$ CFU/g, increasing up to 6.3 and 8 $\log_{10}$ CFU/g at 2 and 21 days, respectively.

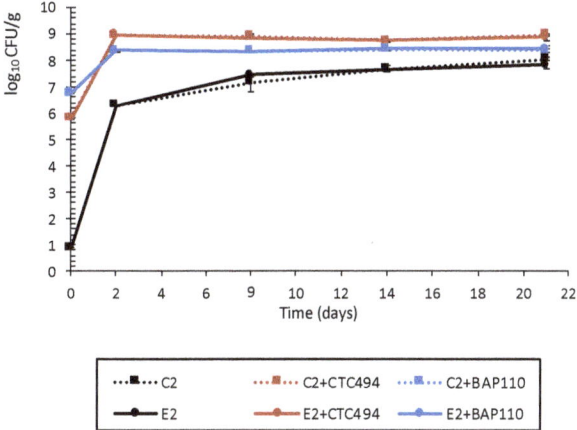

Figure 3. Growth of LAB in pork sausages spontaneously fermented with (E2) and without (C2) the addition of 0.3% LPE or fermented with the starter culture *L. sakei* CTC494 or *L. sakei* BAP110 and with (E2 + CTC494 or E2 + BAP110) or without (C2 + CTC494 or C2 + BAP110) the LPE.

Overall, the addition of the LPE did not affect the LAB counts in any batch. In this context, LAB have been described as highly tolerant to polyphenols in the growth environment [50,60]. Our results agree with those of Wang et al. [61] and Zhang et al. [62], who found that LAB counts were unaltered by the addition of different polyphenols to meat products. Nevertheless, fermented sausages produced with a shiitake by-product extract had higher levels of LAB [16,17]. Ultimately, LAB tolerance of phenolic compounds, and their ability to metabolize them, seems to be strain- or species-specific [50,63].

3.2.2. Impact of the LPE on Pathogenic Bacteria

The effect of the LPE, alone or combined with *L. sakei* CTC494 or *L. sakei* BAP110, against *Salmonella* and *L. monocytogenes* is shown in Figure 4. When no starter culture was added, the addition of the LPE had very little effect on *Salmonella*, whose growth during the first days of fermentation was similar to that of the control (without the LPE and starter culture). However, the mean counts of the pathogen at the end of the ripening were 0.71 $\log_{10}$ lower than in the control sausages. A similar effect was observed for *L. monocytogenes*, although in this case the final counts in the E2 batch were similar to the control sausages (C2).

As expected, *L. sakei*-based starter culture strains exerted a strong antimicrobial effect on *Salmonella* and *L. monocytogenes*, resulting in significantly lower pathogen counts (by 3–4 $\log_{10}$ units) at the end of ripening compared to the spontaneously fermented control sausages. No additional effect was observed when the LPE was added together with the starter cultures. No strain-specific effect was observed against *Salmonella*, which exhibited similar behavior with both starter cultures, in contrast with *L. monocytogenes*, whose behavior differed. The bacteriocinogenic strain *L. sakei* CTC494 not only prevented the growth of *L. monocytogenes* but exerted a listericidal effect from the early stages of fermentation and ripening. The non-bacteriocinogenic *L. sakei* BAP110 reduced but did not prevent the growth of *L. monocytogenes* during fermentation and had an inactivation effect during ripening; at the end of the process, the count was 1 $\log_{10}$ higher than in the batches containing the bacteriocinogenic *L. sakei* CTC494 (C2 + CTC494 and E3 + CTC494). This enhanced lethality can be related to the already reported specific antilisterial effect of *L. sakei* CTC494 in other food matrixes [34,64,65].

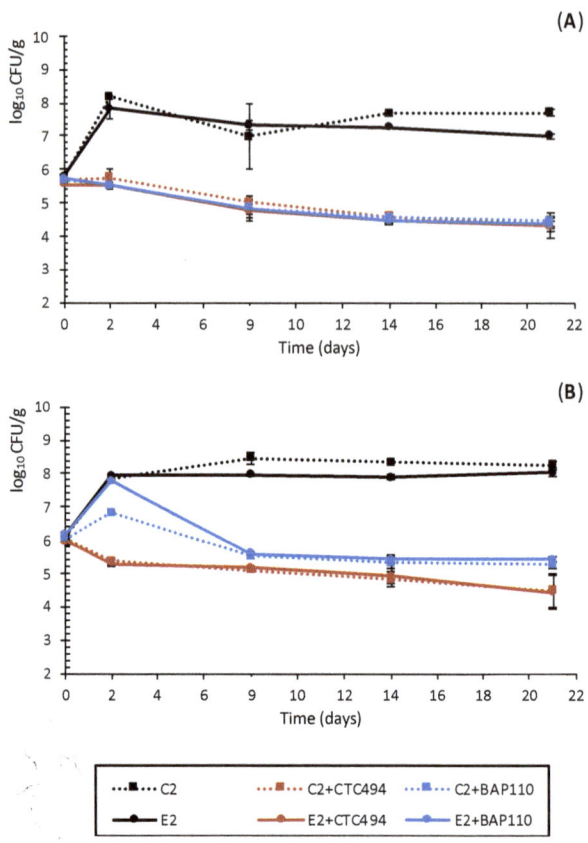

Figure 4. Behavior of *Salmonella* (**A**) and *Listeria monocytogenes* (**B**) strains in sausages spontaneously fermented with (E2) and without (C2) the addition of 0.3% of LPE or fermented with the starter culture *L. sakei* CTC494 or *L. sakei* BAP110 with (E2 + CTC949 or E2 + BAP110) or without (C2 + CTC494 or C2 + BAP110) the LPE.

Polyphenols are widely reported to have an antimicrobial effect against several pathogenic bacteria, including *Staphylococcus aureus*, *Escherichia coli*, *Salmonella* spp. and *L. monocytogenes*, mostly in the context of microbial cultures [20,24,27,29,50,66]. The results obtained here indicate that the anti-*Salmonella* effect of whole cava lees in spontaneously fermented pork sausages may be partially attributed to the phenolic fraction of this by-product (Figure 4A). However, no effect of the LPE was observed against *L. monocytogenes*. Although most reports describe Gram-positive bacteria as far more susceptible to polyphenols than Gram-negative bacteria [20,24,28,67], this trend was not supported by the results of the current study.

The antimicrobial efficacy of polyphenol-rich extracts against pathogenic bacteria varies greatly, depending on both the phenolic structure and the bacterial species [27,63,67]. Cetin-Karaca et al. [29] assessed the antimicrobial potential of different plant phenolic compounds against three *Salmonella* species, reporting that (-)epicatechin, one of the main polyphenols found in cava lees, was the most effective, although with varying degrees of sensitivity according to the species. Conversely, among the range of bacterial species tested by Silva et al. [24], a winery by-product consisting of a grape polyphenol extract showed high antimicrobial activity against two food-borne strains of *L. monocytogenes*, but not *Salmonella*. It seems that the phenolic antimicrobial activity depends not only on the type of bacteria but also on the specific strain or serotype [27].

On the other hand, sausages fermented with *L. sakei* starter cultures exhibited a significant reduction in *Salmonella* and *L. monocytogenes* counts, regardless of the presence of the LPE. Similarly, Tremonte et al. [8] found that the addition of a polyphenol-rich lemon by-product did not enhance the anti-*Listeria* effect of a bioprotective strain of *Lactiplantibacillus plantarum* during the ripening of fermented sausages. Considering that LAB may be able to metabolize plant-derived polyphenols, thus significantly compromising their antimicrobial potential [50,60,63,68], it may be envisaged that in the current study, the endogenous LAB and the inoculated *L. sakei* technological strains could have reduced the polyphenol fraction of both the cava lees and LPE.

Further studies could be designed to elucidate the role of specific components of cava lees in the antimicrobial effect observed in pork fermented sausages, which could at least be attributed to the lower pH and, most probably to acidic compounds (such as tartaric acid) and phenolic compounds (such as caftaric acid, catechin and epicatechin) determining the by-product acidity or other bacteriostatic properties. Furthermore, considering that sensory qualities are essential for consumer acceptance of food, the potential impact of cava lees on the sensorial profile of the final product should also be addressed.

Author Contributions: Conceptualization, M.L.-M., M.d.C.V.-C., A.J. and S.B.-C.; investigation, S.H.-M., N.F.-B., O.C.-B. and M.L.-M.; writing—original draft preparation, S.H.-M., O.C.-B. and M.L.-M.; writing—review and editing, S.H.-M., N.F.-B., O.C.-B., M.L.-M., A.J., S.B.-C. and M.d.C.V.-C.; supervision, M.d.C.V.-C., A.J. and S.B.-C. All authors have read and agreed to the published version of the manuscript.

Funding: This research was funded by the Spanish Government through the Interministerial Commission for Science and Technology (CICYT) project AGL2016-78324-R and the Ministry of Science and Innovation project RTI2018-099195-R-I00 and by the CERCA Programme/Generalitat de Catalunya. Salvador Hernández-Macias and Nuria Ferrer were a recipient of a doctoral fellowships from the University of Guadalajara, Mexico and Ministry of Science and Innovation (PRE2019-087847), respectively.

Acknowledgments: The authors thanks Freixenet S.A. for providing the cava lees used in this study.

Conflicts of Interest: The authors declare no conflict of interest.

References

1. De Iseppi, A.; Lomolino, G.; Marangon, M.; Curioni, A. Current and future strategies for wine yeast lees valorization. *Food Res. Int.* **2020**, *137*, 109352. [CrossRef]
2. Mirabella, N.; Castellani, V.; Sala, S. Current options for the valorization of food manufacturing waste: A review. *J. Clean. Prod.* **2014**, *65*, 28–41. [CrossRef]
3. García, M.; Dominguez, R.; Galvez, M.; Casas, C.; Selgas, M. Utilization of cereal and fruit fibres in low fat dry fermented sausages. *Meat Sci.* **2002**, *60*, 227–236. [CrossRef]
4. Fernández-Ginés, J.M.; Fernández-López, J.; Sayas-Barberá, E.; Sendra, E.; Pérez-Álvarez, J.A. Lemon albedo as a new source of dietary fiber: Application to bologna sausages. *Meat Sci.* **2004**, *67*, 7–13. [CrossRef] [PubMed]
5. Sánchez-Zapata, E.; Zunino, V.; Pérez-Alvarez, J.A.; Fernández-López, J. Effect of tiger nut fibre addition on the quality and safety of a dry-cured pork sausage ('Chorizo') during the dry-curing process. *Meat Sci.* **2013**, *95*, 562–568. [CrossRef]
6. Mendes, A.C.G.; Rettore, D.M.; Ramos, A.d.L.S.; da Cunha, S.d.F.V.; de Oliveira, L.C.; Ramos, E.M. Salames tipo milano elaborados com fibras de subprodutos da produção de vinho tinto. *Cienc. Rural* **2014**, *44*, 1291–1296. [CrossRef]
7. Younis, K.; Ahmad, S. Waste utilization of apple pomace as a source of functional ingredient in buffalo meat sausage. *Cogent Food Agric.* **2015**, *1*, 1119397. [CrossRef]
8. Tremonte, P.; Pannella, G.; Lombardi, S.J.; Iorizzo, M.; Vergalito, F.; Cozzolino, A.; Maiuro, L.; Succi, M.; Sorrentino, E.; Coppola, R. Low-fat and high-quality fermented sausages. *Microorganisms* **2020**, *8*, 1025. [CrossRef] [PubMed]
9. Sayas-Barberá, E.; Viuda-Martos, M.; Fernández-López, F.; Pérez-Alvarez, J.A.; Sendra, E. Combined use of a probiotic culture and citrus fiber in a traditional sausage 'Longaniza de Pascua'. *Food Control* **2012**, *27*, 343–350. [CrossRef]
10. Yalinkiliç, B.; Kaban, G.; Kaya, M. The effects of different levels of orange fiber and fat on microbiological, physical, chemical and sensorial properties of sucuk. *Food Microbiol.* **2012**, *29*, 255–259. [CrossRef]
11. Mikami, N.; Tsukada, Y.; Pelpolage, S.W.; Han, K.H.; Fukushima, M.; Shimada, K. Effects of sake lees (sake-kasu) supplementation on the quality characteristics of fermented dry sausages. *Heliyon* **2020**, *6*, e03379. [CrossRef] [PubMed]

12. Fernández-López, J.; Zhi, N.; Aleson-Carbonell, L.; Pérez-Alvarez, J.A.; Kuri, V. Antioxidant and antibacterial activities of natural extracts: Application in beef meatballs. *Meat Sci.* **2005**, *69*, 371–380. [CrossRef] [PubMed]
13. Lorenzo, J.M.; González-Rodríguez, R.M.; Sánchez, M.; Amado, I.R.; Franco, D. Effects of natural (grape seed and chestnut extract) and synthetic antioxidants (buthylatedhydroxytoluene, BHT) on the physical, chemical, microbiological and sensory characteristics of dry cured sausage "chorizo". *Food Res. Int.* **2013**, *54*, 611–620. [CrossRef]
14. Fasolato, L.; Carraro, L.; Facco, P.; Cardazzo, B.; Balzan, S.; Taticchi, A.; Andreani, N.A.; Montemurro, F.; Martino, M.E.; Di Lecce, G.; et al. Agricultural by-products with bioactive effects: A multivariate approach to evaluate microbial and physicochemical changes in a fresh pork sausage enriched with phenolic compounds from olive vegetation water. *Int. J. Food Microbiol.* **2016**, *228*, 34–43. [CrossRef]
15. Marchiani, R.; Bertolino, M.; Belviso, S.; Giordano, M.; Ghirardello, D.; Torri, L.; Piochi, M.; Zeppa, G. Yogurt enrichment with grape pomace: Effect of grape cultivar on physicochemical, microbiological and sensory properties. *J. Food Qual.* **2016**, *39*, 77–89. [CrossRef]
16. Van Ba, H.; Seo, H.W.; Cho, S.H.; Kim, Y.S.; Kim, J.H.; Ham, J.S.; Park, B.Y.; Nam, S.P. Antioxidant and anti-foodborne bacteria activities of shiitake by-product extract in fermented sausages. *Food Control* **2016**, *70*, 201–209. [CrossRef]
17. Van Ba, H.; Seo, H.W.; Cho, S.H.; Kim, Y.S.; Kim, J.H.; Ham, J.S.; Park, B.Y.; Pil-Nam, S. Effects of extraction methods of shiitake by-products on their antioxidant and antimicrobial activities in fermented sausages during storage. *Food Control* **2017**, *79*, 109–118. [CrossRef]
18. Vázquez-Armenta, F.J.; Silva-Espinoza, B.A.; Cruz-Valenzuela, M.R.; González-Aguilar, G.A.; Nazzaro, F.; Fratianni, F.; Ayala-Zavala, J.F. Antibacterial and antioxidant properties of grape stem extract applied as disinfectant in fresh leafy vegetables. *J. Food Sci. Technol.* **2017**, *54*, 3192–3200. [CrossRef]
19. Menchetti, L.; Taticchi, A.; Esposto, S.; Servili, M.; Ranucci, D.; Branciari, R.; Miraglia, D. The influence of phenolic extract from olive vegetation water and storage temperature on the survival of *Salmonella enteritidis* inoculated on mayonnaise. *LWT* **2020**, *129*, 109648. [CrossRef]
20. Serra, A.T.; Matias, A.A.; Nunes, A.V.M.; Leitão, M.C.; Brito, D.; Bronze, R.; Silva, S.; Pires, A.; Crespo, M.T.; San Romão, M.V.; et al. In vitro evaluation of olive- and grape-based natural extracts as potential preservatives for food. *Innov. Food Sci. Emerg. Technol.* **2008**, *9*, 311–319. [CrossRef]
21. Tsukada, M.; Sheng, H.; Kamachi, T.; Niwano, Y. Microbicidal action of photoirradiated aqueous extracts from wine lees. *J. Food Sci. Technol.* **2016**, *53*, 3020–3027. [CrossRef] [PubMed]
22. Sanz-Puig, M.; Moreno, P.; Pina-Pérez, M.C.; Rodrigo, D.; Martínez, A. Combined effect of high hydrostatic pressure (HHP) and antimicrobial from agro-industrial by-products against *S. Typhimurium*. *LWT* **2017**, *77*, 126–133. [CrossRef]
23. Vazquez-Armenta, F.J.; Bernal-Mercado, A.T.; Lizardi-Mendoza, J.; Silva-Espinoza, B.A.; Cruz-Valenzuela, M.R.; Gonzalez-Aguilar, G.A.; Nazzaro, F.; Fratianni, F.; Ayala-Zavala, J.F. Phenolic extracts from grape stems inhibit *Listeria monocytogenes* motility and adhesion to food contact surfaces. *J. Adhes. Sci. Technol.* **2018**, *32*, 889–907. [CrossRef]
24. Silva, V.; Igrejas, G.; Falco, V.; Santos, T.P.; Torres, C.; Oliveira, A.M.P.; Pereira, J.E.; Amaral, J.S.; Poeta, P. Chemical composition, antioxidant and antimicrobial activity of phenolic compounds extracted from wine industry by-products. *Food Control* **2018**, *92*, 516–522. [CrossRef]
25. Gouvinhas, I.; Santos, R.A.; Queiroz, M.; Leal, C.; Saavedra, M.J.; Domínguez-Perles, R.; Rodrigues, M.; Barros, A.I.R.N.A. Monitoring the antioxidant and antimicrobial power of grape (*Vitis vinifera* L.) stems phenolics over long-term storage. *Ind. Crops Prod.* **2018**, *126*, 83–91. [CrossRef]
26. Ricci, A.; Bernini, V.; Maoloni, A.; Cirlini, M.; Galaverna, G.; Neviani, E.; Lazzi, C. Vegetable by-product lacto-fermentation as a new source of antimicrobial compounds. *Microorganisms* **2019**, *7*, 607. [CrossRef]
27. Taguri, T.; Tanaka, T.; Kouno, I. Antimicrobial activity of 10 different plant polyphenols against bacteria causing food-borne disease. *Biol. Pharm. Bull.* **2004**, *27*, 1965–1969. [CrossRef]
28. Papadopoulou, C.; Soulti, K.; Roussis, I.G. Potential antimicrobial activity of red and white wine phenolic extracts against strains of *Staphylococcus aureus*, *Escherichia coli* and *Candida albicans*. *Food Technol. Biotechnol.* **2005**, *43*, 41–46.
29. Cetin-Karaca, H.; Newman, M.C. Antimicrobial efficacy of plant phenolic compounds against *Salmonella* and *Escherichia coli*. *Food Biosci.* **2015**, *11*, 8–16. [CrossRef]
30. Mattos, G.N.; Tonon, R.V.; Furtado, A.A.L.; Cabral, L.M.C. Grape by-product extracts against microbial proliferation and lipid oxidation: A review. *J. Sci. Food Agric.* **2017**, *97*, 1055–1064. [CrossRef] [PubMed]
31. D.O. Cava Regulatory Board Global Report. 2020. Available online: https://www.cava.wine/documents/258/ENG_KEY_FIGURES_2020.pdf (accessed on 31 May 2021).
32. Lavelli, V.; Torri, L.; Zeppa, G.; Fiori, L.; Spigno, G. Recovery of Winemaking By-Products. *Ital. J. Food Sci.* **2016**, *28*, 542–564.
33. Hernández-Macias, S.; Comas-Basté, O.; Jofré, A.; Bover-Cid, S.; Latorre-Moratalla, M.L.; Vidal-Carou, M.C. Growth-promoting effect of cava lees on lactic acid bacteria strains: A potential revalorization strategy of a winery by-product. *Foods* **2021**, *10*, 1636. [CrossRef]
34. Garriga, M.; Aymerich, M.T.; Costa, S.; Monfort, J.M.; Hugas, M. Bactericidal synergism through bacteriocins and high pressure in a meat model system during storage. *Food Microbiol.* **2002**, *19*, 509–518. [CrossRef]
35. Aguilera, M.Á. Characterization of Recovered Lees from Sparkling Wines. Ph.D. Thesis, University of Barcelona, Barcelona, Spain, 2016.

36. Vallverdú-Queralt, A.; Medina-Remón, A.; Martínez-Huélamo, M.; Jáuregui, O.; Andres-Lacueva, C.; Lamuela-Raventos, R.M. Phenolic profile and hydrophilic antioxidant capacity as chemotaxonomic markers of tomato varieties. *J. Agric. Food Chem.* **2011**, *59*, 3994–4001. [CrossRef]
37. Rubio, R.; Jofré, A.; Aymerich, T.; Guàrdia, M.D.; Garriga, M. Nutritionally enhanced fermented sausages as a vehicle for potential probiotic lactobacilli delivery. *Meat Sci.* **2014**, *96*, 937–942. [CrossRef]
38. Bover-Cid, S.; Izquierdo-Pulido, M.; Vidal-Carou, M.C. Effect of proteolytic starter cultures of *Staphylococcus* spp. on biogenic amine formation during the ripening of dry fermented sausages. *Int. J. Food Microbiol.* **1999**, *46*, 95–104. [CrossRef]
39. Fernández-López, J.; Viuda-Martos, M.; Sendra, E.; Sayas-Barberá, E.; Navarro, C.; Pérez-Alvarez, J.A. Orange fibre as potential functional ingredient for dry-cured sausages. *Eur. Food Res. Technol.* **2007**, *226*, 1–6. [CrossRef]
40. Yang, B.; Prasad, K.N.; Xie, H.; Lin, S.; Jiang, Y. Structural characteristics of oligosaccharides from soy sauce lees and their potential prebiotic effect on lactic acid bacteria. *Food Chem.* **2011**, *126*, 590–594. [CrossRef]
41. Gómez, B.; Peláez, C.; Martínez-Cuesta, M.C.; Parajó, J.C.; Alonso, J.L.; Requena, T. Emerging prebiotics obtained from lemon and sugar beet byproducts: Evaluation of their in vitro fermentability by probiotic bacteria. *LWT* **2019**, *109*, 17–25. [CrossRef]
42. Fernández-López, J.; Sendra, E.; Sayas-Barberá, E.; Navarro, C.; Pérez-Alvarez, J.A. Physico-chemical and microbiological profiles of 'salchichón' (Spanish dry-fermented sausage) enriched with orange fiber. *Meat Sci.* **2008**, *80*, 410–417. [CrossRef]
43. Serra-Castelló, C.; Bover-Cid, S.; Garriga, M.; Hansen, T.B.; Gunvig, A.; Jofré, A. Risk management tool to define a corrective storage to enhance *Salmonella* inactivation in dry fermented sausages. *Int. J. Food Microbiol.* **2021**, *346*, 109160. [CrossRef]
44. Ramachandraiah, K.; Chin, K.B. Antioxidant, antimicrobial, and curing potentials of micronized celery powders added to pork sausages. *Food Sci. Anim. Resour.* **2021**, *41*, 110–121. [CrossRef]
45. Zokaityte, E.; Cernauskas, D.; Klupsaite, D.; Lele, V.; Starkute, V.; Zavistanaviciute, P.; Ruzauskas, M.; Gruzauskas, R.; Juodeikiene, G.; Rocha, J.M.; et al. Bioconversion of milk permeate with selected lactic acid bacteria strains and apple by-products into beverages with antimicrobial properties and enriched with galactooligosaccharides. *Microorganisms* **2020**, *8*, 1182. [CrossRef]
46. Wang, Y.; Li, F.; Zhuang, H.; Li, L.; Chen, X.; Zhang, J. Effects of plant polyphenols and α-tocopherol on lipid oxidation, microbiological characteristics, and biogenic amines formation in dry-cured bacons. *J. Food Sci.* **2015**, *80*, C547–C555. [CrossRef]
47. Leistner, L. Basic aspects of food preservation by hurdle technology. *Int. J. Food Microbiol.* **2000**, *55*, 181–186. [CrossRef]
48. International Commission on Microbiological Specifications for Foods (ICMSF). Microbiological Specifications for Foods (ICMSF). Microbiological specifications of food pathogens. In *Microorganisms in Foods 5*; Roberts, T.A., Baird-Parker, A.C., Tompkin, R.B., Eds.; Springer: New York, NY, USA, 1996; p. 514, ISBN 9780412473500.
49. Jofré, A.; Aymerich, T.; Garriga, M. Improvement of the food safety of low acid fermented sausages by enterocins A and B and high pressure. *Food Control* **2009**, *20*, 179–184. [CrossRef]
50. Piekarska-Radzik, L.; Klewicka, E. Mutual influence of polyphenols and *Lactobacillus* spp. bacteria in food: A review. *Eur. Food Res. Technol.* **2021**, *247*, 9–24. [CrossRef]
51. Romero-Díez, R.; Rodríguez-Rojo, S.; Cocero, M.J.; Duarte, C.M.M.; Matias, A.A.; Bronze, M.R. Phenolic characterization of aging wine lees: Correlation with antioxidant activities. *Food Chem.* **2018**, *259*, 188–195. [CrossRef] [PubMed]
52. Jara-Palacios, M.J. Wine Lees as a Source of Antioxidant Compounds. *Antioxidants* **2019**, *8*, 45. [CrossRef] [PubMed]
53. Razmkhab, S.; Lopez-Toledano, A.; Ortega, J.M.; Mayen, M.; Merida, J.; Medina, M. Adsorption of phenolic compounds and browning products in white wines by yeasts and their cell walls. *J. Agric. Food Chem.* **2002**, *50*, 7432–7437. [CrossRef]
54. Morata, A.; Gómez-Cordovés, M.C.; Colomo, B.; Suárez, J.A. Cell wall anthocyanin adsorption by different *Saccharomyces* strains during the fermentation of *Vitis vinifera* L. cv Graciano grapes. *Eur. Food Res. Technol.* **2005**, *220*, 341–346. [CrossRef]
55. Mazauric, J.-P.P.; Salmon, J.-M.M. Interactions between yeast lees and wine polyphenols during simulation of wine aging. II. Analysis of desorbed polyphenol compounds from yeast lees. *J. Agric. Food Chem.* **2006**, *54*, 3876–3881. [CrossRef]
56. Gallardo-Chacón, J.J.; Vichi, S.; Urpí, P.; López-Tamames, E.; Buxaderas, S. Antioxidant activity of lees cell surface during sparkling wine sur lie aging. *Int. J. Food Microbiol.* **2010**, *143*, 48–53. [CrossRef] [PubMed]
57. Barcia, M.T.; Pertuzatti, P.B.; Gómez-Alonso, S.; Godoy, H.T.; Hermosín-Gutiérrez, I.; Teixeira, H.; Becker, P.; Gómez-Alonso, S.; Teixeira, H.; Hermosín-Gutiérrez, I. Phenolic composition of grape and winemaking by-products of Brazilian hybrid cultivars BRS Violeta and BRS Lorena. *Food Chem.* **2014**, *159*, 95–105. [CrossRef] [PubMed]
58. Marcos, B.; Aymerich, T.; Guardia, M.D.; Garriga, M. Assessment of high hydrostatic pressure and starter culture on the quality properties of low-acid fermented sausages. *Meat Sci.* **2007**, *76*, 46–53. [CrossRef]
59. Latorre-Moratalla, M.L.; Veciana-Nogués, T.; Bover-Cid, S.; Garriga, M.; Aymerich, T.; Zanardi, E.; Ianieri, A.; Fraqueza, M.J.; Patarata, L.; Drosinos, E.H.; et al. Biogenic amines in traditional fermented sausages produced in selected European countries. *Food Chem.* **2008**, *107*, 912–921. [CrossRef]
60. Curiel, J.A.; Rodríguez, H.; Landete, J.M.; de las Rivas, B.; Muñoz, R. Ability of *Lactobacillus brevis* strains to degrade food phenolic acids. *Food Chem.* **2010**, *120*, 225–229. [CrossRef]
61. Wang, Y.; Li, F.; Zhuang, H.; Chen, X.; Li, L.; Qiao, W.; Zhang, J. Effects of plant polyphenols and α-tocopherol on lipid oxidation, residual nitrites, biogenic amines, and N-nitrosamines formation during ripening and storage of dry-cured bacon. *LWT Food Sci. Technol.* **2015**, *60*, 199–206. [CrossRef]
62. Zhang, J.; Wang, Y.; Pan, D.D.; Cao, J.X.; Shao, X.F.; Chen, Y.J.; Sun, Y.Y.; Ou, C.R. Effect of black pepper essential oil on the quality of fresh pork during storage. *Meat Sci.* **2016**, *117*, 130–136. [CrossRef]

63. Sánchez-Maldonado, A.F.; Schieber, A.; Gänzle, M.G. Structure-function relationships of the antibacterial activity of phenolic acids and their metabolism by lactic acid bacteria. *J. Appl. Microbiol.* **2011**, *111*, 1176–1184. [CrossRef]
64. Aymerich, T.; Rodríguez, M.; Garriga, M.; Bover-Cid, S. Assessment of the bioprotective potential of lactic acid bacteria against *Listeria monocytogenes* on vacuum-packed cold-smoked salmon stored at 8 °C. *Food Microbiol.* **2019**, *83*, 64–70. [CrossRef] [PubMed]
65. Serra-Castelló, C.; Costa, J.C.C.P.; Jofré, A.; Bolívar, A.; Pérez-Rodríguez, F.; Bover-Cid, S. A mathematical model to predict the antilisteria bioprotective effect of *Latilactobacillus sakei* CTC494 in vacuum packaged cooked ham. *Int. J. Food Microbiol.* **2021**, in press.
66. Payne, K.D.; Rico-Munoz, E.; Davidson, P.M. The antimicrobial activity of phenolic compounds against *Listeria monocytogenes* and their effectiveness in a model milk system. *J. Food Prot.* **1989**, *52*, 151–153. [CrossRef] [PubMed]
67. Klancnik, A.; Guzej, B.; Kolar, M.H.; Abramovic, H.; Mozina, S.S.; Klančnik, A.; Guzej, B.; Kolar, M.H.; Abramovič, H.; Možina, S.S.; et al. In vitro antimicrobial and antioxidant activity of commercial rosemary extract formulations. *J. Food Prot.* **2009**, *72*, 1744–1752. [CrossRef] [PubMed]
68. Tabasco, R.; Sánchez-Patán, F.; Monagas, M.; Bartolomé, B.; Victoria Moreno-Arribas, M.; Peláez, C.; Requena, T. Effect of grape polyphenols on lactic acid bacteria and bifidobacteria growth: Resistance and metabolism. *Food Microbiol.* **2011**, *28*, 1345–1352. [CrossRef] [PubMed]

Article

RSM Optimization for the Recovery of Technofunctional Protein Extracts from Porcine Hearts

Dolors Parés *, Mònica Toldrà, Estel Camps, Juan Geli, Elena Saguer and Carmen Carretero

Institute for Food and Agricultural Technology (INTEA), University of Girona, Escola Politècnica Superior (EPS-1), C/Maria Aurèlia Capmany 61, 17003 Girona, Spain; monica.toldra@udg.edu (M.T.); estelcamps2@gmail.com (E.C.); juangeli97@gmail.com (J.G.); elena.saguer@udg.edu (E.S.); carmen.carretero@udg.edu (C.C.)
* Correspondence: dolors.pares@udg.edu; Tel.: +34-972-418-347

Received: 2 November 2020; Accepted: 20 November 2020; Published: 25 November 2020

Abstract: Meat byproducts, such as the internal organs from slaughtered animals, are usually underutilized materials with low commercial value. The functional (emulsifying, gelling, and foaming) properties of soluble protein extracts derived from pork hearts were investigated, as well as their molecular weight distribution. A central composite design (CCD) for two process variables (pH and ionic strength of the extraction buffer) was used to foreknow the effects of the process conditions on the physicochemical characteristics and technofunctionality of the protein extracts by means of the response surface methodology (RSM). SDS-PAGE patterns of the heart protein solutions revealed multiple bands with molecular weights ranging from 15 to 220 kDa, mainly corresponding to sarcoplasmic, myofibrillar, as well as blood proteins. The best extraction conditions to obtain protein fractions with good foaming properties would correspond to acid pH (pH ≤ 5) and high salt content (2–4%). On the contrary, solutions recovered at pH > 5 with low NaCl contents were the ones showing better emulsifying properties. Regarding gelation ability, heat-induced gels were obtained from extracts at pH 6.5–8, which showed improved firmness with increasing NaCl content (2–4%). Satisfactory second-order polynomial models were obtained for all the studied response variables, which can be useful in guiding the development of functional ingredients tailored for specific uses to maximize applications.

Keywords: meat byproducts; porcine heart; protein extraction; response surface methodology; technofunctional properties

1. Introduction

Efficient utilization of meat byproducts such as blood and offal is important for the profitability of the meat industry. It has been estimated that significant percentages of the gross income from food-producing animals come from these byproducts, about 11.4% and 7.5% from beef and pork, respectively [1]. Most byproducts have good nutritive value and can be utilized for food, to a greater or lesser degree depending on traditions, culture, and religion, as well as regulatory requirements. Traditional markets for edible meat byproducts have gradually decreased because of quality and health preoccupations; which have led to an increased focus on nonfood uses, such as pet foods, animal feed, pharmaceuticals and cosmetics [1,2], fertilizers, and biodiesel generation sources [3].

Currently, maximizing the use of animal proteins in the food industry is a crucial challenge due to the exponential increase in the meat protein demand and because of environmental concerns related to the meat chain. Thus, it is highly advisable to consider all byproducts as raw materials that can be changed into valuable food products or ingredients [4–6]. Such added value can be obtained in terms of shelf stability, more convenience, better sensory quality, and also improved technological functions, which can lead to the development of flavoring ingredients, water bonding agents, and stabilizers or emulsifiers [3].

Various categories of nonmeat technofunctional ingredients as such are chemicals, like sodium chloride, phosphates, carbonates and citrates; hydrocolloids, starches, flours, and vegetable fibers; or proteins like vegetable, dairy, and egg proteins; are used by meat processors to achieve different technological requirements and to meet consumer expectations. Some of these ingredients can be considered problematic to reach clean-label purposes or even may pose health concerns because they are catalogued as allergens. Thus, the development of alternative functional ingredients of meat origin can be a smart way to solve both aspects.

Meat byproducts can be a valuable source of technofunctional ingredients [2,6–12]. Among them, internal animal organs such are liver, lungs, brain, spleen, or heart, usually are low commercial valued and underutilized materials. Since they contain varying amounts of myofibrillar, sarcoplasmic, and stromal proteins when compared with those of lean meat [2,13,14], differences in the behavior of protein fractions from these byproducts when heated, as well as in the role of salts and pH on water holding capacity and texture, are conceivable. In this context, more information on the functional properties of protein fractions of meat byproducts can be useful for food processors.

The present work focuses on obtaining protein extracts from porcine hearts. Some studies on this topic can be found in the recent literature. Kim et al. (2017) tested the use of ultrasound-assisted extraction methods to extract proteins from myofibrils of porcine myocardium. They reported that sonicated samples produced higher protein extraction rates without the need for high salt concentrations [15]. Tsermoula et al. (2018) compared alkali (pH 11) and acid (pH 2) solubilization followed by isoelectric precipitation to prepare protein-rich extracts from bovine and porcine hearts. Both myofibril extracts showed good heat-induced gelling properties, which allowed for inferring its potential application as functional ingredients for processed meat products [16].

The objective of the present study was gain insight into the knowledge of technofunctional properties of porcine heart soluble proteins, aiming at maximizing their applications in the most efficient way. Both the raw material and the extraction conditions determine the kind and the amount of proteins that can be solubilized, as well as their structure or unfolding degree. All these factors may have a great influence on the functionality of the protein extracts obtained. Thus, we followed the same approach previously used in the obtaining of protein concentrates from porcine spleens [11]. Response surface methodology (RSM) was used to foreknow the effects of the extraction conditions (at different pH and salt concentration of the buffer solution) on the physicochemical characteristics and technofunctional properties of the soluble protein extracts derived from pork heart. The functional properties, such as solubility, gelling, emulsifying, and foaming properties, were determined.

2. Materials and Methods

2.1. Sample Collection and Preparation

Fresh hearts from recently slaughtered white pigs (LargeWhite × Landrace × Pietrain × Duroc commercial crossbred; 6 months old and 100 kg weight) were supplied by a local industrial slaughterhouse (Norfrisa S.A., Riudellots de la Selva, Girona, Spain) and were transported at 5 ± 1 °C to the laboratory. Firstly, the hearts were weighed and their pHs were measured in triplicate using a pH meter with a solid probe (LPG Crison 22, Barcelona, Spain). After being polished through removing fat and valves and blood vessels, hearts were divided into small pieces and kept refrigerated until they were analyzed or processed.

2.2. Experimental Design

The microbiological quality and physicochemical characterization of raw material were carried out on hearts collected in six different days but under the same conditions. Sample units for compositional analysis consisted of a mixture of pieces from at least four different hearts collected in the same day, which were minced in a blade grinder (Moulinex Moulinette MR, France) to obtain representative samples.

In a second stage, protein extraction processes were carried out. Aiming at performing a RSM analysis, a central composite design (CCD) for two process variables (pH and ionic strength of the extraction buffer), five equidistant levels and five replicates at the central point (pH 6.5 and 2% NaCl) resulted in 13 experiments, which were performed randomly (Table 1). A blend of 4–5 hearts were used in every experiment. The appropriate range of each operational factor was selected according to preliminary factorial experiments carried out in our laboratory: the pH conditions were 4.3, 5.0, 6.5, 8.0, and 8.6, and salt concentrations 0%, 0.58%, 2%, 3.42%, and 4% NaCl, the same as for a previous study on porcine spleens [11].

Table 1. Experimental design for the optimization of the recovery technofunctional protein extracts from porcine hearts, according to the central composite design (CCD).

Run	Coded Variables		Order	Experimental Variables	
				pH	NaCl (%)
1	−1	−1	13	5	0.58
2	+1	−1	5	8	0.58
3	−1	+1	7	5	3.42
4	+1	+1	12	8	3.42
5	−α	0	6	4.3	2
6	+α	0	1	8.6	2
7	0	−α	3	6.5	0
8	0	+α	10	6.5	4
9	0	0	4	6.5	2
10	0	0	9	6.5	2
11	0	0	11	6.5	2
12	0	0	8	6.5	2
13	0	0	2	6.5	2

Coded variables: 1 = high factor level, −1 = low factor level, 0 = central point; negative and positive default α-values indicate the low and high axial levels, respectively (α = 1.414).

2.3. Microbiological Analysis

Heart samples were serially diluted in sterile tryptone water (tryptone 10 g (Oxoid, L42 Oxoid Ltd., Basingstoke, UK) and NaCl 5 g L^{-1}), pour-plated in Petri plates with plate count agar (PCA, Oxoid CM361) as culture medium, and incubated aerobically at 30 ± 1 °C for 72 h [17]. Total aerobic mesophilic bacteria were expressed as $\log_{10}$ colony-forming units per g ($\log_{10}$ cfu g^{-1}). Microbiological analyses were performed the same day of sample collection and preparation.

2.4. Protein Extraction

Heart proteins extraction was carried out following the same method as described by Toldrà et al. (2019) [11] with slight modifications. Buffer extraction solutions were prepared with 1:15 M sodium dihydrogen phosphate and 0.1 M citric acid (pH 4.3), 1:15 M potassium dihydrogen phosphate and 1:15 M sodium dihydrogen phosphate (pH 5.0, 6.5, and 8.0), and 0.1 M HCl and 0.1 M Tris (pH 8.6) in distilled water. NaCl was then added to the buffers according to the experimental design.

Five hundred grams of porcine heart, a mixture of pieces from at least four different organs, were introduced into a cutter vessel Sammic CKE-5 (Sammic S.L., Barcelona, Spain) and minced at 2100 rpm for 45 s. Subsequently, the ground sample was mixed, at 1800 rpm for 2 min, with 1000 mL of the corresponding buffer solution according to the CCD, and kept agitated at 300 rpm for 30 min at room temperature. The suspensions were then centrifuged at 20,000× g for 15 min at 20 °C in a Sorvall RC-SC plus centrifuge (Dupont Co, Newton, CT, USA), and the soluble fractions obtained by decanting were stored at 5 °C until analysis.

2.5. Physicochemical Characterization

2.5.1. Proximate Analysis

Standard methods were used to analyze the proximate composition of porcine hearts and the soluble fractions of each extraction process. Each product was analyzed in triplicate. Moisture and ash contents were determined according to the Association of Official Analytical Chemists (AOAC) methods [18]. Protein content was estimated from the total Kjeldahl nitrogen (TKN × 6.25) [19] by using a digestion block (C. Gerhardt GmbH & Co., Königswinter, Germany) and a distillation unit Büchi K-314 (Büchi Labortechnik AG, Flawil, Switzerland). Total collagen in heart samples was calculated from the hydroxyproline content (Hyp × 8), which was determined through the NMKL–AOAC colorimetric method described by Kolar [20]. It consisted of protein hydrolysis with sulfuric acid, oxidation of hydroxyproline with chloramine-T, and formation of a red-purple complex with p-dimethylaminobenzaldehyde, which was measured at 560 nm by using a Cecil CE 1021 UV-Vis spectrophotometer (Cecil Instruments, Cambridge, UK). Total fat content was determined gravimetrically by Sohxlet extraction [21]. The fat was extracted for 5 h from the previously hydrolyzed and dried sample, using diethyl ether. The extraction solvent was removed through evaporation, and the residue was dried and subsequently weighed after cooling. Atomic emission spectrophotometry (SpectrAA Varian 50B; Agilent Technologies, Palo Alto, CA, USA) with a multielement lamp at a wavelength of 248.3 nm was used to analyze the iron content in aqueous solutions of previously ashed samples.

2.5.2. SDS–Polyacrylamide Gel Electrophoresis (SDS–PAGE)

SDS-PAGE was performed using acrylamide gels with stacking (T 3.94%, C 2.66%) and separating gel (T 15%, C 2.66%) zones, which were prepared with acrylamide:bisacrylamide 40% (37.5:1) solution (Bio-Rad), according to the Laemmli method as described by Fort et al. [22], using a Mini-Protean® 3 electrophoresis system (Bio-Rad Laboratories Inc., Hercules, CA, USA). Gels were run at 70 V for 30 min and then at a constant voltage of 120 V for 45–60 min. The approximate molecular weights were estimated using molecular weights markers from 10 to 220 kDa (BenchMark™ Protein Ladder, Invitrogen, Carlsbad, CA, USA). The gels were fixed with 2.5% gluteraldehyde solution, then stained with Coomassie Blue (0.1% PhastGel Blue R solution in 30% ethanol and 10% acetic acid), distained with 30% ethanol and 10% acetic acid, and preserved in 10% acetic acid and 10% glycerol.

Protein extracts were diluted at a ratio of 1:4 in pH 6.8, 10 mM TrisHCl, 1 mM EDTA, 1% sodium dodecyl sulfate (SDS) and 1% β-mercaptoethanol buffer, and subsequently heated at 100 °C for 5 min and centrifuged at 8000 rpm. Immediately before to the electrophoresis performance, these solutions were diluted to 50% with a buffer containing 1.25 mL Tris HCl 0.05 M pH 8.8, 1% SDS, 2 mL glycerol, and 1.75 mL alcoholic solution of bromophenol blue (0.01%) as tracking dye.

2.6. Technofunctional Properties

2.6.1. Protein Solubility

Protein solubility was calculated from the protein content and the corresponding yield of soluble fraction obtained after protein extraction procedure as explained in "Protein Extraction" (Section 2.4), relative to the total protein content of heart.

2.6.2. Foaming Properties

The foaming properties were determined as described in Davila et al. [23]. Three aliquots of 200 mL of protein solutions (5 g L^{-1}) of every fraction were prepared in distilled water, and then transferred to 1000 mL volumetric flasks. Solutions were whipped in a Braun Multimix M700 mixer (Braun Española S.A., Barcelona, Spain) with two whisks (Ø = 5 cm) at 1000 rpm for 10 min. The flasks were placed on a rotational plate during mixing to form homogeneous foams. Afterward, the foaming capacity (FC) was determined as the volume (mL) of foam after 2 min at rest. Foam stability was determined using a gravimetric method as follows: measured quantities of foam were carefully placed in three dry stainless steel sieves to let the released liquid drain, and the remaining foam was weighted every 10 min for a period of 60 min. The percentage of remaining foam versus time was plotted, and relative foam stability (RFS), defined as the time (min) needed for the disappearance of 50% of the initial foam, was calculated by fitting data to an exponential decay function $y = y_0 + a \times 10^{(-bx)}$.

2.6.3. Emulsifying Properties

The turbidimetric method reported by Pearce and Kinsella [24] with slight modifications [25] was used to determine the emulsifying properties. Solutions of protein extracts in distilled water were prepared at 5 g L^{-1} of protein. One hundred fifty milliliters of each solution was homogenized along with 50 mL of commercial corn oil using a hand-operated laboratory piston-type homogenizer (MFC MicrofluidizerTM Series 5000, Microfluidics Corporation, Newton, MA, USA) at 12 MPa and 40 L h^{-1} output flow, with recirculation for 90 s. Triplicate preparations were carried out for each sample. The emulsions were diluted 2500-fold with 0.1% sodium dodecyl sulfate (SDS), immediately after homogenization (t = 0) and after 10 min of emulsion rest (t = 10). The absorbance of the diluted emulsions was then determined at 500 nm in a Cecil CE 7400 spectrophotometer (Cecil Instruments Ltd., Cambridge, UK). Each determination was performed in duplicate. Results were reported as emulsifying activity index (EAI) and emulsion stability index (ESI), which were calculated as follows: EAI (m^2 g^{-1} protein) = $2 \cdot T / \phi \cdot C$; where T is turbidity, ϕ is the volume fraction of the dispersed phase, and C is the weight of protein per unit volume of aqueous phase before the emulsion is formed; ESI (min) = $T \cdot (\Delta t / \Delta T)$; where ΔT is the change in turbidity (T) occurring during the interval Δt (between t = 0 and t = 10 min).

2.6.4. Heat-Induced Gelation

The heat-induced gelation capacity of the soluble fractions was tested. Five aliquots of 20 mL of every sample were poured into twist-off glass containers (50 mm in diameter and 20 mm in depth), hermetically closed, and heated for 40 min in a water bath at 80 ± 1 °C. After that, samples were immediately cooled at room temperature. A uniaxial compression test with a TA-XT2 texture analyzer (Stable Micro Systems Ltd., Surrey, UK) was undertaken on the samples in the same receptacle where the gelation was produced. A cylindrical aluminum plunger of 25 mm diameter was used to compress samples until 50% deformation at a rate of 1 mm s^{-1}. The strain/distance curve was recorded and the total work involved in performing the test was considered as a measure of gel firmness (N mm). Mean values (n = 5) were used for sample comparison.

2.7. Statistical Analysis

The SPSS software package version 23.0 (2015) for Windows (IBM Corporation, Armonk, NY, USA) was used for statistical analysis. The experimental data of the protein extracts were fitted to a second-degree polynomial regression model, which included the coefficients of linear, quadratic, and the two-factor interaction effects. Parameters of the second-order polynomial model were estimated using the linear regression procedure with the backward method (inclusion and exclusion criteria at $p < 0.05$ and $p > 0.1$, respectively). Analysis of variance (ANOVA) was performed for the models calculated from the linear regression and the criteria used to evaluate the models were the R^2 adj value, and significance of the model and the estimated coefficients. Significance was attained for $p < 0.05$. The 3D response graphs of predicted values through the RSM models were plotted using SigmaPlot for Windows v. 11 (2008) (Systat Software Inc., San Jose, CA, USA).

3. Results and Discussion

3.1. Physicochemical and Microbiological Characterization of Porcine Hearts

Proximate composition, collagen and iron content, weight, pH, color, and microbiological counts from porcine hearts are shown in Table 2. The average weight of the hearts was 407.67 ± 36.14 g and the pH 5.89 ± 0.10, which is in the normal pH range of pork meat (5.6–6.2). Overall proximate composition did not differ much from that reported by Tsermoula et al. [16], except for the slightly higher humidity in our samples and the consequent somewhat lower values of the other components, but leading to similar moisture:protein ratios [16]. Moisture in the heart samples analyzed was also 3.5% higher than that reported by Seong et al. [26], who also found fat contents that doubled our values. Such variability could be explained not only by the different crossbred used in both studies but also by the higher life weights of the pigs, which in the referred work were reported to be about 130–140 kg. The total mesophilic aerobic bacterial counts were in the interval between 3 and 4 log units, in agreement with the values reported elsewhere [27], and within the range of acceptability of the hygienic processing criteria for meat products (Commission Regulation (EC) No. 2073/2005) [28].

Table 2. Physicochemical characteristics and microbiological quality of raw porcine hearts (means ± SD, $n = 6$).

Weight (g)	407.67 ± 36.14
pH	5.89 ± 0.10
Moisture (%)	79.31 ± 0.34
Protein (%)	16.70 ± 0.18
Collagen (%)	1.13 ± 0.30
Fat (%)	2.46 ± 0.54
Ashes (%)	1.04 ± 0.07
Iron (ppm)	60.33 ± 20.19
Color properties	
Chroma	19.79 ± 0.59
Hue (°)	25.60 ± 1.39
Lightness (L*)	32.55 ± 1.16
Bacterial counts (log cfu g^{-1})	3.49 ± 0.46

3.2. Extractability and Compositional Characteristics of Heart Protein Fractions

Table 3 shows the yield of every extraction process and the proximate composition of the soluble extracts as a function of pH and salt concentration of the extraction buffer. A significant increase in viscosity was observed in all solutions after refrigeration at 5 °C, in some cases (all extracts at pH ≥ 6.5) becoming very dense solutions with lumpy gelled appearance.

Concerning the proximate composition of the protein extracts, average contents of 94.9 ± 1.1% moisture, 2.91 ± 0.5% protein, and 1.8 ± 0.75% ashes, were obtained. The NaCl added to the extraction buffer influenced the moisture content of the soluble extracts. As expected, low NaCl entailed a lower ash content and consequently higher relative moisture in the liquid extracts. At any pH, the protein content of the extracts was higher at increasing ionic strength. Although there was not as much clear effect of the pH, for the same NaCl content the highest protein percentage was always found in the extracts obtained at higher pH values.

As can be observed, the average yield of soluble fraction for all extractions was 68.03 ± 1.74; the most distant from the mean value were the extracts obtained at pH 5 and 3.42% NaCl, which showed the minimum yield (59.5%), and at pH 8.6 and 2% NaCl, which reached the maximum (72.82%). The highest values corresponded to the pH axial points, pH 8.6 and 4.3, both containing the salt content corresponding to the central point (2% NaCl). Overall, all processes at pH 6.5 resulted in a yield close to the mean value, regardless of their ionic strength. Nevertheless, the same behavior was found when combining acid pH with low salt or alkaline pH with a considerable amount of NaCl. It is worth noting that the extract obtained by using a buffer at pH 8.6 and 2% NaCl was the one that showed both the best yield and the highest protein content.

Table 3. Yield and proximate composition of soluble protein fractions from porcine heart as influenced by pH and NaCl content (%) of the extraction buffer.

pH	NaCl	Yield (%)	Proximate Composition (%)		
			Moisture	Protein	Ash
5	0.58	67.2	96.63	2.33	0.89
8	0.58	63.2	95.57	2.85	1.11
5	3.42	59.5	93.78	3.03	2.74
8	3.42	68.6	94.21	3.35	2.00
4.3	2	71.3	95.63	1.79	2.01
8.6	2	72.8	94.61	3.59	1.62
6.5	0	68.6	96.82	2.24	0.37
6.5	4	67.7	92.84	3.35	3.30
6.5	2	68.0 ± 1.7	94.64 ± 0.39	1.88 ± 0.20	3.05 ± 0.01

Central point (pH 6.5 and 2% NaCl); mean ± SD ($n = 5$).

Soluble protein recovery was calculated from the protein content of the extracts referred to the corresponding yield of the extraction process and it was reported as protein solubility, that is, percentage of protein in the solutions with respect to the total protein content of porcine heart (Table 4). Solubility ranged from 22.4% (at pH 4.3 and 2% NaCl) to 44.3% (at pH 8.6 and 2% NaCl), confirming the influence of pH in this property. In our heart extracts, the protein contents were somewhat higher while the solubility was slightly lower, as compared to protein extracts from porcine spleen obtained in a previous study, due to the different ratio of solvent to material used in the extraction processes, 10:1 for the spleen proteins and 2:1 for the heart proteins, which result in a higher protein content solution but with a lower solubilization yield [11]. Protein yields around 50% (w/w) were obtained in several studies on the extraction of water and salt soluble proteins from bovine coproducts [14,29]. According to Selmane et al., yields of protein recovery from pork and beef lungs were between 48 and 55% (w/w) [30]. Moreover, Tsermoula et al. reported 51.53–55.74% recovery of the total protein from bovine and porcine hearts through alkali and acid solubilization [16].

Table 4. Technofunctional properties of soluble protein fractions from porcine heart as influenced by pH and NaCl content (%) of the extraction buffer.

pH	NaCl	Solubility (%)	Foaming		Emulsifying		Gelling
			FC	RFS	EAI	ESI	GS
5	0.58	26.9	596.5	8.53	236.44	20.13	1.31
8	0.58	29.9	375.9	12.37	297.32	170.62	2.18
5	3.42	31.5	572.0	12.89	278.89	41.10	1.19
8	3.42	38.9	343.2	8.03	150.39	22.47	11.47
4.3	2	22.4	738.1	22.04	259.09	25.65	0.82
8.6	2	44.3	283.3	16.17	251.79	19.94	2.42
6.5	0	26.3	365.0	5.55	354.74	35.19	2.15
6.5	4	39.2	324.1	11.64	206.35	26.15	18.77
6.5	2	35.8 ± 1.1	326.8 ± 41.6	11.15 ± 1.78	308.00 ± 39.01	25.32 ± 6.50	5.38 ± 0.89

FC: foaming capacity (mL). RFS: relative foam stability (min). EAI: emulsifying activity index ($m^2\ g^{-1}$). ESI: emulsifying stability index (min). GS: gel firmness (N mm). Central point (pH 6.5 and 2% NaCl); mean ± SD ($n = 5$).

The quadratic model for solubility was found to be significant ($p = 0.000$) and revealed an adjusted coefficient of determination (R^2 adj) of 0.978 (Table 5), indicating that close to 98% of experimental data variation is explained by the estimated equation. The proposed quadratic model includes pH and pH^2, NaCl concentration, and three pH–NaCl interactions. It is widely understood that the solubility of a protein is highly dependent on pH and salt concentration. As shown in the response surface (Figure 3a), higher solubility values correspond to the pH range from 6.5 to 8.5, and solubility improves with salt addition from 1.5 to 3% NaCl, due to an enhancement of myofibrillar proteins solubilization by salt. These results agree with studies on porcine cardiac proteins [15] or proteins extracted from other sources, i.e., porcine spleen [11] or mechanically separated turkey meat [31]. Additionally, ultrasonic treatments were claimed as useful to increase the effective extraction of protein from normal residual meat byproducts such as porcine myocardium without the need for high salt concentrations, as required in conventional extraction methods [15].

Table 5. Polynomial models for the response surface methodology (RSM) optimization of functionality of protein extracts from porcine heart as a function of pH and salt concentration of the extraction buffer.

	Protein Solubility (%)	Foaming		Emulsifying		Gelling
		FC	RFS	EAI	ESI	GS
Coefficients						
Constant	14.762	2732.309	64.900	−385.954	−110.123	1.629
pH	6.090	−647.477	−18.880	201.037	ns	ns
NaCl	−16.740	ns	4.396	ns	55.987	ns
pH*NaCl	ns	ns	ns	ns	ns	ns
pH^2	−0.669	43.023	1.502	−13.923	3.971	ns
NaCl^2	ns	ns	ns	28.200	ns	−11.310
pH^2*NaCl	0.550	ns	ns	ns	−1.576	ns
pH*NaCl^2	1.408	ns	ns	−5.456	ns	3.544
pH^2*NaCl^2	−0.236	ns	−0.021	ns	ns	−0.253
Adjusted R^2	0.978	0.901	0.614	0.655	0.490	0.946
Significance	0.000	0.000	0.017	0.011	0.028	0.000

FC: foaming capacity (mL); RFS: relative foam stability (min); EAI: emulsifying activity index (m^2 g^{-1}); ESI: emulsifying stability index (min); GS: gel firmness (N mm); ns: not significant (variable not included in the model).

3.3. SDS-PAGE Profiles

SDS-PAGE electrophoretograms of soluble protein fractions from porcine hearts are shown in Figure 1. All samples showed broadly similar electrophoretic profiles with bands corresponding to the MW of the major sarcoplasmic and myofibrillar myocardial proteins, as well as blood proteins.

Soluble protein fractions are composed mainly of myofibrillar proteins, actin, and myosin [14,32]. All samples exhibited bands corresponding to heavy (H) and light (L) myosin chains (with a MW of 200–220 and 15–25 kDa, respectively) and bands around 40–50 kDa that could correspond to the globular monomeric form of actin (G-actin) (MW of 42 kDa) [15,33]. Bands in the range between 30 and 40 kDa may be assigned to the regulatory proteins associated with actin, troponin complex, and tropomyosin. Sarcoplasmic proteins are water-soluble globular proteins of relatively low MW (15–20 kDa), consisting mainly of enzymes and hem pigments [29]. Other bands in the range from 20 to 70 kDa can be observed in all samples, probably corresponding to blood proteins, albumin (MW 68–70 kDa), which is the most abundant globular protein in plasma; globulins, a heterogeneous group of globular proteins that include a variety of enzymes, carrier, and antigenic proteins with MW ranging from few to hundreds kDa; and proteins associated to red cell membranes (MW ~ 30–240 kDa) [34,35].

From the main differences among samples, presumably attributable to the pH and ionic strength of the extraction buffer, we can make the following remarks:

Focusing on myosin, the solubility is favored by extraction conditions combining high pH, far from the isoelectric point, together with high ionic strength. This is likely due to alkaline pH loosening the tissues and facilitating structural changes [15]. Samples at pH 8.6 (lane 7) and pH 8 (lane 5) showed the most intense bands of heavy myosin (220 kDa) along with the presence of a band that matches the MW of light myosin (20 kDa). At pH 4.3 and 5 the light chain is not observed and the intensity of the myosin H band is weaker. Increasing the ionic strength of the buffer by the addition of NaCl aids myosin extraction, as confirmed by the fading of the myosin H band and the absence of the myosin L band in samples with low salt content (0.58% NaCl), despite being at pH 8 (lane 3). This effect is also noticeable even when comparing samples at pH 6.5 (lanes 8, 9, and 10), since a slight increase in myosin recovery can be observed as salt concentration increases, in agreement with the protein solubility results. On the other hand, the bands corresponding to the molecular weight of G-actin also show greater intensity in soluble extracts at pH ≥ 5 and seems to be less influenced by NaCl contents.

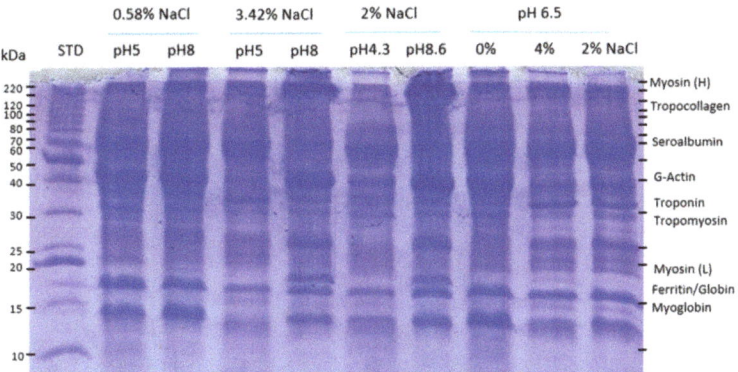

Figure 1. SDS-PAGE (12.5% PA) patterns of soluble proteins extracted from porcine heart as a function of pH and salt concentration of the extraction buffer. STD: molecular weight standard from 10 to 200 kDa.

One band in the range 25–30 kDa can be observed only in some extracts, those corresponding to pH ≥ 6.5 and salt content ≥ 2%. The band could putatively be assigned to low MW salt soluble globulins or several sarcoplasmic enzymes (myokinase, triosephosphate isomerase, phosphoglycerate mutase) [36].

Moreover, in most protein extracts there were two bands in the range of 15–20 kDa MW that follow the same pattern of intensity among the different samples. Looking at their MW, the band of lower MW may correspond to the hemoproteins, myoglobin, and the globin monomer of hemoglobin (16–18 kDa MW approximately), and the other band, with a slightly higher MW (19–21 kDa) to ferritin subunits [37]. The effect of ionic strength on these proteins can be easily observed by comparing the three samples at pH 6.5 (lanes 8–10), the two samples at pH 8 (lanes 3 and 5) or at pH 5 (lanes 2 and 4). The intensities of the bands are always greater in samples with low salt content (0–0.58%) for all pHs. The effect of pH is evidenced from the comparison of 2% NaCl samples at pH 4.3, 8.6, and 6.5 (lanes 6, 7, and 10, respectively). The bands become paler as pH decreases; the highest intensity corresponding to samples at alkaline pH.

Although SDS-PAGE, as carried out in this study, cannot be considered as a quantitative method, it has been shown that differences in polypeptide profiles may be related to the functional properties of protein samples [38].

3.4. Effects of the Extraction Conditions on the Functional Properties of Soluble Protein Extracts

The technofunctionality of protein extracts obtained at different extraction conditions are shown in Table 4. Significant second-order polynomial models were obtained for all the response variables ($p < 0.05$) (Table 5). The 3D response surface graphs and contour plots can be found in Figure 3.

3.4.1. Foaming Properties

Proteins are known to enhance and to stabilize foams by adsorbing at the air–liquid interface after undergoing unfolding and molecular rearrangement. Looking at the experimental results on foamability (Table 4), highest foaming capacity (FC) and stability (RFS) corresponded to the solutions extracted with acid buffers. FC values ranged from 283 to 738 mL of foam. It showed low values for solutions at pH ≥ 6.5 and reached higher values at pH 5 (572 mL) and at pH 4.3 (738 mL). At pH 6.5–8.3, a drop of up to about half the volume of foam as compared to the acid extracts was observed. This behavior is in agreement with the one described for other proteins, e.g., globulins from porcine plasma [23]. Concerning the foam stability (Figure 2, RFS values Table 4), the more unstable foam corresponded to the extract recovered at pH 6.5 without added NaCl, and the one showing better stability was again the one produced with the solution of the acid extract (pH 4.3). At pH 5 and 6.5 and using a buffer with increased ionic strength (2–4% NaCl) improved the foam stability of the protein extracts. RFS was also good for solutions at pH 8.6, but this feature is of little relevance due to the fact that it was the extract with the lowest foaming capacity (283.3 mL).

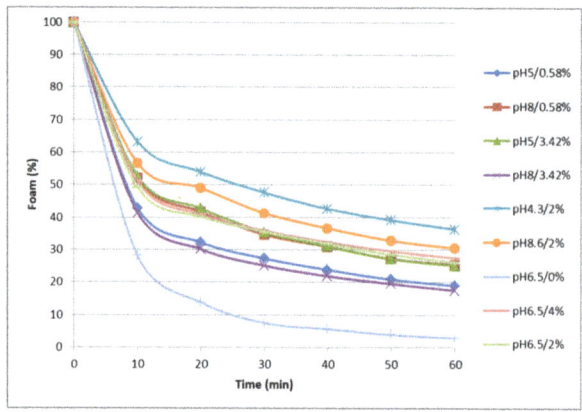

Figure 2. Vanishing kinetics of foam from porcine heart protein extracts (5 g L^{-1}), obtained with extraction buffers at different pH and NaCl contents. Points refer to the relative percentage of the initial foam remaining at 10 min intervals.

Since the foaming properties of proteins are related to their hydrophobicity and their charge [39], our results could be explained by increased protein surface hydrophobicity at acid pH. The unfolding of protein molecules exposes hydrophobic groups, resulting in increased interaction at the air–water interface. According to Yang et al., the hydrophobicity of myofibrillar proteins decrease as pH increases from 5.0 to 8.0 [40].

Table 5 shows the models that better fit to our experimental data on FC and RFS. The response surfaces of both variables are shown in Figure 3c,d, respectively. Regarding to the second-order polynomial functions obtained, the FC response can be explained by the pH of the extraction buffer itself and shows a good fit to the experimental data (R^2 = 0.901). Conversely, a more complex model that includes NaCl concentration and pH–NaCl interaction was obtained for the RFS variable, although it shows a worse fit to the experimental data (R^2 = 0.614). As can be seen in Figure 3c, foaming ability reaches a maximum at acidic pH values and worsens with increasing pH, this behavior being irrespective of the NaCl concentration. The model proposes that the FC of 5 g L^{-1} solutions (200 mL) of the water-soluble heart proteins after the whipping process would be good at pH lower than 5.5, with specific foam volumes of 500–750 mL. Surprisingly, the behavior of the FC variable for porcine heart extracts showed to be opposite to the extracts coming from porcine spleen, as reported in

Toldrà et al. [11], thus confirming the influence of both the raw material and the extraction process in the technofunctionality of the soluble extracts obtained from different organs [30,41]. Only the heart extracts obtained at acidic pH showed foaming capacity comparable to the extracts from spleen proteins, which had shown to be good at any pH. This behavior can also be related to the solvent:material proportions used to obtain heart extracts; with a ratio of 2:1, a solubilization yield lower than that obtained with the 10:1 ratio used for extractions of soluble spleen proteins. It is possible that the availability of solvent acted as a limiting factor and the proteins with the best surface properties remained in the insoluble residue, even at the most favorable pH and ionic strength conditions for their extraction.

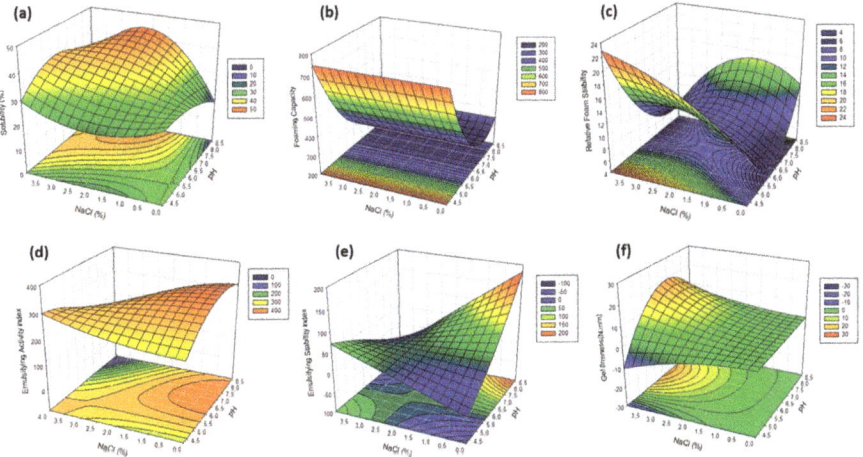

Figure 3. Response surface graphs and contour plot of the protein extracts from porcine heart as a function of pH and salt concentration of the extraction buffer. (**a**) Protein solubility (%) and surface properties (5 g L^{-1} protein); (**b**) foaming capacity (mL); (**c**) relative foam stability (min); (**d**) emulsifying activity index (m^2 g^{-1}); (**e**) emulsion stability index (min); and (**f**) gel firmness after heating (80 °C for 40 min) (N mm).

According to the response surface for the variable RFS (Figure 3d), the foam stability can also be considered acceptable at acid conditions and improves with increasing NaCl concentration. The foams show to reach maximum stability at low pH values and high ionic strength, with a linear increase from 0 to 4% NaCl when using an extraction buffer at pH 4.5. The predicted minimum foam stability corresponds to pH near the isoelectric point (*pI*) of myofibrillar proteins (pH 5–5.5). This fact could be a consequence of the protein precipitation, so that there would not be enough protein able to adsorb at the surfaces. Far from this *pI*, proteins with flexible structure form denser and thicker adsorption layers, thus ensuring better stabilization. Added NaCl at acid pH increases further the adsorption and the repulsion between the surfaces (probably by steric and/or osmotic mechanism).

From our results, the best extraction conditions to obtain a soluble protein fraction from porcine hearts with good foaming properties would correspond to acid pH and high salt content.

3.4.2. Emulsifying Properties

The emulsification properties of acid and alkaline extracted proteins were evaluated by their ability to form and stabilize oil-in-water emulsions. EAI indicates the ability of protein to adsorb at the interface during the formation of emulsion, avoiding flocculation and coalescence of the small fat droplets. ESI estimates the decreasing rate of the emulsion turbidity due to fat droplet coalescence and creaming. Therefore, EAI and ESI increase when proteins favor emulsion formation and stabilization, respectively [30].

Experimental results concerning emulsifying properties are shown in Table 4. The mean EAI value for all the protein extracts was 260.3 ± 59.7 $m^2 g^{-1}$, ranging from 150.4 (pH 8 and 3.42% NaCl) to 354.7 (pH 6.5 without added NaCl). Overall, solutions at pH > 6.5 with low NaCl contents were the ones leading to higher EAI. Steen et al. also found that an increase in salt concentration decreased the emulsifying properties of water and salt soluble proteins from pork liver [29]. Nevertheless, in our results the low emulsifying activity of acid solutions (pH 5) seemed to be somewhat compensated by increasing NaCl. In fact, the impact of salt on the emulsifying properties of proteins, both of animal and vegetable origin, which can be found in the literature leads to contradictory results and diverse mechanisms are proposed to support each finding [29]. Looking at ESI, significantly more stable emulsions were obtained at pH 8 and 0.58% NaCl (170.62 min) as compared to the rest of extracts, which on the whole showed an average ESI value of about 26.43 ± 7.17 min. Dàvila et al. [23] also reported ESI values for porcine plasma globulins and albumin from four- to five-fold higher at pH 7.5 than at lower pH (pH 6.5 and 4.5), which were attributed to the effect of protein aggregation phenomena induced by acidification. However, none of the other alkaline extracts with higher ionic strength allowed for obtaining such stable emulsions.

Regardless of the fractions that provided emulsions with higher indexes, it should be noted that the heart extracts at any condition displayed rather poorer emulsifying properties than extracts from porcine spleen, which were reported to achieve EAI values from 244 to 500 $m^2 g^{-1}$ and ESI values from 57 to 381 min [11]

Mathematical models for both the emulsifying activity and emulsion stability indexes are shown in Table 5. Emulsion indexes show to be influenced by both pH and ionic strength, but it is worth noting that the models merely account for 65.5% and less than 45% of the data variability for EAI and ESI, respectively. As can be observed from the response surface in Figure 3d, heart protein extracts (5 g L^{-1}) would exhibit relatively low emulsifying activity at any pH/NaCl combination, up to 300–400 $m^2 g^{-1}$ at best, which is much lower than the capacity to form o:w emulsions of blood proteins [23]. In spite of this, it is important to highlight that the EAI values obtained were similar to those displayed by palm seed protein extracts [42], and even higher than those from ovine whey proteins [43], fish protein extracts [44], yeast proteins, and egg albumins [24].

3.4.3. Gelling Properties

Gelation is as a multistage process involving the initial denaturation of native protein structure and unfolding of the protein molecules, followed by aggregation and cross-linking between proteins. Since heating induces protein denaturation, aggregation, and eventually the formation of a gel that gives consistency to the food product, many proteins act as gelling or thickening agents when a thermal treatment is applied [45]. Indeed, one of the important functional properties of muscle proteins is their ability to form a gel. Different contribution to both gel formation and gel characteristics have been reported for sarcoplasmic and myofibrillar proteins. For myofibrillar individual proteins such as myosin and protein complexes such as actomyosin, pH significantly influences gel formation. Optimal pH for gelation depends on the concentration of myosin and actin in solution. The state of meat protein molecules, which is influenced by the ionic strength, is also important for gelling capacity since gel formation results from protein–protein and protein–solvent interactions [46].

Although salt-soluble muscle proteins are known to contribute primarily to gel formation, their recovery from some organs is difficult due to the high content of connective tissue [7]. All extracts contain myofibrillar proteins, as shown by SDS-PAGE, but their relatively low protein content has probably impaired the heat-induced gel forming ability. Nonetheless, it is interesting to realize (Table 4 and Figure 3f) that some of them displayed heat-induced gelling capacity in spite of their relatively low protein content, which would probably improve after a gentle concentration process.

Protein extracts at pH ≤ 5 and pH 8.6 showed some aggregation after the heat treatment, but they were not able to form consistent gels. On the other hand, self-supporting gels were obtained on heating extracts at pH 6.5–8, which showed improved firmness with increasing NaCl concentration. It is worth noting that all heart protein gels obtained in the present work showed high syneresis. Addition of NaCl has been described as a key requirement for good gelling properties for myofibrillar protein-rich extracts prepared from bovine hearts using a surimi-like process [47]. Tsermoula et al. also reported an increase in gel hardness when 2% NaCl was added to the extracts from porcine heart standardized at 8% protein content [16]. Nevertheless, gel forming ability of water soluble pork liver proteins without salt showed to be higher compared to high salt concentrations [29]. It seems interesting to highlight that extracts able to form firm gels somewhat coincide with the ones showing the band in the range 25–30 kDa of their electrophoretic profile, which supposedly could correspond to several blood globulins. Gelation studies on blood proteins reported that the development of strong gels in a wide range of pH was mainly attributed to the globulins fraction that were also responsible for the strength of gels when mixed with albumin [23,48]. Likewise, Steen et al. observed that higher salt concentrations shifted onset gelation temperature to lower temperatures [29].

The mathematical model obtained (Table 5) fits well to the experimental data ($R^2 = 0.946$) and confirms the strong dependence of the gelling capacity on the ionic strength, as well as its interaction with the pH. As can be seen in Figure 3f, the effect of pH is only important when the extraction process is carried out with buffers with high NaCl content. Higher gel firmness values correspond to solutions at pH 6–7.5, showing maximum values at 3–4% NaCl. At lower NaCl contents, the firmness is reduced in half and becomes practically independent of the pH factor.

4. Conclusions

The protein functional properties strongly depend on the extraction conditions, which affect each property in a different way. Protein extracts showing diverse functionality were obtained from porcine hearts by using different extraction buffers. The conditions that showed both the best extraction yield and the highest protein content were pH 8.6 and 2% NaCl. Highest foaming capacity and stability corresponded to the solutions extracted with acid buffers. The foam stability improved with increased ionic strength. The highest emulsifying activity was found in solutions at pH > 6.5 with low NaCl, whilst the more stable emulsions were obtained at pH 8 and 0.58% NaCl; nevertheless, heart protein extracts showed rather poorer emulsifying properties than extracts from other porcine organs. Although protein extracts at pH ≤ 5 and pH 8.6 were not able to form consistent gels, self-supporting gels were obtained from extracts at pH 6.5–8, all showing enhanced firmness as NaCl concentration increases.

Response surface methodology has been used successfully to systematically study the effects of extraction conditions on the recovery and functionality of soluble proteins from porcine hearts. The mathematical models obtained allow for knowing the variables that influence every functional property and their interactions, and also to determine the best conditions to obtain ingredients with a specific functionality. Therefore, these models can be used to foreknow the most suitable operating conditions according to the intended use of the protein extracts.

Recovering technofunctional proteins for food applications from industrial byproducts would contribute to improving the sustainability of the meat industry.

Author Contributions: Each author has made substantial contributions to the present work and has approved the submitted version. Their individual contributions follow: conceptualization, D.P., M.T., and C.C.; methodology, D.P., M.T., and E.S.; investigation, D.P., E.C., and J.G.; data curation and interpretation, D.P., M.T., E.C., J.G., E.S., and C.C.; writing, D.P. (original draft preparation), M.T., E.S., and C.C. (review and editing); supervision and project administration, C.C. All authors have read and agreed to the published version of the manuscript.

Funding: This research was funded by Government of Catalonia, grant number 56.21.031.2016 3A.

Acknowledgments: This work was supported by the industrial abattoirs: Patel SAU, Olot Meats SA, Friselva SA, Frigorifics del Nordeste SA, and Frigoríficos Costa Brava SA. We acknowledge A.M. Aymerich and NORFRISA (Girona, Spain) for the helpful technical assistance and for kindly donation of heart samples, respectively.

Conflicts of Interest: The authors declare no conflict of interest.

References

1. Jayathilakan, K.; Sultana, K.; Radhakrishna, K.; Bawa, A.S. Utilization of byproducts and waste materials from meat, poultry and fish processing industries: A review. *J. Food Sci. Technol.* **2012**, *49*, 278–293. [CrossRef]
2. Rivera, J.A.; Sebranek, J.G.; Rust, R.E. Functional properties of meat by-products and mechanically separated chicken (MSC) in a high-moisture model petfood system. *Meat Sci.* **2000**, *55*, 61–66. [CrossRef]
3. Toldrá, F.; Aristoy, M.C.; Mora, L.; Reig, M. Innovations in value-addition of edible meat by-products. *Meat Sci.* **2012**, *92*, 290–296. [CrossRef] [PubMed]
4. Toldrá, F.; Mora, L.; Reig, M. New insights into meat by-product utilization. *Meat Sci.* **2016**, *120*, 54–59. [CrossRef] [PubMed]
5. Toldrá, F.; Reig, M. Innovations for healthier processed meats. *Trends Food Sci. Technol.* **2011**, *22*, 517–522. [CrossRef]
6. Zhang, W.; Xiao, S.; Samaraweera, H.; Lee, E.J.; Ahn, D.U. Improving functional value of meat products. *Meat Sci.* **2010**, *86*, 15–31. [CrossRef]
7. Lynch, S.A.; Mullen, A.M.; O'Neill, E.; Drummond, L.; Álvarez, C. Opportunities and perspectives for utilisation of co-products in the meat industry. *Meat Sci.* **2018**, *144*, 62–73. [CrossRef]
8. Matak, K.E.; Tahergorabi, R.; Jaczynski, J. A review: Protein isolates recovered by isoelectric solubilization/precipitation processing from muscle food by-products as a component of nutraceutical foods. *Food Res. Int.* **2015**, *77*, 697–703. [CrossRef]
9. Mullen, A.M.; Álvarez, C.; Zeugolis, D.I.; Henchion, M.; O'Neill, E.; Drummond, L. Alternative uses for co-products: Harnessing the potential of valuable compounds from meat processing chains. *Meat Sci.* **2017**, *132*, 90–98. [CrossRef]
10. Papier, K.; Ahmed, F.; Lee, P.; Wiseman, J. Stress and dietary behaviour among first-year university students in Australia: Sex differences. *Nutrition* **2015**, *31*, 324–330. [CrossRef]
11. Toldrà, M.; Parés, D.; Saguer, E.; Carretero, C. Recovery and Extraction of Technofunctional Proteins from Porcine Spleen Using Response Surface Methodology. *Food Bioprocess Technol.* **2019**, *12*, 298–312. [CrossRef]
12. Zouari, N.; Fakhfakh, N.; Amara-Dali, W.B.; Sellami, M.; Msaddak, L.; Ayadi, M.A. Turkey liver: Physicochemical characteristics and functional properties of protein fractions. *Food Bioprod. Process.* **2011**, *89*, 142–148. [CrossRef]
13. Kim, Y.H.; Cheong, J.K.; Yang Sungnam, S.Y.; Lee Suwon, M.H. Functional properties of the porcine variety meats. *Korean J. Anim. Sci.* **1991**, *33*, 507–514.
14. Nuckles, R.O.; Smith, D.M.; Merkel, R.A. Meat By-product protein composition and functional properties in model systems. *J. Food Sci.* **1990**, *55*, 640–643. [CrossRef]
15. Kim, H.K.; Ha, S.J.; Kim, Y.H.; Hong, S.P.; Kim, Y.U.; Song, K.M.; Lee, N.H.; Jung, S.K. Protein extraction from porcine myocardium using ultrasonication. *J. Food Sci.* **2017**, *82*, 1059–1065. [CrossRef]
16. Tsermoula, P.; Virgili, C.; Ortega, R.G.; Mullen, A.M.; Álvarez, C.; O'Brien, N.M.; O'Flaherty, E.A.A.; O'Neill, E.E. Functional protein rich extracts from bovine and porcine hearts using acid or alkali solubilisation and isoelectric precipitation. *Int. J. Food Sci. Technol.* **2019**, *54*, 1292–1298. [CrossRef]
17. International Organization for Standardization. *Microbiology of the Food Chain. Horizontal Method for the Enumeration of Microorganisms—Part 1: Colony Count at 30 °C by the Pour Plate Technique*; Standard No. 4833-1:2013; International Organization for Standardization: Geneva, Switzerland, 2013.
18. Association of Official Analytical Chemists (AOAC). *Official Methods of Analysis of AOAC*; Association of Analytical Communities: Gaithersburg, MD, USA, 2000.
19. International Organization for Standardization. *Determination of Nitrogen Content. International Standards Meat and Meat Products*; Standard No. 93:1978; International Organization for Standardization: Geneva, Switzerland, 1978.
20. Kolar, K. Colorimetric determination of hydroxyproline as measure of collagen content in meat and meat products: NMKL collaborative study. *J. AOAC* **1990**, *73*, 54–57. [CrossRef]
21. International Organization for Standardization. *Determination of Total Fat Content. International Standards Meat and Meat Products*; Standard No. 1443:1973; International Organization for Standardization: Geneva, Switzerland, 1973.

22. Fort, N.; Kerry, J.P.; Carretero, C.; Kelly, A.L.; Saguer, E. Cold storage of porcine plasma treated with microbial transglutaminase under high pressure. Effects on its heat-induced gel properties. *Food Chem.* **2009**, *115*, 602–608. [CrossRef]
23. Dàvila, E.; Saguer, E.; Toldrà, M.; Carretero, C.; Parés, D. Surface functional properties of blood plasma protein fractions. *Eur. Food Res. Technol.* **2007**, *226*, 207–214. [CrossRef]
24. Pearce, K.N.; Kinsella, J.E. Emulsifying properties of proteins—Evaluation of a turbidimetric technique. *J. Agric. Food Chem.* **1978**, *26*, 716–723. [CrossRef]
25. Parés, D.; Ledward, D.A. Emulsifying and gelling properties of porcine blood plasma as influenced by high-pressure processing. *Food Chem.* **2001**, *74*, 139–145. [CrossRef]
26. Seong, P.N.; Park, K.M.; Cho, S.H.; Kang, S.M.; Kang, G.H.; Park, B.Y.; Moon, S.S.; Van Ba, H. Characterization of edible pork by-products by means of yield and nutritional composition. *Korean J. Food Sci. Anim. Res.* **2014**, *34*, 297–306. [CrossRef] [PubMed]
27. Kang, G.; Seong, P.; Moon, S.; Cho, S.; Ham, H.; Park, K.; Kang, S.; Park, B. Distribution Channel and Microbial Characteristics of Pig By-products in Korea. *Korean J. Food Sci.* **2014**, *34*, 792–798. [CrossRef] [PubMed]
28. European Food Safety Authority (EFSA). Commission regulation (EC) no 2073/2005 of 15th November 2005 on microbiological criteria for foodstuffs. *Off. J. Eur. Union* **2005**, *338*, 1–26.
29. Steen, L.; Glorieux, S.; Goemaere, O.; Brijs, K.; Paelinck, H.; Foubert, I.; Fraeye, I. Functional properties of pork liver protein fractions. *Food Bioprocess Technol.* **2016**, *9*, 970–980. [CrossRef]
30. Selmane, D.; Christophe, V.; Gholamreza, D. Extraction of proteins from slaughterhouse by-products: Influence of operating conditions on functional properties. *Meat Sci.* **2008**, *79*, 640–647. [CrossRef] [PubMed]
31. Hrynets, Y.; Omana, D.A.; Xu, Y.; Betti, M. Effect of acid- and alkaline-aided extractions on functional and rheological properties of proteins recovered from mechanically separated turkey meat (MSTM). *J. Food Sci.* **2010**, *75*, 477–486. [CrossRef]
32. Krasnowska, G.; Gorska, I.; Gergont, J. Evaluation of functional properties of offal proteins. *Meat Sci.* **1995**, *39*, 149–155. [CrossRef]
33. Pérez-Chabela, M.L.; Soriano-Santos, J.; Ponce-Alquicira, E.; Díaz-Tenorio, L.M. Electroforesis en gel de poliacrilamida-SDS como herramienta en el estudio de las proteínas miofibrilares. *Nacameh* **2015**, *9*, 77–96.
34. Howell, N.K.; Lawrie, R.A. Functional aspects of blood plasma proteins I. Separation and characterization. *J. Food Technol.* **1983**, *18*, 747–762. [CrossRef]
35. Luna, E.J.; Hitt, A.L. Cytoskeleton-plasma membrane interactions. *Science* **1992**, *258*, 955–964. [CrossRef] [PubMed]
36. Grujić, R.; Savanović, D. Analysis of myofibrillar and sarcoplasmic proteins in pork meat by capillary gel electrophoresis. *Foods Raw Mater.* **2018**, *6*, 421–428. [CrossRef]
37. Toldrà, M.; Parés, D.; Saguer, E.; Carretero, C. Hemoglobin hydrolysates from porcine blood obtained through enzymatic hydrolysis assisted by high hydrostatic pressure processing. *Innov. Food Sci. Emerg. Technol.* **2011**, *12*, 435–442. [CrossRef]
38. Aluko, R.E.; McIntosh, T. Polypeptide profile and functional properties of defatted meals and protein isolates of canola seeds. *J. Sci. Food Agric.* **2001**, *81*, 391–396. [CrossRef]
39. Indrawati, L.; Wang, Z.; Narsimhan, G.; Gonzalez, J. Effect of processing parameters on foam formation using a continuous system with a mechanical whipper. *J. Food Eng.* **2008**, *88*, 65–74. [CrossRef]
40. Yang, Q.L.; Lou, X.W.; Wang, Y.; Pan, D.D.; Sun, Y.Y.; Cao, J.X. Effect of pH on the interaction of volatile compounds with the myofibrillar proteins of duck meat. *Poultry Sci.* **2017**, *96*, 1963–1969. [CrossRef]
41. Lynch, S.A.; Álvarez, C.; O'Neill, E.E.; Keenan, D.F.; Mullen, A.M. Optimization of protein recovery from bovine lung by pH shift process using response surface methodology. *J. Sci. Food Agric.* **2018**, *98*, 1951–1960. [CrossRef]
42. Akasha, I.; Campbell, L.; Lonchamp, J.; Euston, S.R. The major proteins of the seed of the fruit of the date palm (Phoenix dactylifera L.): Characterisation and emulsifying properties. *Food Chem.* **2016**, *197*, 799–806. [CrossRef]
43. Díaz, O.; Pereira, C.D.; Cobos, A. Functional properties of ovine whey protein concentrates produced by membrane technology after clarification of cheese manufacture by-products. *Food Hydrocoll.* **2004**, *18*, 601–610. [CrossRef]
44. Liceaga-Gesualdo, A.M.; Li-Chan, E.C.-Y. Functional Properties of Fish Protein Hydrolysate from Herring (Clupea harengus). *J. Food Sci.* **1999**, *64*, 1000–1004. [CrossRef]

45. Ziegler, G.R.; Foegeding, E.A. The gelation of proteins. *Adv. Food Nutr. Res.* **1990**, *34*, 203–298.
46. Zayas, J.F. Gelling Properties of proteins. In *Functionality of Proteins in Food*; Springer: Berlin/Heidelberg, Germany, 1997; pp. 310–366.
47. James, J.M.; Mireles DeWitt, C.A. Gel Attributes of Beef Heart When Treated by Acid Solubilization Isoelectric Precipitation. *J. Food Sci.* **2004**, *69*, 473–480. [CrossRef]
48. Parés, D.; Saguer, E.; Carretero, C. Blood by-products as ingredients in processed meat. In *Processed Meats: Improving Safety, Nutrition and Quality*; Kerry, J.P., Kerry, J.F., Eds.; Woodhead Publishing Ltd.: Cambridge, UK, 2011; pp. 218–242.

Publisher's Note: MDPI stays neutral with regard to jurisdictional claims in published maps and institutional affiliations.

© 2020 by the authors. Licensee MDPI, Basel, Switzerland. This article is an open access article distributed under the terms and conditions of the Creative Commons Attribution (CC BY) license (http://creativecommons.org/licenses/by/4.0/).

Article

Development of an Accelerated Stability Model to Estimate Purple Corn Cob Extract Powder (Moradyn) Shelf-Life

Lucia Ferron [1,2], Chiara Milanese [3], Raffaella Colombo [1] and Adele Papetti [1,*]

[1] Department of Drug Sciences, University of Pavia, Viale Taramelli 12, 27100 Pavia, Milan, Italy; lucia.ferron01@universitadipavia.it (L.F.); raffaella.colombo@unipv.it (R.C.)
[2] FlaNat Research Italia Srl, Via Giuseppe di Vittorio 1, 20017 Rho, Milan, Italy
[3] Consorzio interuniversitario per i Sistemi a Grande Interfase & Department of Chemistry, Physical Chemistry Section, University of Pavia, Viale Taramelli 12, 27100 Pavia, Milan, Italy; chiara.milanese@unipv.it
* Correspondence: adele.papetti@unipv.it; Tel.: +39-0382-98-7863

Abstract: Moradyn is an Italian purple corn variety whose cobs represent a rich source of polyphenols. At the industrial level, they are used to produce a dried extract (MCE) by the addition of 20% Arabic gum. In order to evaluate the extract solid-state stability, an innovative accelerated stress protocol was developed following the isoconversion approach. The degradation kinetics of cyanidin-3-*O*-glucoside (C3G), the most suitable marker to monitor the overall MCE degradation status, was monitored under five temperature–humidity (RH) combinations. These data were used to build a mathematical model, able to estimate the C3G stability at 25 °C and 30% RH, whose predictiveness was further assessed by comparing the predicted vs. experimental C3G isoconversion time. Finally, by applying this model, the expiry date of the extract was calculated to be within 26–33 days, confirming that the addition of 20% Arabic gum is insufficient to stabilize MCE and highlighting the need of a new formula in order to prolong MCE shelf-life.

Keywords: purple corn cob; anthocyanins; Arabic gum; accelerated stress protocol; forced degradation; moisture-modified Arrhenius equation

Citation: Ferron, L.; Milanese, C.; Colombo, R.; Papetti, A. Development of an Accelerated Stability Model to Estimate Purple Corn Cob Extract Powder (Moradyn) Shelf-Life. *Foods* **2021**, *10*, 1617. https://doi.org/foods10071617

Academic Editors: Simona Grasso, Konstantinos Papoutsis, Claudia Ruiz-Capillas and Ana Herrero Herranz

Received: 11 June 2021
Accepted: 9 July 2021
Published: 13 July 2021

Publisher's Note: MDPI stays neutral with regard to jurisdictional claims in published maps and institutional affiliations.

Copyright: © 2021 by the authors. Licensee MDPI, Basel, Switzerland. This article is an open access article distributed under the terms and conditions of the Creative Commons Attribution (CC BY) license (https://creativecommons.org/licenses/by/4.0/).

1. Introduction

Today people are more aware of the potential health benefits derived from the consumption of nutritionally valuable foods than in previous decades; these are clearly recognized as the primary source of bioactive compounds, claimed not only to have a nutritional function but also to improve health condition by inhibiting typical chronic disease risk factors, such as hyperglycemia, hypertension, obesity, and oxidative stress [1]. Nowadays, a huge number of both epidemiological and in vitro studies support the correlation between the consumption of plant secondary metabolites—such as alkaloids, terpenoids, flavonoids, and phenolic acids—and their bio-protective effects [2]. In particular, flavonoids are known to exert a wide range of different health benefits, including hypoglycemic, hypolipidemic, anti-inflammatory, antimicrobial, and antioxidant effects [3].

In this context, plant-based food supplements (botanicals) represent an interesting approach in the prevention and management of the diseases strictly related to aging and life-style, such as metabolic syndrome [4]. Although researchers and pharmaceutical companies have demonstrated a renewed interest in investigating plants and crops as a source of bioactive phytocomplexes, their application is still limited since the physico-chemical properties and chemical stability of these complex matrices are strongly affected by environmental factors such as pH, temperature, and light [5,6]. Exposure to these factors triggers the degradation pathways of many bioactive metabolites, especially anthocyanins, strongly affecting their storage stability [6].

Over the last five years, interest in the identification of new and effective strategies to prolong the shelf-life of polyphenols and preserve their health value has grown. In particular, research has focused on the use of polysaccharides such as Arabic gum (AG), maltodextrins, or β-cyclodextrin as stabilizing agents. In fact, soluble fibers are able to bind and entrap both hydrophilic and hydrophobic bioactive metabolites (such as lutein, anthocyanins, or piperine) which become less sensitive to environmental factors [7–13]. Moreover, encapsulation protocols have been designed to optimize stability and bioavailability of molecules present in natural extracts. Maltodextrins and AG are often used as common coating agents, thanks to their high solubility, lack of color and odor, and low cost. AG belongs to a group of water soluble and undigestible polysaccharides which are known to inhibit oxidation reactions and, therefore, are able to protect sensitive compounds when encapsulated [9].

Among flavonoids, anthocyanins have been thoroughly investigated in stability studies because of their extremely low storage stability, especially when affected by high temperatures and pH values higher than 5.0 [5,8]. Different stress testing procedures, based on guidelines reported by the International Council for Harmonisation of Technical Requirements for Pharmaceuticals for Human Use (ICH), the World Health Organization (WHO), and the European Medicines Agency (EMA) [14–16], can be applied to verify the stability of a molecule. Stress testing's main goal is to identify the degradation fate of an active principle ingredient (API), its intrinsic stability, through a validated analytical procedure. Guidance on stability testing of active pharmaceutical ingredients and finished pharmaceutical products, as reported by EMA, states: "The objective of stress testing is to identify primary degradation products and not to completely degrade the API. The conditions studied should cause degradation to occur to a small extent, typically 10–30% loss of API as determined by assay when compared with non-degraded API." This test should be carried out by submitting a single batch of the API to different temperature (T) and relative humidity (RH) conditions, in order to monitor the effects of these factors on active molecules over time (European Medicines Agency). However, guidelines typically refer to a single bioactive compound, an API, or drugs at the final preparation stage, and not to a natural phytocomplex. Currently, there is a general lack of commitment to pass worldwide effective legislation concerning the evaluation of natural extract quality, chemical stability, and efficacy; however, these features are already mandatory in the production of drugs [11,16–19]. Compared with well-defined synthetic drugs, plant extract standardization and stability assessment are hard tasks; the high variability of a phytocomplex composition, in addition to all factors influencing the secondary metabolite profile such as the plant's developmental stage, environmental factors, and post-harvest processing, make standardization feasible, but not easy.

The biological activity of plant-based preparations cannot be attributed to a small number of compounds, but is usually related to the synergistic action of a complex pattern of molecules. Thus, in order to evaluate the quality of the extract, a target group of compounds should be selected and monitored during standardization, as well as during stability studies, following the above-mentioned guidelines [20]. However, the application of these standardized protocols to evaluate a natural extract's chemical stability is an extremely complex, expensive, and time-consuming procedure.

Recently, our group investigated the chemical composition of an anthocyanin-enriched extract obtained from a new Italian purple corn cob variety, Moradyn (MCE). The phytocomplex of MCE differs in composition from the typical Peruvian purple corn cob in its greater variety of flavonols, such as quercetin, myricetin, isorhamnetin, and kaempferol derivatives, and its lack of malonylated anthocyanins, which are typically detected in purple corn varieties. MCE demonstrated good hypoglycemic and antiglycative activities in several in vitro experiments, attributed to the synergistic action of all polyphenols present in the extract [21]. Therefore, MCE could be a good candidate for a healthy formulation, but only after stability studies.

Therefore, the aim of this work was to assess MCE storage stability, when AG is used as a carrier agent, by applying an innovative statistical approach. In the current study, stress testing was carried out using a new solid-state stability model based on the accelerated stability assessment program (ASAP) which allowed for evaluation of the effect of temperature and humidity on MCE and prediction of its storage stability [22,23].

2. Materials and Methods

2.1. Chemicals

Ethanol, magnesium chloride, sodium chloride, HPLC-grade formic acid, and acetonitrile were purchased from Carlo Erba (Milan, Italy). Water was obtained from a Millipore Direct-QTM system (Merck-Millipore, Milan, Italy). Arabic gum was purchased from Merck Life Science (Milan, Italy). Kuromanin chloride (cyanidin-3-O-glucoside) was purchased from Extrasynthese (Genay, Rhone, France).

2.2. Moradyn Corn Cob Extract (MCE) Preparation

Chopped Moradyn corn cobs (Community Plant Variety Office Registration-Examination Ref. 4067062) were provided by FlaNat Research Italia S.r.l. (Milan, Italy), and extracted following the procedure described in [21]. MCE dry matter was then suspended in AG aqueous solution (4:1, w/w), and dried using a vacuum drying oven at 40 °C for 48 h.

2.3. Stress Conditions and Accelerated Stability Model Validation

Stress tests were performed by submitting MCE-AG dry matter to five different storage conditions: 25 °C at 75% RH, 45 °C at 30 and 75% RH, and 70 °C at 30 and 75% RH. Twenty milligrams of dried MCE-AG were weighed in a 15 mL open glass vial, inserted into a sealed vessel containing saturated salt solutions, in order to maintain a controlled relative humidity value in the chamber, and then stored in a controlled temperature oven. Saturated magnesium chloride and sodium chloride solutions were used to maintain the environment at 30% RH and 75% RH, respectively [24].

In order to validate the mathematical model, three MCE-AG samples from three different batches were prepared following the above-mentioned protocol. Three different samples obtained from each batch were submitted to stress testing at 45 °C-30% RH, 45 °C-75% RH, and 70 °C-75% RH, and their degradation was assessed by monitoring cyanidin-3-O-glucoside (C3G) concentration by HPLC at three different times (4, 5, and 7 days for 45 °C-30% RH; 4 h, 1 day, and 4 days for 45 °C-75% RH; 2 h, 5 h, and 1 day for 70 °C-75% RH). All chromatograms were recorded both at 520 nm and 370 nm in order to monitor the changes occurring in anthocyanin and flavonoid concentrations.

2.4. Chemical Characterization by RP-HPLC-WVD

An Agilent Technologies 1260 Infinity high-performance liquid chromatography system (Santa Clara, CA, USA), which included a quaternary gradient pump, an autosampler, a degasser, and a variable wavelength detector (VWD), was used. The chromatographic separation was performed using a Gemini C18 analytical column (150 × 2.0 mm i.d., 5 µm, Phenomenex, Torrance, CA, USA) thermostatted at 25.0 ± 0.5 °C, operating at a constant flow rate of 0.3 mL/min, injection volume 20 µL. The mobile phase consisted of 0.1% formic acid aqueous solution (solvent A) and acetonitrile acidified with 0.01% formic acid (solvent B); the gradient elution was: 0–3 min, 2–15% B; 3–45 min, 15–25% B; 45–48 min, 25–35% B; 48–58 min, 2% B, followed by a column reconditioning step of 10 min. Chromatograms were recorded at 370 and 520 nm. The HPLC-VWD system was controlled by Agilent OpenLab CDS ChemStation software (Windows 10, Agilent Technologies, Santa Clara, CA, USA).

RP-HPLC-WVD Method Validation

The validation tests were carried out using the external standard method, following ICH guidelines on bioanalytical method validation [25]. C3G was used as standard and chromatograms were registered at 520 nm. In order to verify a putative matrix effect, C3G was dissolved both in MCE and in acidulate aqueous-acetonitrile solution at 100 µg/mL. Both solutions were used to construct two different five-point calibration curves in the

5–20 µg/mL concentration range; each point was analyzed in triplicate and R^2 values were compared.

Specificity was assessed by overlapping chromatograms obtained for C3G and blank solution (mobile phase) and no peak was detected in the latter; selectivity was verified by comparing the retention time of standard reference C3G in sample solvent to the retention time of C3G in MCE, recorded at 520 nm.

As regards linearity, three calibration curves (in three non-consecutive days) in the range 5–20 µg/mL were constructed for C3G by plotting the integrated peak area (y) at five different concentration levels vs. the theoretical concentration.

Limit of detection (LOD) and limit of quantification (LOQ) were determined by the signal-to-noise ratio (S/N) and defined as the concentration levels at S/N of about 3 and 10, respectively. The precision of the method was evaluated by intra- and inter-day estimation, repeating the analysis of the highest tested concentration (20 µg/mL) nine times within a single day (intra-day) and, in triplicate, three different concentrations (5, 10, and 20 µg/mL) for three consecutive days (inter-day). Finally, the precision was measured by computing the relative standard deviation (RSD%) on replicate analyses, while accuracy was assessed by recovery studies of C3G at three different concentration levels (5, 10, and 20 µg/mL), performed in triplicate.

2.5. Solid-State Characterization

Differential scanning calorimetry (DSC) analyses were performed by heating the samples (about 5 mg) from -80 °C to 300 °C at 5 °C/min under N_2 atmosphere in open Al crucibles in a Q2000 instrument interfaced with a TA 5000 data station (TA Instruments, USA) and subsequently cooling them back to room temperature.

Fourier-transform infrared (FT-IR) spectra were acquired using a Nicolet FT-IR iS10 Spectrometer (Nicolet, Madison, WI, USA) equipped with an attenuated total reflectance (ATR) sampling accessory (Smart iTR with diamond plate) by co-adding 32 scans in the 4000 cm^{-1}–650 cm^{-1} range with resolution set at 4 cm^{-1}.

2.6. Statistical Analysis

Matlab statistical software version R2019B was used to build the solid-state stability model, based on moisture-corrected Arrhenius equation, and to perform ANOVA tests.

3. Results and Discussion

MCE is a rich source of polyphenols and its anthocyanin fraction represents 28% (w/w) of its total phenolic content [21]. Anthocyanins are well known for their health benefits; however, their use is still limited by their sensitivity to pH, light, oxygen, and heat, which lead to weak storage stability [26]. In order to increase anthocyanins' storage stability, different formulations developed by coupling these compounds with different wall materials, such as maltodextrins [27] or Arabic gum, are present in the literature. In particular, Arabic gum could be a suitable anthocyanin stabilizer when drying techniques are used to obtain the final ingredient [28]. Therefore, in this study, MCE was formulated with 20% AG (w/w) (MCE-AG), which is a concentration generally used at the industrial manufacturing level to produce a dried extract.

3.1. HPLC Validation

The RP-HPLC method was validated according to ICH guidelines on bioanalytical method validation [25]. Specificity, selectivity, linearity, limit of detection (LOD), limit of quantification (LOQ), precision (intra- and inter-day), and accuracy were evaluated.

Preliminary experiments demonstrated that no matrix effect was registered, thus all the analyses were carried out using C3G dissolved in 0.1% formic acid aqueous solution—acetonitrile acidified with 0.01% formic acid, 80:20 (v/v).

Method specificity was assessed by overlapping chromatograms obtained for C3G at the concentration of 20 µg/mL and blank solution (mobile phase) and no peak was detected in the latter, while selectivity was verified by comparing the retention time of the standard with the retention time of C3G detected in MCE, recorded at 520 nm. Over three different days, a calibration curve (5–20 µg/mL) was freshly prepared, analyzed, and linear regression analysis was performed using the least-squares method. The resulting coefficient of correlation mean value of these three curves was higher than 0.998, indicating a good linearity.

The method was shown to be accurate, with values ranging from 99.648 to 101.604%, and precise, since all standard deviation values were lower than 0.3%. Finally, the sensitivity was assessed by calculating LOD and LOQ values, which were 0.925 µg/mL and 3.082 µg/mL, respectively.

Therefore, overall, the data demonstrated that this analytical method is suitable to quantify C3G in MCE.

3.2. Preliminary Physico-Chemical Characterization of the Raw Materials and Their Mixtures

Preliminary thermal analysis (DSC) and IR measurements were performed on MCE, C3G, AG, and their mixture to evaluate if 20% AG (w/w) could modify MCE and C3G when exposed to high temperatures.

The calorimetric profiles acquired for the different substances are reported in Figure 1. In the range 80–300 °C, MCE presented three large endothermic peaks (Figure 1a), centered, respectively, at 90 °C, 160 °C, and 225 °C, due to the decomposition of some of its different constituents, as proven by the fact that the cooling curve showed no peaks. The first peak appeared as the superimposition of two processes with different kinetics, the first one, faster, starting at around 0 °C and the second one, slower, starting at about 50 °C. C3G (Figure 1b) showed only one sharper endothermic peak, starting at about 0 °C and centered at 40 °C, under which the substance decomposed. This peak resembled the features of the first signal in MCE, indicating that C3G could be the first component decomposing and hence the most instable of the MCE constituents. For this reason, C3G was chosen as a marker of the stability of the MCE mixtures, and its decomposition temperatures as the best temperature range to study the behavior of the mixtures under thermal stress. The Arabic gum (Figure 1c) showed a very large and asymmetric signal starting at −35 °C, centered at 60 °C and not ended by 110 °C, due to irreversible processes happening in the matrix. The calorimetric curves of the mixtures MCE–20% AG (w/w) (Figure 1d) and C3G-20% AG (w/w) (Figure 1e) were superimposable in shape and characteristic temperatures, up to 150 °C, to those of the main components (Figure 1a,b, respectively), with no effect and by-side reactions attributable to the gum presence in the examined temperature range. To further demonstrate the compatibility between C3G and Arabic gum, IR spectra of the pure component and of the 20% mixture were acquired. As evident in Figure 2a,b, the two graphs were superimposable and all the bands and peaks resulted attributable to the C3G functional groups.

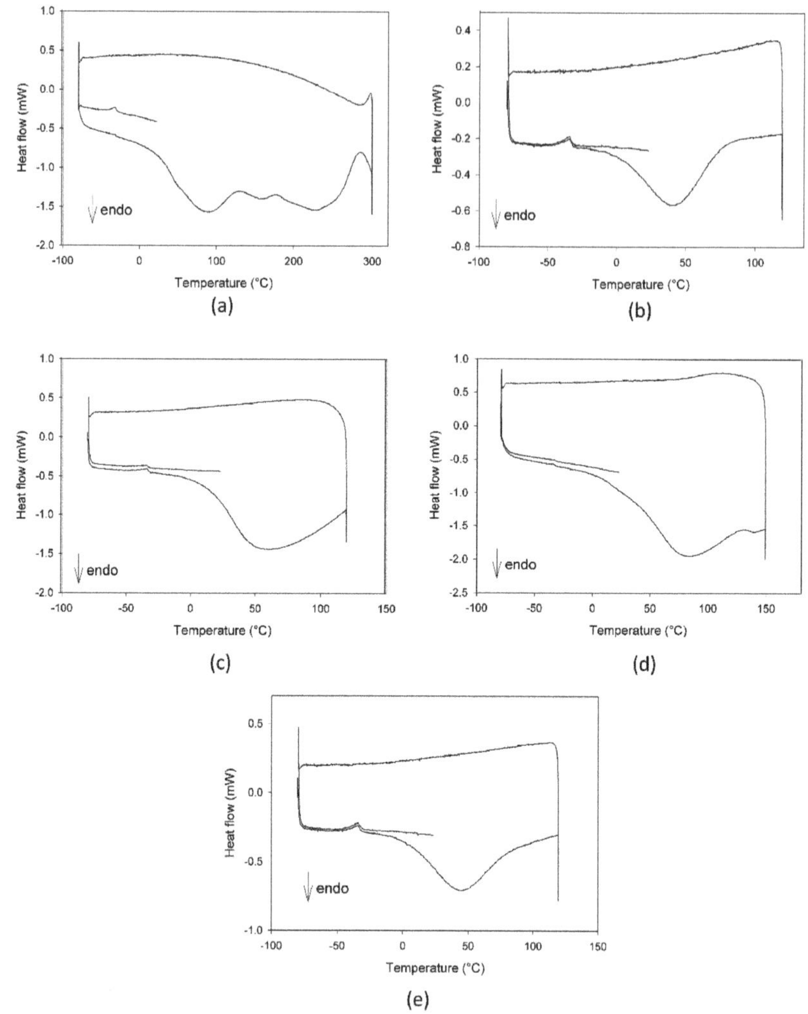

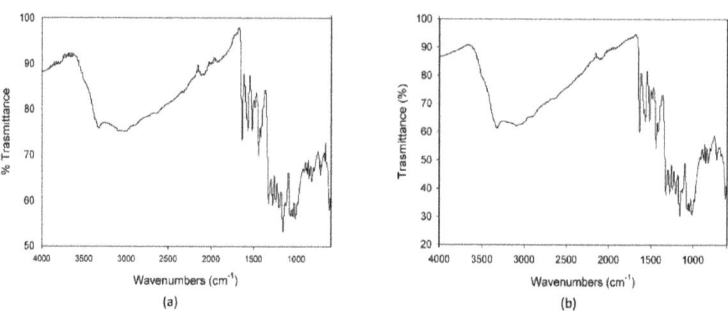

Figure 1. DSC profiles of Moradyn corn cob extract-MCE (**a**), cyanidi-3-O-glucoside-C3G (**b**), Arabic gum-AG (**c**), mixture MCE-20% AG, w/w (**d**), and mixture C3G-20% AG, w/w (**e**).

Figure 2. IR spectra of pure cyanidyn-3-O-glucoside-C3G (**a**) and of the mixture C3G-20% AG, w/w (**b**).

3.3. Stability Assessment Model

MCE stability was estimated by conducting a new isoconversion approach based on the accelerated stability assessment program [22,23], usually applied to drugs and never, to the best of the authors' knowledge, to a natural product consisting of many compounds.

Bioactive molecules' degradation behavior, when submitted to high temperature and humidity levels, is typically described by a moisture-corrected Arrhenius equation [23],

$$\ln K = \ln A + [E_a/(R \times T)] + [(B \times \%RH)] \quad (1)$$

where K is the degradation rate (degradation %/day), A is the collision frequency factor, T is temperature (K), R is the gas constant, RH is relative humidity, and Ea and B are the activation energy and humidity sensitivity constant related to the studied compounds, respectively.

Temperature and relative humidity are the main drivers of instability, however K is also related to time, as the registered reaction rates typically decrease with the extent of the conversion of reacting compounds. This is due to the fact that all of the compounds present in a mixture, at the solid state, coexist in multiple states, each characterized by a different sensitivity to environmental conditions. When complex mixtures are exposed to stressful conditions, each state reacts with its own kinetics, based on T or RH levels. During the degradation process, quantities and ratios between these different states are continuously changing, influencing the overall kinetics and degradation rate constant. Therefore, if the degradation is monitored at fixed time points independently from T and RH values, the degradation rates extrapolated from the so obtained dataset cannot be predictive of degradation rates occurring under different environmental conditions. In fact, the overall composition of each collected sample at a fixed time point is characterized by a unique pattern of states, based on T or RH, which cannot be compared with the one obtained at the same time point but under different conditions.

Following the isoconversion approach, degradation state is no longer monitored at multiple, fixed times, but is instead determined considering only the isoconversion time, i.e., the period required by the selected marker compound or API to reach the specification limit of degradation when exposed to a specific storage condition, which varies based on temperature and humidity values.

During stress testing, the selected specification limit should be within the range of 5–20% degradation of the selected markers [14,29], as it would lie in the first degradation stage, which generally follows a linear behavior [30]. Moreover, this method allowed for the mathematical prediction of the isoconversion time at any T and RH combination from Equation (1), knowing the three terms, A, Ea, and B, related to the marker compound. In fact, when the sample reached the specification limit of degradation, the K constant received the same contribution from each state, independently of T and RH levels [22].

Therefore, in order to extrapolate A, Ea, and B terms related to MCE, this extract was submitted to five different combinations of T and % RH (Table 1). Its degradation was monitored by assessing the gradual decrease of C3G, the marker selected as it is the most abundant anthocyanin in MCE [21], at four different time points, as reported in Table 1, and the specification limit was selected considering the highest degradation value lies on the first linear stage of the reaction.

Table 1. Conditions used for forced degradation tests and experimental scheme applied.

Storage Condition	Monitoring Times
25 °C-75% RH	0 d, 1 d, 3 d, 7 d, 14 d
45 °C-30% RH	0 h, 4 h, 8 h, 3 d, 7 d
45 °C-75% RH	0 h, 4 h, 8 h, 3 d, 7 d
70 °C-30% RH	0 h, 4 h, 8 h, 1 d, 3 d
70 °C-75% RH	0 h, 4 h, 8 h, 1 d, 3 d

Then, the experimental degradation rate for each T and RH combination, expressed as lnK value, was extrapolated from the curve obtained by plotting % C3G reduction vs. time, and it corresponded to the slope of the straight line passing from the origin and the point related to the chosen specification limit [29]. By using these experimental lnK values, a multiple regression model was built and the three terms of the Arrhenius equation were extrapolated. These extrapolated terms were used to realize the construction of a predictive mathematical model, based on the Arrhenius equation [22], which was applied to calculate MCE isoconversion time at 25 °C and 30% RH, aiming to approximate its expiry date.

3.3.1. Degradation Kinetics

Stress tests were performed by submitting MCE to five different storage conditions and monitoring the C3G reduction percentage (quantified by RP-HPLC-WV) at fixed time points, as reported in Table 1. The gradual C3G reduction was followed for each stress condition and the order of reaction kinetics was graphically determined by choosing the fitting which gave the best correlation coefficient (Figure 3) [31].

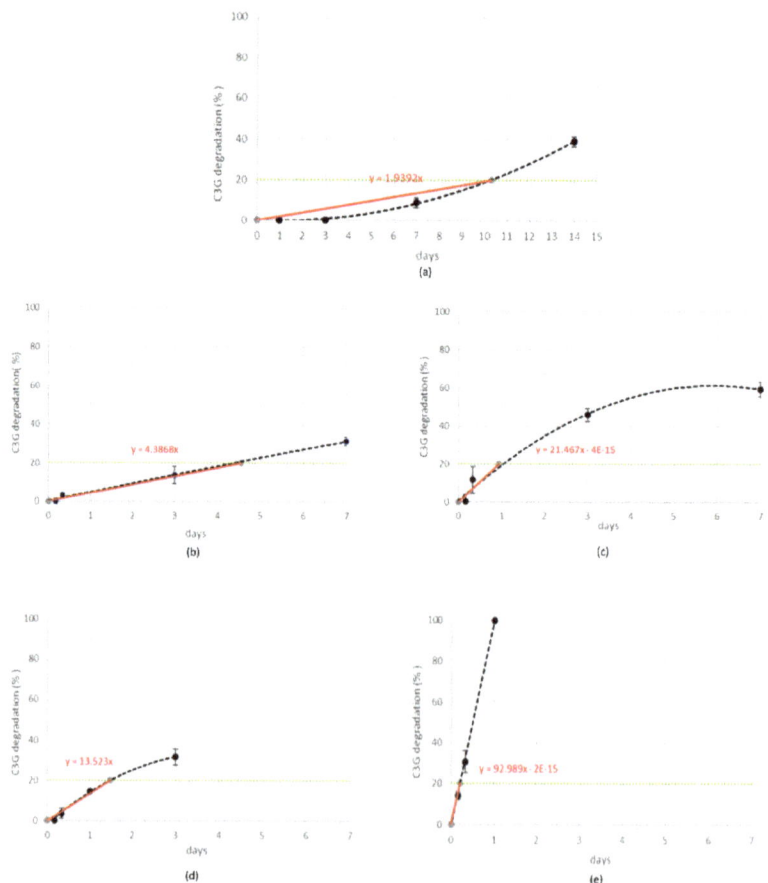

Figure 3. Cyanidin-3-O-glucoside (C3G) in MCE degradation kinetics (black line) when exposed to 25 °C-75% RH (**a**), 45 °C-30% RH (**b**), 45 °C-75% RH (**c**), 70 °C-30% RH (**d**), or 70 °C-75% RH (**e**). Experimental rate constants (ExpK) for each treatment were determined from the slope of the straight line (red line) passing from the origin and the specification limit (green line).

When MCE was exposed to the strongest condition, (70 °C-75% RH), C3G totally degraded during the first 24 h, following zero-order kinetics, and therefore its degradation was not dependent on its initial concentration (Figure 3e). This is demonstrated also by the IR spectroscopy; the acquired spectrum for this sample was totally different from those before treatment (Figures 2b and 4a). On the other hand, when samples were submitted to mild conditions (45 °C-75% RH and 70 °C-30% RH), two different steps were identified: in the first part of the reaction the degradation rate was faster, while it decreased step by step in the last phase (Figure 3c,d), following the classical degradation kinetic already stated by Waterman [22]. This behavior could be due to an initial faster degradation of free C3G which was highly exposed to oxygen action, followed by a slower degradation of bounded anthocyanin fraction, in accordance with the results reported by Tonon for açai juice [13]. The IR spectra for the samples treated for 4 h and 3 d at 70 °C-30% RH are reported in Figure 4b,c; after only 4 h of treatment the characteristic features of the functional groups of C3G were almost totally lost and after 3 days the spectrum was almost superimposable to the spectrum of a totally degraded sample (Figure 4a). This trend was not observed at 45 °C-30% RH (Figure 3b), but a longer monitoring period is likely required and more time points should be evaluated to better investigate degradation kinetics under this condition. When the degradation rate was monitored in the presence of high humidity at room temperature (25 °C and 75% RH), the reaction had an opposite trend (Figure 3a). In fact, anthocyanins are known to be extremely sensitive to high temperatures, but high humidity levels could also contribute to increases in their instability by accelerating chemical and physical degradations [9,13]. Thus, at room temperature and high humidity values, C3G degradation is only delayed and gets faster after three days of exposure to such conditions (see IR spectra in Figure 4b,e for 3 days and 7 days aging).

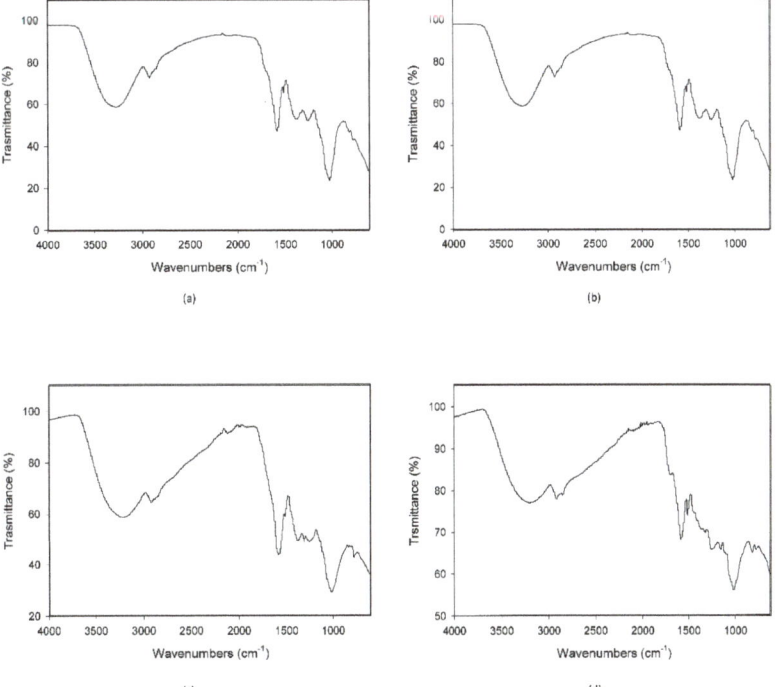

Figure 4. *Cont.*

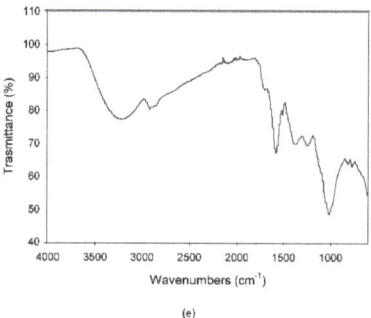

Figure 4. IR spectra of the MCE-20% AG (*w/w*) after 24 h of degradation at 70 °C-75% RH (**a**), after 4 h (**b**) and 3 days of degradation at 70 °C-30% RH (**c**), and after 3 (**d**) and 7 days of degradation at 25 °C-75% RH (**e**).

In Figure 5, the DSC profiles of the same samples discussed for the IR results are reported. As evident in all of the plots, no signals attributable to C3G are present, with the exception of a small shoulder in Figure 5b, confirming its degradation due to ageing.

Both the DSC and IR plots also demonstrated the gradual degradation of the different components of the mixtures, with a different evolution depending on the ageing conditions.

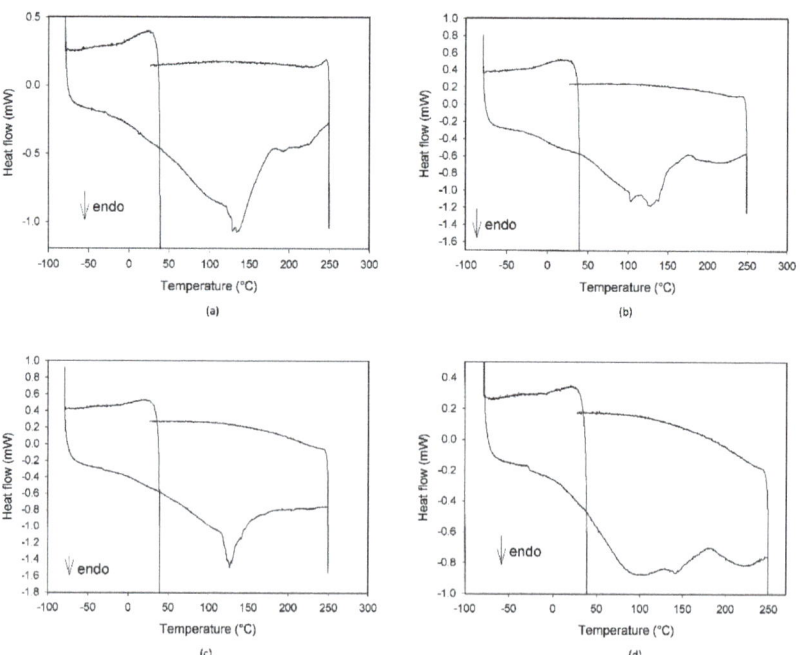

Figure 5. *Cont.*

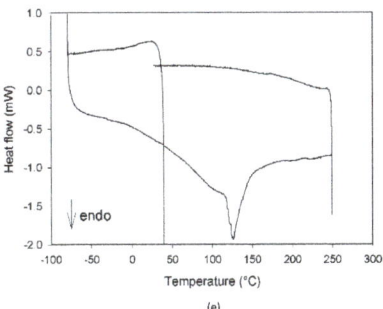

Figure 5. DSC profiles of the mixture MCE-20% AG (w/w) after 24 h of degradation at 70 °C-75% RH (**a**), after 4 h (**b**) and 3 days of degradation at 70 °C-30% RH (**c**), and after 3 (**d**) and 7 days of degradation at 25 °C-75% RH (**e**).

3.3.2. Determination of Ea, lnA, and B Arrhenius Equation Terms and Development of a Mathematical Model

As mentioned above, during stress testing it is recommended to select the specification limit of degradation within the range of 5–20% degradation of the monitored marker [14,29], as it would lie in the first degradation stage which typically follows a linear behavior [30].

During preliminary stress testing, MCE was exposed to high temperature values (both in the presence and absence of high humidity values), and C3G degradation kinetics between 0% and 20% could always be simplified and approximated as zero order (Figure 3), therefore 20% was selected as the specification limit to be used during the ASAP protocol. Thus, based on the isoconversion concept, the experimental rate constant (exp K), expressed as degradation % per day, was derived for each tested condition from the slope of straight lines passing from the origin of graphs and the point corresponding to 20% degradation, calculated by interpolation of each curve (Figure 3a–e, red lines). The natural logarithms of all five exp K (exp lnK) were fit with the reciprocal of temperature (1/T (K)) and relative humidity (RH (%)), using the Matlab multiple regression function, to extrapolate Ea, lnA, and B terms related to C3G. Then, the Matlab curve-fitting tool was applied to graphically represent this model (Figure 6a,b). Following ASAP assumption, the surface slope represented lnA, while angles determined by the surface and 1/T and RH % axes corresponded to Ea and B, respectively.

Finally, a mathematical model was constructed by fitting Equation (1) to the experimental data:

$$\ln K = 24.209 - [(7626.6 \times 1/T) + (0.033 \times RH\%)] \qquad (2)$$

This model was described by an adjusted-R squared of 0.94, and all residuals were properly distributed within the 95% confidence interval, as shown in Figure 7. The accuracy of the regression was estimated by calculating the coefficient of variation of the root mean squared error (which is a not-scaled dependent factor) and it was found to be 3.1% [32,33].

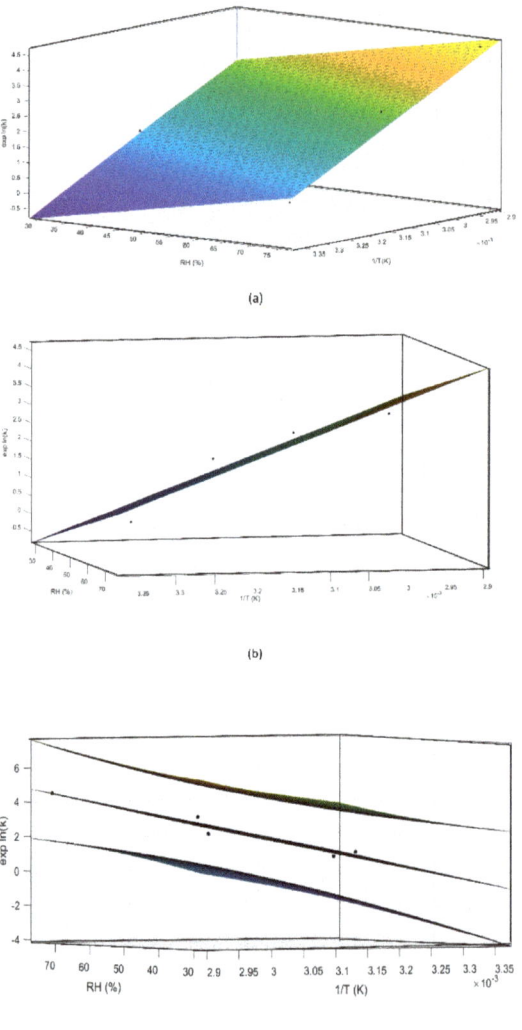

Figure 6. Different perspectives of a three-dimensional surface plot of the model generated by the Matlab curve-fitting tool (**a,b**). Using the same tool the confidence bounds for this fitting were calculated with a level of certainty of 95%, represented by the upper and lower planes (**c**).

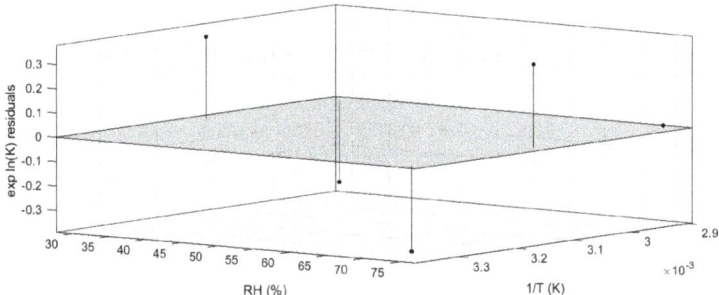

Figure 7. Residual plot generated by Matlab curve-fitting tool.

The quality of this model was further evaluated by the application of ANOVA analysis (α = 0.05), which highlighted the significance of the regression with a p-value of 0.0063.

However, the effect of T and RH could be modeled on lnK following different approaches in order to find the best fit with experimental data (e.g., adopting other degradation shape parameters, or introducing new interaction terms to Equation (1), especially when no linear relation between the observed degradation rate and time is present.) Thus, in order to investigate the possibility of further improvement of the fit, the effects of the two predictors, 1/T and RH%, on lnK values and their interactions were also graphically estimated using the interaction plot analysis (Figure 8). Temperature had a positive effect on lnK values, which increased with lower 1/T values, as shown in Figure 8. Moreover, humidity exerted an additive effect on temperature, as evident from a positive slope of the three curves. However, the interactions between these two predictors were not meaningful, confirming the proper use of the original moisture-corrected Arrhenius equation.

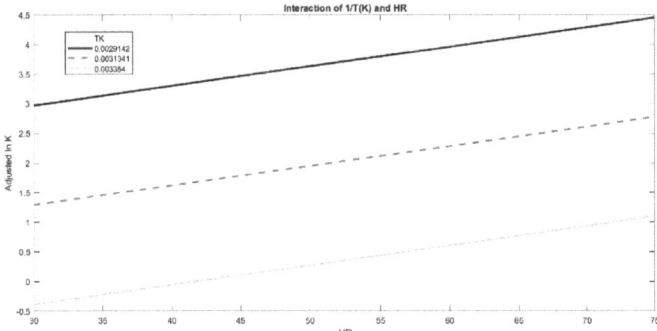

Figure 8. Interaction plot of the effect exerted by growing RH % on lnK at fixed T values (1/T (K)), represented by the different lines.

The so-constructed mathematical model described C3G behavior when exposed to stressful conditions.

Moreover, this model can be further used to extrapolate C3G lnK values and predict the isoconversion time for each temperature–humidity combination. Hence, applying this model, it is possible to calculate the C3G isoconversion time at 25 °C-30% RH, the condition used to mimic typical storage conditions [1]. By applying the model (Equation (2)), it was predicted that C3G could decrease to its specification limit after a period of between 26 and 33 days of storage.

3.4. Mathematical Model Validation

Further experiments were performed in order to assess the predictiveness of this mathematical model and thus validate it. The model (Equation (2)) was applied to extrapolate the C3G isoconversion times (using 20% degradation as a specification limit) for an intermediate temperature with high or low humidity values (45 °C-75% RH, 45 °C-30% RH, respectively) or the strongest condition tested (70 °C-75% RH).

Then, three different samples obtained from different independent batches of MCE were submitted to these selected stress conditions. Furthermore, degradation state was predicted at two additional time points for each tested condition, respectively before and after the isoconversion time, to better investigate the predictive potential of this model. At the end of each treatment, the degradation rate was assessed and the deviation of the predicted degradation from experimental values were calculated for all time points and conditions tested (Table 2). The main disagreement deviance (24%) was observed at 45 °C-75% RH after 1 day of storage, when the estimated degradation was 20%, but the experimental degradation was 28.44% ± 0.25. This difference was probably due to the fact that, at mild temperatures, degradation is strongly driven by humidity, whose effect cannot be fully described by a linear fit, as previously mentioned.

Table 2. Cyanidin-3-O-glucoside (C3G) relative degradations observed after each treatment at the monitored times and their deviation percentage (Dev %) from the predicted values. Average deviations (Ave dev) were calculated for each condition. Exp Degradation was calculated as the mean of all degradation values registered for the three tested batches.

Condition	Time (h)	Exp Degradation (%) ± DS	Predicted Degradation (%)	Dev (%)	Ave Dev (%)
70 °C-75% RH	2	5.311 ± 1.67	7.10	11.51	
	5	18.953 ± 7.38	20.46	5.41	6.27
	24	87.553 ± 0.54	85.25	1.89	
45 °C-75% RH	4	4.194 ± 0.825	3.35	15.86	
	24	28.444 ± 0.255	20.09	24.34	13.72
	96	79.292 ± 3.35	80.36	0.95	
45 °C-30% RH	96	15.841 ± 5.67	15.52	1.45	
	120	18.237 ± 6,66	19.40	4.37	3.2
	168	28.651 ± 0.896	27.16	3.78	

However, considering the overall result, the model demonstrated a satisfactory predictive capability with an average fitting deviation of 11.87% (calculated as the mean of the deviations observed at each tested condition).

3.5. MCE Quality Assessment

Considering that MCE is a phytocomplex characterized by the presence of a great variety of flavonols such as quercetin, myricetin, isorhamnetin, and kaempferol derivatives, and that its healthy properties are correlated to the whole phytocomplex [21], any stability assessment study of MCE has to also include flavonols. This became mandatory in order to perform a comprehensive quality control study. However, application of the ASAP approach to each polyphenol present in MCE would require too long a time and was thus unsuitable for industrial scale-up. Therefore, a smart approach could be to monitor the flavonols' degradation states at C3G degradation isoconversion time when exposed to different T and RH% conditions. The reduction percentage of the five most abundant flavonols (myricetin-3,7-di-O-hexoside, quercetin-7-O-glucoside, kaempferol-7-O-(6"-O-malonyl)-hexoside, isorhamnetin-7-O-rutinoside, and isorhamnetin-3-O-hexoside) and the other two anthocyanins (perlagonidin-3-O-glucoside and peonidin-3-O-glucoside) present in MCE was monitored at the predicted C3G isoconversion times (120 h, 24 h, 5 h) under the three selected T and RH% conditions (45 °C-30% RH, 45 °C-75% RH, and 70 °C-75% RH) for evaluation of the stability quality of MCE.

The results obtained by monitoring the reduction percentage of the eight markers from three different MCE batches at 520 and 370 nm are reported in Table 3. Perlagonidin-3-O-glucoside followed a degradation trend similar to that of C3G, while peonidin-3-O-glucoside reached a higher degradation level at all monitoring conditions. On the other hand, the degradation of each flavonol never reached 20%, with the exception of those stored at the most extreme conditions applied (70 °C–75% RH). However, under these conditions the overall degradation percentage was very similar to that reached by C3G.

Table 3. Peak area relative reduction (%) of selected marker compounds present in MCE, monitored at the estimated isoconversion times (120 h, 24 h, and 5 h) for each treatment (45 °C-30% RH, 45 °C-75% RH, and 70 °C-75% RH). All values were calculated as the mean of the relative reductions observed for the three tested batches.

Compound	Reduction (%)		
	45 °C-30% RH (120 h)	45 °C-75% RH (24 h)	70 °C-75% RH (5 h)
cyanidin-3-O-glucoside	21.70 ± 1.66	28.72 ± 0.16	19.13 ± 6.08
perlagonidin-3-O-glucoside	20.78 ± 2.77	25.59 ± 6.99	14.57 ± 5.7
peonidin-3-O-glucoside	30.70 ± 3.9	30 ± 1.07	34.45 ± 4.49
myricetin-3,7-di-O-hexoside	0	8.44 ± 1.89	3.02 ± 2,66
quercetin-7-O-glucoside	0	0	0
kaempferol-7-O-(6″-O-malonyl)-hexoside	0	1.59 ± 0.99	27.15 ± 4.71
isorhamnetin-7-O-rutinoside	0	0	12.96 ± 1.14
isorhamnetin-3-O-hexoside	7.41 ± 3.83	0	29.18 ± 10.63

4. Conclusions

Before now, to our knowledge, stability studies of natural extracts have never been conducted using the isoconversion approach, even though this methodology allows for estimation of the degradation entity in a shorter time, and with better accuracy, than the classical protocol.

Our approach allowed us to estimate the storage stability of a natural extract, consisting of a large number of compounds, without generating an enormous dataset. In fact, by selecting the marker compound which was the most sensitive to degradation parameters and extrapolating its isoconversion time, it was possible to predict the overall phytocomplex shelf-life. In the current study, C3G was shown to be one of the most sensitive compounds in MCE as it reached the specification limit (20%) faster than the other compounds. For this reason, it was selected as a marker in order to monitor MCE degradation. Therefore, considering the isoconversion time of C3G at 25 °C-30% RH (which mimics ambient storage conditions) as a reference for the prediction of the storage stability of the entire phytocomplex, the shelf-life range of MCE-20% Arabic gum (w/w) formulation was estimated to be between 26 and 33 days.

In conclusion, the application of this new approach to estimate the stability of anthocyanin enriched-extracts could represent an interesting strategy and may be particularly useful for rapidly screening different formulations with the aim of optimizing the solid-state stability of such products. As a result of this work, the MCE-20% Arabic gum (w/w) formulation's storage stability was found to be inappropriate for market purposes, and thus a future perspective will include further stress tests performed using different Arabic gum concentrations and considering the effect of oxygen and light in order to prolong MCE shelf-life.

Author Contributions: Conceptualization, A.P., C.M. and L.F.; methodology, C.M. and L.F.; software, L.F.; validation, L.F., C.M., R.C. and A.P.; formal analysis, L.F. and C.M.; investigation, L.F., C.M. and A.P.; data curation, L.F. and C.M.; writing—original draft preparation, L.F. and C.M.; writing—review and editing, A.P.; visualization, R.C.; supervision, A.P.; project administration, A.P. All authors have read and agreed to the published version of the manuscript.

Funding: This research received no external funding.

Institutional Review Board Statement: Not applicable.

Informed Consent Statement: Not applicable.

Data Availability Statement: All data used in the study are available in the present article.

Conflicts of Interest: The authors declare no conflict of interest.

References

1. Melse-Boonstra, A. Bioavailability of Micronutrients From Nutrient-Dense Whole Foods: Zooming in on Dairy, Vegetables, and Fruits. *Front. Nutr.* **2020**, *7*, 101. [CrossRef] [PubMed]
2. Panickar, K.S.; Jewell, D.E. The beneficial role of anti-inflammatory dietary ingredients in attenuating markers of chronic low-grade inflammation in aging. *Horm. Mol. Biol. Clin. Investig.* **2015**, *23*, 59–70. [CrossRef] [PubMed]
3. Panche, A.N.; Diwan, A.D.; Chandra, S.R. Flavonoids: An overview. *J. Nutr. Sci.* **2016**, *5*, e47. [CrossRef] [PubMed]
4. Sirtori, C.R.; Pavanello, C.; Calabresi, L.; Ruscica, M. Nutraceutical approaches to metabolic syndrome. *Ann. Med.* **2017**, *49*, 678–697. [CrossRef]
5. Zhang, L.; McClements, D.J.; Wei, Z.; Wang, G.; Liu, X.; Liu, F. Delivery of synergistic polyphenol combinations using biopolymer-based systems: Advances in physicochemical properties, stability and bioavailability. *Crit. Rev. Food Sci. Nutr.* **2019**, *60*, 2083–2209. [CrossRef]
6. Kapcum, C.; Uriyapongson, J. Effects of storage conditions on phytochemical and stability of purple corn cob extract powder. *Food Sci. Technol.* **2018**, *38*, 301–305. [CrossRef]
7. Ahmadian, Z.; Niazmand, R.; Pourfarzad, A. Microencapsulation of Saffron Petal Phenolic Extract: Their Characterization, In Vitro Gastrointestinal Digestion, and Storage Stability. *J. Food Sci.* **2019**, *84*, 2745–2757. [CrossRef]
8. Dutta, S.; Bhattacharjee, P. Nanoliposomal encapsulates of piperine-rich black pepper extract obtained by enzyme-assisted supercritical carbon dioxide extraction. *J. Food Eng.* **2017**, *201*, 49–56. [CrossRef]
9. Jafari, S.M.; Mahdavi-Khazaeia, K.; Hemmati-Kakhki, A. Microencapsulation of saffron petal anthocyanins with cress seedgum compared with Arabic gum through freeze drying. *Carbohydr. Polym.* **2016**, *140*, 20–25. [CrossRef]
10. Pal, S.; Bhattacharjee, P. Spray dried powder of lutein-rich supercritical carbon dioxide extract of gamma-irradiated marigold flowers: Process optimization, characterization and food application. *Powder Technol.* **2018**, *327*, 512–523. [CrossRef]
11. Patel, L.J.; Raval, M.; Patel, S.G.; Patel, A.J. Development and Validation of Stability Indicating High-Performance Thin-Layer Chromatographic (HPTLC) Method for Quantification of Asiaticoside from *Centella asiatica* L. and its Marketed Formulation. *J. AOAC Int.* **2019**, *102*, 1014–1020. [CrossRef]
12. Pieczykolan, E.; Kurek, M.A. Use of guar gum, gum arabic, pectin, beta-glucan and inulin for microencapsulation of anthocyanins from chokeberry. *Int. J. Biol. Macromol.* **2019**, *129*, 665–671. [CrossRef]
13. Tonon, R.V.; Brabet, C.; Hubinger, M.D. Anthocyanin stability and antioxidant activity of spray-dried açai (*Euterpe oleracea* Mart.) juice produced with different carrier agents. *Food Res. Int.* **2010**, *43*, 907–914. [CrossRef]
14. ICH Steering Committee. Ich Harmonised Tripartite Guideline, Stability Testing of New Drug Substances and Products. *Int. Conf. Harmon.* **2003**, *4*, 1–24.
15. European Medicines Agency. CPMP/QWP/122/02-Guide Line on Stability Testing: Stability Testing on Existing Active Substances and Related Finished Products. 2003. Available online: https://www.ema.europa.eu/en/stability-testing-existing-active-ingredients-related-finished-products (accessed on 4 June 2020).
16. World Health Organization. Guidelines on Safety Monitoring of Herbal Medicines in Pharmacovigilance Systems. 2004. Available online: https://apps.who.int/iris/handle/10665/43034 (accessed on 10 October 2019).
17. Calixto, J. Efficacy, safety, quality control, marketing and regulatory guidelines for herbal medicines (phytotherapeutic agents). *Braz. J. Med Biol. Res.* **2000**, *33*, 179–189. [CrossRef]
18. Fibigr, J.; Satínský, D.; Solich, P. Current trends in the analysis and quality control of food supplements based on plant extracts. *Anal. Chim. Acta* **2018**, *1036*, 1–15. [CrossRef]
19. Xin, W.; Hongbingc, Z.; Shanshana, F.; Yidana, Z.; Zhena, Y.; Simiao, F.; Pengweia, Z.; Yanjuna, Z. Quality markers based on biological activity: A new strategy for the quality control of traditional Chinese medicine. *Phytomedicine* **2018**, *44*, 103–108.
20. Pferschy-Wenzig, E.-M.; Bauer, R. The relevance of pharmacognosy in pharmacological research on herbal medicinal products. *Epilepsy Behav.* **2015**, *52*, 344–362. [CrossRef]
21. Ferron, L.; Colombo, R.; Mannucci, B.; Papetti, A. A New Italian Purple Corn Variety (Moradyn) Byproduct Extract: Antiglycative and Hypoglycemic In Vitro Activities and Preliminary Bioaccessibility Studies. *Molecules* **2020**, *25*, 1958. [CrossRef]
22. Waterman, K.C. The Application of the Accelerated Stability Assessment Program (ASAP) to Quality by Design (QbD) for Drug Product Stability. *AAPS PharmSciTech* **2011**, *12*, 932–937. [CrossRef]
23. Waterman, K.C.; Adami, R.C. Accelerated aging: Prediction of chemical stability of pharmaceuticals. *Int. J. Pharm.* **2005**, *293*, 101–125. [CrossRef]
24. Grenspan, L. Humidity fixed points of binary saturated aqueous solutions. *J. Res. Natl. Bur. Stand. A Phys. Chem.* **1976**, *81*, 89–96. [CrossRef]
25. ICH Guideline Q2(R1). *Validation of Analytical Procedures: Text and Methodology*; Somatek Inc.: San Diego, CA, USA, 2005.
26. Guo, J.; Giusti, M.; Kaletunça, G. Encapsulation of purple corn and blueberry extracts in alginate-pectin hydrogel particles: Impact of processing and storage parameters on encapsulation efficiency. *Food. Res. Int.* **2018**, *107*, 414–422. [CrossRef]

27. Lao, F.; Giusti, M.M. The effect of pigment matrix, temperature and amount of carrier on the yield and final color properties of spray dried purple corn (*Zea mays* L.) cob anthocyanin powders. *Food Chem.* **2017**, *227*, 376–382. [CrossRef]
28. Khazaeia, K.M.; Jafaria, S.M.; Ghorbania, M.; Kakhkib, A.H. Application of maltodextrin and gum Arabic in microencapsulation of saffron petal's anthocyanins and evaluating their storage stability and color. *Carbohydr. Polym.* **2014**, *105*, 57–62. [CrossRef]
29. Alsante, K.M.; Ando, A.; Brown, R.; Ensing, J.; Hatajik, T.D.; Kong, W.; Tsuda, Y. The role of degradant profiling in active pharmaceutical ingredients and drug products. *Adv. Drug Deliv. Rev.* **2007**, *59*, 29–37. [CrossRef]
30. Fan, Z.; Zhang, L. One- and Two-Stage Arrhenius Models for Pharmaceutical Shelf Life Prediction. *J. Biopharm. Stat.* **2015**, *25*, 307–316. [CrossRef]
31. Khalid, H.; Zhari, I.; Amirin, S.; Pazilah, I. Accelerated Stability and Chemical Kinetics of Ethanol Extracts of Fruit of Piper sarmentosum Using High Performance Liquid Chromatography. *Iran. J. Pharm. Res. IJPR* **2011**, *10*, 403–413.
32. Bezzerra, M.A.; Santelli, R.E.; Oliveira, E.P.; Villar, L.S.; Escaleira, L.A. Response surface methodology (RS31. M) as a tool for optimization in analytical chemistry. *Talanta* **2008**, *76*, 965–977. [CrossRef]
33. Fu, M.; Perlman, M.; Lu, Q.; Varga, C. Pharmaceutical solid-state kinetic stability investigation by using moisture-modified Arrhenius equation and JMP statistical software. *J. Pharm. Biomed. Anal.* **2015**, *107*, 370–377. [CrossRef]

Article

Defatted Seeds of *Oenothera biennis* as a Potential Functional Food Ingredient for Diabetes

Zhiqiang Wang [1,2,*], Zhaoyang Wu [1], Guanglei Zuo [3], Soon Sung Lim [3] and Hongyuan Yan [1,2,*]

- [1] Key Laboratory of Public Health Safety of Hebei Province, College of Public Health, Hebei University, Baoding 071002, China; wzy19970202@163.com
- [2] Key Laboratory of Medicinal Chemistry and Molecular Diagnosis of Ministry of Education, Institute of Life Science and Green Development, Hebei University, Baoding 071002, China
- [3] Department of Food Science and Nutrition, Hallym University, 1 Hallymdeahak-gil, Chuncheon 24252, Korea; B16504@hallym.ac.kr (G.Z.); limss@hallym.ac.kr (S.S.L.)
- * Correspondence: wangzq2017@hbu.edu.cn (Z.W.); yanhy@hbu.edu.cn (H.Y.); Tel.: +86-312-5079010 (Z.W.); +86-312-5078507 (H.Y.)

Citation: Wang, Z.; Wu, Z.; Zuo, G.; Lim, S.S.; Yan, H. Defatted Seeds of *Oenothera biennis* as a Potential Functional Food Ingredient for Diabetes. *Foods* **2021**, *10*, 538. https://doi.org/10.3390/foods 10030538

Academic Editors: Simona Grasso, Konstantinos Papoutsis, Claudia Ruiz-Capillas and Ana Herrero Herranz

Received: 2 February 2021
Accepted: 2 March 2021
Published: 5 March 2021

Publisher's Note: MDPI stays neutral with regard to jurisdictional claims in published maps and institutional affiliations.

Copyright: © 2021 by the authors. Licensee MDPI, Basel, Switzerland. This article is an open access article distributed under the terms and conditions of the Creative Commons Attribution (CC BY) license (https://creativecommons.org/licenses/by/4.0/).

Abstract: The defatted seeds of *Oenothera biennis* (DSOB) are a by-product of evening primrose oil production that are currently not effectively used. In this study, α-glucosidase inhibition, aldose reductase inhibition, antioxidant capacity, polyphenol composition, and nutritional value (carbohydrates, proteins, minerals, fat, organic acid, and tocopherols) of DSOB were evaluated using the seeds of *Oenothera biennis* (SOB) as a reference. DSOB was an excellent inhibitor of α-glucosidase (IC_{50} = 3.31 µg/mL) and aldose reductase (IC_{50} = 2.56 µg/mL). DSOB also showed considerable antioxidant capacities (scavenging of 2,2-diphenyl-1-picrylhydrazyl, 2,2′-azino-bis(3-ethylbenzothiazoline-6-sulfonic acid, nitric oxide, peroxynitrite, and hydroxyl radicals). DSOB was a reservoir of polyphenols, and 25 compounds in DSOB were temporarily identified by liquid chromatography coupled with electrospray ionization–quadrupole time of flight–mass spectrometry analysis. Moreover, the carbohydrate, protein, and mineral content of DSOB were increased compared to that of SOB. DSOB contained large amounts of fiber and low levels of sugars, and was rich in calcium and iron. These results imply that DSOB may be a potential functional food ingredient for diabetes, providing excellent economic and environmental benefits.

Keywords: defatted seeds of *Oenothera biennis*; α-glucosidase; aldose reductase; antioxidant; polyphenols; nutrients

1. Introduction

Diabetes mellitus, characterized by abnormal hyperglycemia, affects 463 million patients globally [1], of which 60% have one or more complications, such as diabetic retinopathy, nephropathy, and neuropathy [2]. The results of epidemiological and clinical research show that effective blood glucose control postpones the occurrence of diabetic complications; however, its development in patients with diabetes is certain to happen [3,4]. Furthermore, changes in diet or medication of patients with diabetes often results in nutritional deficiencies, dramatically damaging the health and lowering quality of life [5]. Thus, the health management of patients with diabetes should be concerned with the control of hyperglycemia, prevention of complications, and supplementation of nutrition. α-Glucosidases are a group of intestinal enzymes involved in carbohydrate digestion and are critical targets for the amelioration of postprandial hyperglycemia [6]. Aldose reductase is a pivotal target for preventing the onset and progression of diabetic complications [7]. Moreover, oxidative stress is strongly related to the onset of diabetes and exacerbation of complications [8]. Currently, the increasingly mindful attitudes of consumers to their diet and health has led to the development of new trends, including the widespread use of functional foods. Hence, there is growing interest in developing functional food ingredients that possess effective inhibitory activities against α-glucosidase, aldose reductase, and free radicals, particularly those with nutritional supplementation that may be useful for promoting the health of patients with diabetes.

Oenothera biennis (OB), commonly known as evening primrose, is an herbaceous plant of the family *Onagraceae* [9]. In recent decades, oil extracted from the seeds of OB (SOB) has drawn attention because of its high polyunsaturated fatty acid content (γ-linolenic acid) and excellent bioactive properties [10]. Thus, OB is globally cultivated as an industrial oil crop for the production of evening primrose oil (EPO) used in the development of pharmaceuticals, cosmetics, food products, bakery products, and confectionery [11]. The annual global production of SOB has increased 20-fold in the last 20 years, producing kilotons of seeds annually, and is expected to experience robust growth in the foreseeable future [12]. However, 50–55% of SOB generated after EPO manufacture are residues (defatted SOB (DSOB)), resulting in a large waste of resources. The disposal of DSOB seed residues generated by oil extraction is an important issue for related industries, and its use may lead to great economic and environmental benefits, if utilized properly.

SOB is a rich source of not only a valuable oil, containing essential fatty acids, but also of polyphenols, which have shown remarkable bioactivity [13]. Moreover, SOB is also rich in nutrients containing proteins, carbohydrates, minerals, and vitamins, in addition to the oil [14]. However, to date, few studies have evaluated the nutritional value and polyphenol composition of DSOB, nor the biological properties related to diabetes, which limits the potential reuse of DSOB. In this study, to evaluate the potential of DSOB as a functional food ingredient in diabetes, we investigated the inhibitory effects of DSOB on α-glucosidase and aldose reductase for the first time, and evaluated the antioxidant capacities of DSOB by electron transfer assays (1,1-diphenyl-2-picrylhydrazyl (DPPH) and 2,2-azinobis(3-ethylbenzothiazoline-6-sulfonic acid) diammonium salt (ABTS)) and reactive oxygen species (ROS) scavenging assays (hydroxyl radical (HO•), nitric oxide (NO), and peroxynitrite ($ONOO^-$)). Moreover, the polyphenol composition and nutritional value of DSOB were analyzed compared to SOB as a reference.

2. Materials and Methods

2.1. Chemicals

α-Glucosidase from *Saccharomyces cerevisiae*, ascorbic acid, acarbose, ammonium sulfate, *p*-nitrophenyl-α-D-glucopyranoside, Trolox, β-nicotinamide adenine dinucleotide 2-phosphate reduced tetrasodium salt hydrate (NADPH), DL-glyceraldehyde, epalrestat, quercetin, oxalic acid, citric acid, glucose, fructose, galactose, lactose, maltose, sucrose, tocopherols (α-, β-, and γ-), hydrogen peroxide (H_2O_2), ABTS, DPPH, Folin and Ciocalteu's phenol reagent, Griess reagent, and 2,6-di-tert-butyl-4-methylphenol (BHT) were purchased from Sigma-Aldrich (St. Louis, MO, USA). The standards used for the determination of mineral contents, including sodium (Na), potassium (K), calcium (Ca), magnesium (Mg), iron (Fe), manganese (Mn), copper (Cu), and zinc (Zn), were obtained from the Central Iron and Steel Research Institute (Beijing, China). $ONOO^-$ was obtained from Cayman Chemical (Ann Arbor, MI, USA). All organic solvents were purchased from Concord Technology (Tianjin, China). Ultrapure water (18.2 MΩ cm) was prepared using a water purification system (Milli-Q; Millipore, Billerica, MA, USA). Unless otherwise stated, all other reagents were purchased from Sigma-Aldrich.

2.2. Plant Samples

SOB was purchased from Yiyuan Chinese Herb Medicine Co. Ltd. (Anguo, Hebei, China) in August 2018. Dry ground SOB (1000.21 g) was defatted twice with *n*-hexane over a period of 48 h to obtain DSOB (788.69 g). Voucher DSOB (WLL-2018-01) and SOB (WLL-2018-02) were deposited at the College of Public Health, Hebei University.

2.3. Macronutrients, Energy, and Sugars

Macronutrients (carbohydrates, proteins, fats, and ash), energy, and sugars (glucose, fructose, galactose, lactose, maltose, and sucrose) were determined using Association of Official Agricultural Chemists methods [15]. Specifically, the amount of protein was assessed using the micro-Kjeldahl method (N × 6.25). Ash was evaluated by calcining at 550 ± 10 °C for 5 h. Fats were measured using a Soxhlet apparatus. Carbohydrates were estimated using difference analysis. Glucose, fructose, lactose, maltose, and sucrose contents were determined using ion chromatography, and the energy value was calculated as follows:

$$\text{Energy (kcal/100 g dry weight [dw])} = 4 \times (\text{g/100 g dw proteins} + \text{g/100 g dw carbohydrates}) + 9 \times (\text{g/100 g dw fats}) \quad (1)$$

2.4. Amino Acids

Proteins extracted from SOB or DSOB were hydrolyzed with hydrochloric acid at 110 °C for 21 h, and then filtered using a 0.22 μm membrane for injection into a high-performance liquid chromatography (HPLC) (UltiMate 3000; Thermo Fisher Scientific, Waltham, MA, USA) instrument equipped with a fluorescence detector (UltiMate 3000; Thermo Fisher Scientific). Separation was achieved using a ZORBAX Eclipse-AAA column (150 mm length, 3.0 mm i.d., and 3.5 μm particle size; Agilent, Santa Clara, CA, USA) at 35 °C. The samples (10 μL) were eluted with sodium dihydrogen phosphate solution (40 mM in water, pH 7.8, (A) and acetonitrile–methanol–water solution (45:45:10, $v/v/v$, (B) at 0.7 mL/min: 0% B at 0–1.9 min, 0–57% B at 1.9–18.1 min, 57–100% B at 18.1–18.6 min, and 100% B at 18.6–22.3 min. The excitation and emission wavelengths of the fluorescence detector were 340 nm and 450 nm, respectively. Amino acids were identified by comparing the retention times with those of the standards and were further quantified using calibration curves. Moreover, the chemical score (CS), essential amino acid index (EAAI), and biological value (BV) were calculated using the following equations:

$$\text{CS} = \text{EAA in sample protein} / \text{EAA in egg protein} \times 100 \quad (2)$$

$$\text{EAAI} = [(\text{Lysine}_s \times \text{Leucine}_s \times \cdots \times \text{Histidine}_s) / (\text{Lysine}_r \times \text{Leucine}_r \times \cdots \times \text{Histidine}_r)]^{1/n} \quad (3)$$

where "EAA" is essential amino acids, "s" is essential amino acid in sample, "r" is essential amino acid in whole egg, and "n" is amount of amino acids (assuming "Phenylalanine + Tyrosine," and "Methionine + Cysteine" are together).

2.5. Minerals

All samples were digested using a MARS6 microwave digester (CEM Corporation, Charlotte, NC, USA). The digested solutions were then filtered using 0.22 μm membranes. The filtrate was subsequently analyzed using an inductively coupled plasma-atomic emission spectrometer (Prodigy7; Teledyne Leeman Labs, Mason, OH, USA). The response signals were tracked at 766.49 nm (K), 589.59 nm (Na), 324.75 nm (Cu), 317.93 nm (Ca), 279.55 nm (Mg), 259.94 nm (Fe), 257.61 nm (Mn), and 206.20 nm (Zn). Each element was quantitatively analyzed in accordance with its standard calibrated curves.

2.6. Tocopherols

Tocopherols in SOB and DSOB were determined using the UltiMate 3000 HPLC system with a UV detector (UltiMate 3000), as previously described with some modifications [16]. Briefly, 2 g of the samples were mixed with methanol (4 mL), n-hexane (4 mL), and saturated sodium chloride water solution (2 mL). The clear upper layer was collected and filtered for further HPLC analysis. The Eclipse Plus C18 column (150 × 4.6 mm, 3.5 μm; Agilent) was used for separation. Samples were eluted with methanol at 0.7 mL/min and detected at 300 nm. Tocopherols (α-, β-, and γ-isoforms) were confirmed and quantified using standard compounds with calibrated curves.

2.7. Organic Acids

Organic acids (oxalic acid and citric acid) in SOB and DSOB were extracted by metaphosphoric acid and subsequently analyzed by HPLC-UV (UltiMate 3000) at 215 nm and 210 nm, respectively. Sodium dihydrogen phosphate solution (0.01 mol/L, pH 3.5) was used as the mobile phase. The eluent was monitored using an Ultimate AQ-C18 column (250 × 4.6 mm, 5 µm; Welch Materials, Shanghai, China) that was used for separation at 35 °C. Compounds were identified by comparison with standards and further quantified using calibration curves [16].

2.8. Total Phenolic Content (TPC) and Total Flavonoid Content (TFC)

SOB and DSOB were extracted twice by sonication at 30 °C with an ethanol aqueous solution (70%, v/v) for 30 min. The obtained extracts were concentrated at 37 °C using a rotary evaporator (RE-2000A; YARONG, Shanghai, China) under vacuum, lyophilized (FreeZone 4.5; Labconco, Kansas City, MO, USA), and stored at −20 °C for the determination of TPC, TFC, polyphenols, and antioxidant capacities. TPC was determined using the Folin–Ciocalteu colorimetric method and expressed as gallic acid equivalents (GAE) [17]. TFC was evaluated using the method reported by Kainama et al. [18] and expressed as (+)-catechin equivalents (CE).

2.9. Characterization of Polyphenols by Liquid Chromatography Coupled with Electrospray Ionization–Quadrupole Time of Flight–Mass Spectrometry Analysis (LC–ESI–QTOF/MS)

The DSOB and SOB polyphenols were characterized using an LC–ESI–QTOF/MS instrument (XEVO G2, Waters, Milford, MA, USA) for tentative identification. Separation was accomplished at 30 °C using a CORTECS C18 column (50 × 2.1 mm, 1.6 µm; Waters). SOB or DSOB (4 µL) were eluted with acidified water (0.1% formic acid, (A) and organic solutions (acetonitrile/methanol = 3:1, v/v, (B) at a flow rate of 0.3 mL/min. The following were used as the optimized gradient chromatography conditions: 5–20% B at 0–2 min, 20–30% B at 2–5 min, 30–40% B at 5–7 min, 40–100% B at 7–10 min. Peak identification was performed in both negative and positive modes, and mass spectra in the m/z range of 50–1000 were obtained. The mass spectrometry conditions were as follows: source temperature, 120 °C; desolvation temperature, 450 °C; gas flow rate, 800 L/h; and nebulizer gas pressure, 6.5 bar. The capillary voltage was set to 2.0 kV. Data acquisition and analysis were performed using MassLynx V4.1 SCN 901 (Waters). The polyphenols were temporarily identified by comparing the literature and searching the library using UNIFI (Waters). The content ratio of polyphenols in DSOB and SOB were calculated by peak areas in TIC.

2.10. Antioxidant

The antioxidant capacities of SOB and DSOB were assessed using electron transfer assays (DPPH and ABTS) and ROS scavenging assays (HO•, NO, and ONOO$^-$). The DPPH, ABTS, HO•, and NO scavenging assays were performed as reported by Kwon et al. [19]. The ONOO$^-$ scavenging assay was carried out as reported by Hazra et al. [20]. BHT and Trolox were used as positive controls for the DPPH and ABTS assays, respectively. Ascorbic acid was used as the positive control for HO•, NO, and ONOO$^-$ scavenging assays. The antioxidant activities were expressed as the percentage of radical elimination (%) and the sample concentration for 50% inhibition (IC_{50}).

2.11. α-Glucosidase Inhibition Assay

The α-glucosidase inhibition assay was performed as previously reported [6]. Specifically, a mixture of phosphate-buffered saline (PBS, pH 7.4, 90 µL), sample solution (20 µL), p-nitrophenyl-α-D-glucopyranoside solution (30 µL, 1 mM in PBS), and α-glucosidase solution (60 µL, 15 µg/mL in PBS) were incubated at 37 °C for 30 min. Acabose was used as a positive control. Then, the absorbance of the incubated mixture was measured at 405 nm using a microplate reader (Synergy HTX, BioTek Instruments, Winooski, VT, USA). The inhibitory activity was expressed as the percentage inhibition (%) and the IC_{50}.

2.12. Rat Lens Aldose Reductase Inhibition Assay

The rat lens aldose reductase inhibition assay of SOB and DSOB was performed as described previously [21]. In brief, the eye lens of Wistar rats (10 weeks old, weight 250–280 g) were collected and homogenized for further centrifugation. The rat lens aldose reductase remained in the supernatant. Then, 900 µL of a total mixture of NADPH (0.16 mM), ammonium sulfate (2.5 mM), DL-glyceraldehyde (2.5 mM), rat lens aldose reductase, and samples were incubated, and the activity of aldose reductase was determined by measuring the decrease in NADPH absorbance at 340 nm for 3 min using a spectrophotometer. Epalrestat was used as a positive control. The inhibitory activity was expressed as the percentage inhibition (%) and the IC_{50}.

2.13. Animal Care

All animal experimental procedures were conducted in accordance with the guidelines and approval of the Institutional Animal Care and Use Committee (IACUC) of Hebei University (IACUC-20180051). Prior to the experiment, rats were placed in a standardized laboratory environment under a 12 h light/12 h dark cycle at temperature of 20–26 °C and humidity of 40–70%. All animals had access to water and food.

2.14. Statistical Analysis

All experiments were repeated at least in triplicate. Results are expressed as the mean ± standard deviation. Data were analyzed using SPSS Statistics software version 19.0 (IBM, Armonk, NY, USA). The comparison of mean values was performed using Student's unpaired *t*-test or one-way analysis of variance, as required. $p < 0.05$ was considered significant.

3. Results and Discussion

3.1. α-Glucosidase Inhibition, Aldose Reductase Inhibition, and Antioxidant Capacity of DSOB

α-Glucosidase inhibitors from natural sources are often used as functional foods for intervening in postprandial hyperglycemia in patients with diabetes. Thus, the effect of DSOB on α-glucosidase inhibition was first investigated to assess its potential as a functional food ingredient in diabetes. As shown in Figure 1A, DSOB inhibited α-glucosidase activity by 1.47%, 72.41%, and 88.52% at concentrations of 1, 5, and 10 µg/mL, respectively. SOB decreased α-glucosidase activity by 2.52%, 69.33%, and 96.22% at 1, 5, and 10 µg/mL, respectively. The α-glucosidase inhibitory capacity of DSOB was lower than that of SOB at 10 µg/mL and showed no significant differences at 1 and 5 µg/mL. As shown in Table 1, DSOB had an IC_{50} of 3.31 µg/mL for α-glucosidase inhibition, which showed no significant difference compared to the IC_{50} of SOB (3.18 µg/mL). Interestingly, both DSOB and SOB showed dramatically higher α-glucosidase inhibitory capacities than acarbose. Grape seed and green tea are popular functional food ingredients that possess excellent α-glucosidase inhibitory activities. In a previous study, Yilmazer-Musa et al. reported that grape seed has an IC_{50} of 1.2 µg/mL for α-glucosidase inhibition, green tea has an IC_{50} of 0.5 µg/mL for α-glucosidase inhibition, and acarbose has an IC_{50} of 91 µg/mL for α-glucosidase inhibition [22]. Another study reported that fruiting body of *Phellinus merrillii* has an IC_{50} of 13.73 µg/mL for α-glucosidase inhibition [23]. These results implied that DSOB was a potent α-glucosidase inhibitor, in which the α-glucosidase inhibitory activity of DSOB was not affected by the defatting process from SOB. This work is the first report to indicate the α-glucosidase inhibitory activities of DSOB and SOB.

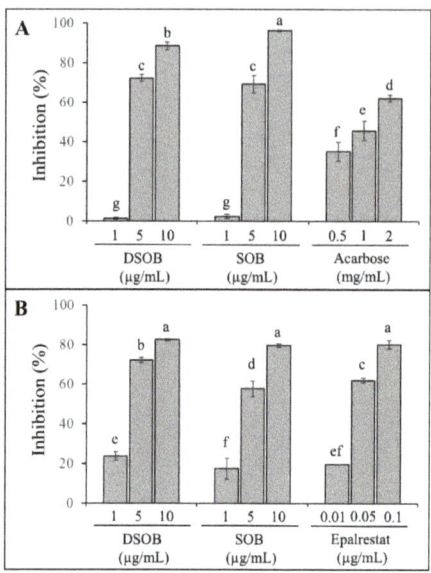

Figure 1. The inhibitory effects of defatted seeds of *Oenothera biennis* (DSOB) and seeds of *Oenothera biennis* (SOB) on α-glucosidase (**A**) and rat lens aldose reductase (**B**). Data in the bar graph with different letters are significantly different from each other ($p < 0.05$).

Table 1. The IC_{50} values of defatted seeds of *Oenothera biennis* (DSOB) and seeds of *Oenothera biennis* (SOB).

Sample	IC_{50} (μg/mL) [1]	
	α-Glucosidase	Rat Lens Aldose Reductase
DSOB	3.31 ± 0.09 [a,2]	2.56 ± 0.06 [a]
SOB	3.18 ± 0.13 [a]	3.44 ± 0.39 [b]
Positive control [3]	1265.7 ± 29.4 [b]	0.031 ± 0.001 [c]

[1] IC_{50} is the sample concentration providing 50% inhibition. [2] Results are presented as the mean ± standard deviation ($n = 3$). Values within a column of the table marked with different letters are significantly different from each other ($p < 0.05$). [3] Acarbose was used as positive control in α-glucosidase assay; epalrestat was used as positive control in rat lens aldose reductase assay.

Recently, the development of naturally derived aldose reductase inhibitors with less toxicity as functional foods has attracted much attention. Our previous research has reported that SOB is a potent aldose reductase inhibitor [24], but the aldose reductase inhibitory activity of DSOB is unclear. Therefore, in the present study, the inhibitory effect of DSOB on aldose reductase was also investigated using rat lens to assess its potential as a functional food ingredient in diabetes. As shown in Figure 1B, DSOB decreases the catalyzing activities of aldose reductase by 23.79%, 72.33%, and 82.63% at concentrations of 1, 5, and 10 μg/mL, respectively. SOB decreased rat lens aldose reductase activities by 17.56%, 57.89%, and 79.70% at 1, 5, and 10 μg/mL, respectively. The inhibitory effects of DSOB at 1 and 5 μg/mL on rat lens aldose reductase were significantly higher than those of SOB. Moreover, DSOB has a significantly lower IC_{50} (2.56 μg/mL) than that of SOB (3.44 μg/mL), as shown in Table 1. Huang et al. reported that fruiting body of *Phellinus merrillii* has an IC_{50} of 12.55 μg/mL for aldose reductase inhibition [23]. Although the aldose reductase inhibitory activity of DSOB was much lower than that of epalrestat, these results indicated that DSOB was an excellent natural origin aldose reductase inhibitor, and the defatting process from SOB seemed to improve its inhibition.

The antioxidant capacities of DSOB were further assessed using five assays, namely, O•, NO, ONOO⁻, DPPH, and ABTS, on the basis of the antioxidant mechanisms of electron transfer and ROS scavenging [19]. HO• is the main ROS in the body, inducing oxidative stress that breaks down DNA strands and damages proteins [25]. As shown in Table 2 and Figure 2, DSOB scavenges 45.18% HO• at 3.33 mg/mL. Although the HO• scavenging capacity of DSOB was significantly lower than that of SOB and ascorbic acid, these results implied that DSOB contained potent HO• scavengers. NO provides vascular protection; nevertheless, the highly reactive radical, ONOO⁻, is formed when NO interacts with superoxide radicals, leading to a series of harmful events [26]. DSOB showed good antioxidant activities against NO (IC_{50} = 1.08 μg/mL) and ONOO⁻ (IC_{50} = 512.49 μg/mL), with no significant difference with SOB. DSOB also showed excellent scavenging capacities for ABTS (IC_{50} = 1.22 μg/mL) and DPPH (IC_{50} = 15.62 μg/mL) radicals, with no significant difference compared to SOB. These findings demonstrated that DSOB is a good natural antioxidant.

Table 2. IC_{50} of defatted seeds of *Oenothera biennis* (DSOB) and seeds of *Oenothera biennis* (SOB) on radical scavenging.

Samples	IC_{50} [1] (μg/mL)				
	DPPH [2]	ABTS [3]	HO• [4]	NO [5]	ONOO⁻ [6]
DSOB	15.62 ± 4.79 [a,7]	1.12 ± 0.11 [a]	Na [8]	1.08 ± 0.20 [a]	512.49 ± 9.64 [a]
SOB	18.04 ± 5.20 [ab]	1.43 ± 0.30 [a]	1784.01 ± 475.14 [a]	1.01 ± 0.22 [a]	470.54 ± 49.68 [b]
Positive control	33.12 ± 11.93 [b]	1.34 ± 0.07 [a]	110.06 ± 11.38 [b]	0.039 ± 0.019 [b]	94.99 ± 0.23 [c]

[1] IC_{50} is the sample concentration providing 50% radical scavenging. [2] DPPH is 1,1-diphenyl-2-picrylhydrazyl. 2,6-di-tert-Butyl-4-methylphenol was used as positive control. [3] ABTS is 2,2-azinobis(3-ethylbenzothiazoline-6-sulfonic acid) diammonium salt. Trolox was used as positive control. [4] HO• is hydroxyl radical. Ascorbic acid was used as positive control. [5] NO is nitric oxide radical. Ascorbic acid was used as positive control. [6] ONOO⁻ is peroxynitrite. Ascorbic acid was used as positive control. [7] Results are presented as the mean ± standard deviation (n = 3). Values within a column superscripted with different letters are significantly different from each other ($p < 0.05$). [8] "Na" indicates not active.

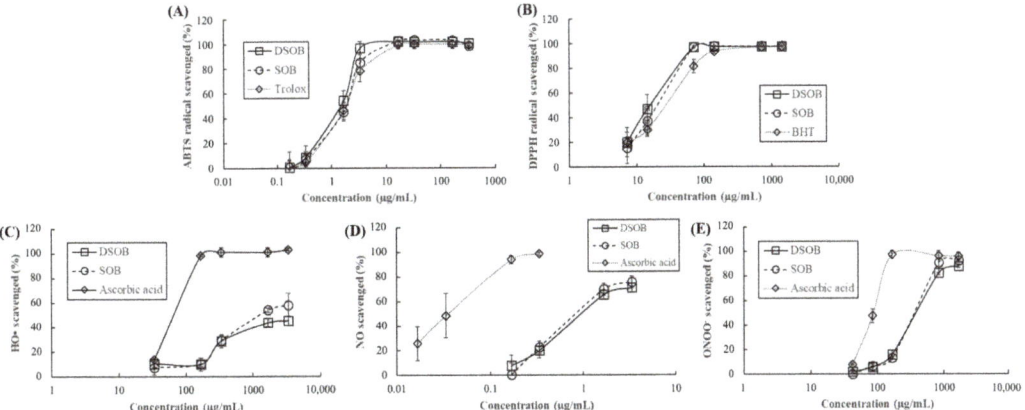

Figure 2. Antioxidant capacities of defatted seeds of *Oenothera biennis* (DSOB) and seeds of *Oenothera biennis* (SOB). (**A**) 2,2-Azinobis(3-ethylbenzothiazoline-6-sulfonic acid) diammonium salt (ABTS) radical scavenging activity of DSOB, SOB, and Trolox (positive control). (**B**) 1,1-Diphenyl-2-picrylhydrazyl (DPPH) radical scavenging activity of DSOB, SOB, and 2,6-di-tert-butyl-4-methylphenol (BHT, positive control). (**C**) Hydroxyl radical (HO•) scavenging activity of DSOB, SOB, and ascorbic acid (positive control). (**D**) Nitric oxide (NO) scavenging activity of DSOB, SOB, and ascorbic acid (positive control). (**E**) Peroxynitrite (ONOO⁻) scavenging of DSOB, SOB, and ascorbic acid (positive control).

3.2. Polyphenol Composition of DSOB

In recent years, an increase in the incidence of diabetes associated with lifestyle has led to a search for effective and safe methods of prophylaxis and/or therapy. The results of epidemiological studies indicate that diabetes may be prevented by enriching diets with plant foods, since these foods are rich in polyphenols with pharmacological activity [27,28]. Over the last few years, SOB has attracted wide attention because of its high polyphenol content. However, understanding of the polyphenols in DSOB remains limited. Therefore, the DSOB TPC and TFC were tested using SOB as a reference, and the results are listed in Table 3. The TPC of DSOB was 48.91 mg GAE/g dw, significantly higher than that of SOB (33.10 mg GAE/g dw). The TPC of grape seed and green tea were 86 and 74 mg GAE/g dw, respectively [22]. The amount of TFC in DSOB (33.93 mg CE/g dw) was also significantly higher than that in SOB (21.97 mg CE/g dw). These findings indicated that DSOB was a bulky reservoir of polyphenols. Therefore, the aforementioned remarkable bioactivities of SOB may be associated with its high phenolic and flavonoid content; the polyphenol composition of DSOB needs to be further illustrated.

Table 3. Total phenolic content (TPC) and total flavonoid content (TFC) of defatted seeds of *Oenothera biennis* (DSOB) and seeds of *Oenothera biennis* (SOB).

	DSOB	SOB	p-Value [4]
TPC (mg GAE/g dw) [1]	48.91 ± 2.08 [3]	33.10 ± 2.27	<0.001
TFC (mg CE/g dw) [2]	33.93 ± 0.60	21.97 ± 0.56	<0.001

[1] Results of TPC are expressed as milligrams of gallic acid equivalents (GAE) per gram dry weight of DSOB or SOB; "dw" is short for dry weight. [2] Results of TFC are expressed as milligrams of (+)-catechin equivalents (CE) per gram dry weight of DSOB or SOB. [3] Results are presented as the mean ± standard deviation (n = 3). [4] p-value was calculated by Student's unpaired t-test. p < 0.05 is considered significant.

To better understand the polyphenol composition of DSOB, we performed LC–ESI–QTOF/MS because it provides high-resolution mass data containing affluent structural information of the analyte. In this work, 25 components in DSOB have been identified, including (+)-catechin, ellagic acid, gallic acid, procyanidins, quercetin, syringic acid, and their derivatives, as listed in Table 4. Among the 25 compounds, 14 compounds were reported to be α-glucosidase inhibitors, 13 compounds were reported to be aldose reductase inhibitors, and 21 compounds were reported to be antioxidants. Interestingly, although the profiles of the polyphenol composition in DSOB and SOB were similar (Figure 3), the quantities of each component in DSOB and SOB were quite different (Table 4). Among the 25 compounds, the amounts of syringic anhydride, trimethylenglykol-digalloat, monogalloylglucose, ellagic acid glycoside, ellagic acid xyloside, syringic acid, ellagic acid, quercetin glucuronide, and kaempferol glucuronide were higher in DSOB than those in SOB. The amounts of digalloylglucose, galloylxy trihydroxyflavanone, (+)-catechin, catechin gallate, methyl ellagic acid, procyanidins, and procyanidin B gallate were lower in DSOB than those in SOB. The differences in biological activities between DSOB and SOB may be caused by differences in the content of these polyphenols.

Table 4. Polyphenol compounds detected and tentatively characterized in defatted seeds of *Oenothera biennis* (DSOB) and seeds of *Oenothera biennis* (SOB) using liquid chromatography coupled with electrospray ionization and quadrupole time of flight mass spectrometry (LC–ESI–QTOF/MS) in positive and negative ion modes.

	t_R (min) [1]	[M+H]$^+$ m/z	[M−H]$^-$ m/z	Predicted Formula	Observed (m/z)	Theroetical (m/z)	Mass Error (mDa)	DBE [2]	Temporarily Identified	Activity	Content Ratio (DSOB/SOB)
1	0.70	−[3]	377.0824	$C_{18}H_{18}O_9$	377.0824	377.0873	−4.9	10.5	Syringic anhydride	-	4.47
2	0.72	381.0837	-	$C_{17}H_{16}O_{10}$	381.0837	381.0822	1.5	9.5	Trimethylenglykol digalloat	-	3.05
3	0.93	-	331.0713	$C_{13}H_{16}O_{10}$	331.0713	331.0665	4.8	6.5	Monogalloylglucose	α-Glucosidase inhibitor [29]; antioxidant [30]	1.06
4	1.13	-	331.0634	$C_{13}H_{16}O_{10}$	331.0634	331.0665	−3.1	6.5	Monogalloylglucose	α-Glucosidase inhibitor [29]; antioxidant [30]	3.38
5	1.40	171.0356	169.0145	$C_7H_6O_5$	169.0145	169.0137	0.8	5.5	Gallic acid	α-Glucosidase inhibitor [31]; aldose reductase inhibitor [24]; antioxidant [32]	1.04
6	1.71	449.0734	-	$C_{20}H_{16}O_{12}$	449.0734	449.0722	1.4	12.5	Methyl ellagic acid xyloside	-	1.01
7	1.98	-	483.0811	$C_{20}H_{20}O_{14}$	483.0811	483.0775	3.6	11.5	Digalloylglucose	α-Glucosidase inhibitor [33]; aldose reductase inhibitor [34]; antioxidant [35]	0.88
8	2.07	-	153.0180	$C_7H_6O_4$	153.0180	153.0188	−0.8	5.5	Protocatechuic acid	α-Glucosidase inhibitor [36]; aldose reductase inhibitor [37]; antioxidant [38]	1.03
9	2.29	579.1470	577.1298	$C_{30}H_{26}O_{12}$	579.1470	579.1503	−3.3	17.5	Procyanidin B	α-Glucosidase inhibitor [39]; aldose reductase inhibitor [24]; antioxidant [40]	0.87

Table 4. Cont.

	t_R (min)[1]	[M+H]$^+$ m/z	[M-H]$^-$ m/z	Predicted Formula	Observed (m/z)	Theroetical (m/z)	Mass Error (mDa)	DBE[2]	Temporarily Identified	Activity	Content Ratio (DSOB/SOB)
10	2.44	579.1470	577.1403	$C_{30}H_{26}O_{12}$	579.1470	579.1503	−3.3	17.5	Procyanidin B	α-Glucosidase inhibitor [39]; aldose reductase inhibitor [24]; antioxidant [40]	0.71
11	2.52	867.2014	865.1972	$C_{45}H_{38}H_{18}$	865.1972	865.1980	−0.8	27.5	Procyanidin trimer	α-Glucosidase inhibitor [41]; antioxidant [42]	0.60
12	2.62	291.0899	289.0694	$C_{15}H_{14}O_6$	289.0694	289.0712	−1.8	9.5	(+)-Catechin	α-Glucosidase inhibitor [43]; aldose reductase inhibitor [44]; antioxidant [45]	0.86
13	2.72	185.0494	183.0301	$C_8H_8O_5$	183.0301	183.0293	0.8	5.5	Methyl gallate	α-Glucosidase inhibitor [46]; aldose reductase inhibitor [24]; antioxidant [47]	0.92
14	2.90	731.1622	729.1451	$C_{37}H_{30}O_{16}$	729.1451	729.1456	−0.5	23.5	Procyanidin B gallate	Antioxidant [48]	0.82
15	3.04	465.0714	463.0554	$C_{20}H_{16}O_{13}$	463.0554	463.0513	4.1	13.5	Ellagic acid glycoside	Aldose reductase inhibitor [49]; antioxidant [50]	1.67
16	3.39	443.0980	441.0837	$C_{22}H_{18}O_{10}$	443.0980	443.0978	0.2	13.5	Catechin gallate	Aldose reductase inhibitor [51]; antioxidant [52]	0.79
17	3.63	435.0565	433.0388	$C_{19}H_{14}O_{12}$	435.0565	435.0564	0.1	12.5	Ellagic acid xyloside	Antioxidant [53]	1.37
18	3.79	199.0663	197.0456	$C_9H_{10}O_5$	197.0456	197.0450	0.6	5.5	Syringic acid	α-Glucosidase inhibitor [54]; aldose reductase inhibitor [55]; antioxidant [56]	1.87
19	4.09	303.0162	300.9980	$C_{14}H_6O_8$	300.9980	300.9984	−0.4	12.5	Ellagic acid	α-Glucosidase inhibitor [57]; aldose reductase inhibitor [58]; antioxidant [59]	1.13

Table 4. Cont.

	t_R (min)[1]	[M+H]$^+$ m/z	[M-H]$^-$ m/z	Predicted Formula	Observed (m/z)	Theroetical (m/z)	Mass Error (mDa)	DBE[2]	Temporarily Identified	Activity	Content Ratio (DSOB/SOB)
20	4.23	479.0820	477.0708	$C_{21}H_{18}O_{13}$	479.0820	479.0826	−0.6	12.5	Quercetin glucuronide	α-Glucosidase inhibitor [60]; antioxidant [61]	1.18
21	4.67	441.0876	439.0679	$C_{22}H_{16}O_{10}$	439.0679	439.0665	1.4	15.5	Galloylxy trihydroxyflavanone	-	0.05
22	4.79	-	433.0752	$C_{20}H_{18}O_{11}$	433.0752	433.0771	−1.9	12.5	Quercetin xylopyranoside	Antioxidant [62]	1.15
23	4.92	463.0877	461.0743	$C_{21}H_{18}O_{12}$	463.0877	463.0877	−0.8	12.5	Kaempferol glucuronide	Antioxidant [63]	2.37
24	5.54	-	315.0179	$C_{15}H_8O_8$	315.0179	315.0141	3.8	12.5	Methyl ellagic acid	α-Glucosidase inhibitor [64]; aldose reductase inhibitor [2]; antioxidant [65]	0.88
25	6.99	303.0542	301.0359	$C_{15}H_{10}O_7$	301.0359	301.0348	1.1	11.5	Quercetin	α-Glucosidase inhibitor [66]; aldose reductase inhibitor [67]; antioxidant [68]	1.08

[1] t_R, retention time. [2] DBE, double bond equivalency. [3] "-" indicates not detected.

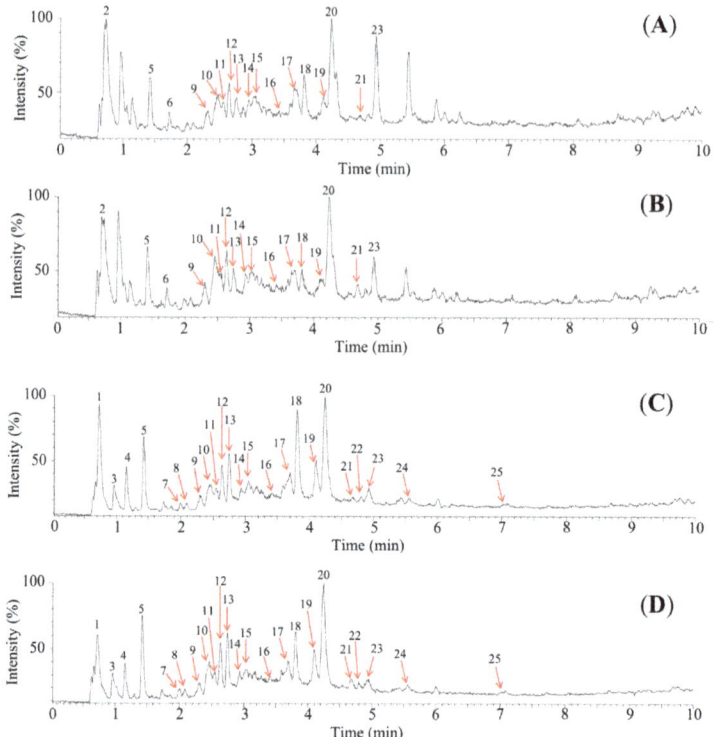

Figure 3. Total ion chromatography of 70% ethanol extracts from defatted seeds of *Oenothera biennis* ((**A**) in positive mode, (**C**) in negative mode) and seeds of *Oenothera biennis* ((**B**) in positive mode, (**D**) in negative mode).

3.3. Nutritional Value of DSOB

The nutritional and energy values of DSOB are presented as dw in Table 5, using SOB as a reference. Carbohydrates (72.60 g/100 g dw) were the main nutrient in DSOB, followed by protein (15.86 g/100 g dw), ash (8.86 g/100 g dw), and fats (2.68 g/100 g dw). However, carbohydrate, protein, ash, and fat content in SOB were 56.34, 12.80, 7.00, and 23.86 g/100 g dw, respectively. All nutritional values, except fats, were higher in DSOB than those in SOB. Seeds are normally a fat reservoir; thus, fats are the major macronutrients in SOB, ranking only second to carbohydrates. Compared to SOB, the proportion of fats in DSOB was greatly reduced by oil extraction, which led to decreased biomass and thus an increased nutritional content (carbohydrates, proteins, and ash) in disguise. The energy of SOB was 491.33 Kcal/100 g dw, calculated on the basis of its carbohydrate, protein, and fat content. A lower energy was observed for DSOB (378.00 Kcal/100 g dw) due to its lower fat content, since fats are the main contributor to energy.

The amount and type of carbohydrates to eat is an important consideration for patients with diabetes when planning their diet. As shown in Table 6, the predominant sugars in DSOB were found to be sucrose (1.27 g/100 g dw) and glucose (0.16 g/100 g dw), whereas other sugars were not detected. Briefly, there was 0.13 g/100 g dw maltose in SOB, and the contents of sucrose (0.99 g/100 g dw) and glucose (0.13 g/100 g dw) in SOB were significantly lower than those in DSOB. However, both DSOB and SOB presented low levels of monosaccharides and disaccharides. Furthermore, a higher fiber content was observed in DSOB (34.78 g/100 g dw) compared to that in SOB (29.40 g/100 g dw). These results demonstrated that DSOB had a lower glycemic index, thereby reducing the health risks induced by sugars.

Table 5. Nutritional value and energetic value of defatted seeds of *Oenothera biennis* (DSOB) and seeds of *Oenothera biennis* (SOB).

Samples	Carbohydrates (g/100 g dw [1])	Proteins (g/100 g dw)	Ash (g/100 g dw)	Fat (g/100 g dw)	Energy (Kcal/100 g dw)
DSOB	72.60 ± 1.12 [2]	15.86 ± 0.95	8.86 ± 0.24	2.68 ± 0.89	378.00 ± 4.74
SOB	56.34 ± 1.96	12.80 ± 0.77	7.00 ± 0.03	23.86 ± 1.16	491.33 ± 5.68
p-Value [3]	0.002	0.013	0.001	<0.001	<0.001

[1] "dw" is short for dry weight. [2] Results are presented as the mean ± standard deviation (n = 3). [3] p-value was calculated by Student's unpaired t-test. $p < 0.05$ was considered significant.

Table 6. The contents of fibers and sugars in defatted seeds of *Oenothera biennis* (DSOB) and seeds of *Oenothera biennis* (SOB).

Sample	DSOB	SOB	p-Value [1]
Glucose (g/100 g dw [2])	0.16 ± 0.010 [3]	0.13 ± 0.006	0.034
Fructose (g/100 g dw)	ND [4]	ND	_ [5]
Galactose (g/100 g dw)	ND	ND	-
Sucrose (g/100 g dw)	1.27 ± 0.08	0.99 ± 0.04	0.042
Maltose (g/100 g dw)	ND	0.13 ± 0.005	<0.001
Lactose (g/100 g dw)	ND	ND	-
Fibers (g/100 g dw)	34.78 ± 2.50	29.40 ± 1.28	0.03

[1] p-value was calculated by Student's unpaired t-test. $p < 0.05$ was considered significant. [2] "dw" is short for dry weight. [3] Results are presented as the mean ± standard deviation (n = 3). [4] "ND" indicates not detected. [5] "-" indicates not available.

As mentioned previously, SOB is a good source of protein and, interestingly, DSOB has a higher protein content. As shown in Table 7, non-essential amino acids in DSOB proteins are completely dominant. In particular, glutamic acid (14.24 g/100 g protein) was highest in DSOB, followed by arginine (6.72 g/100 g protein), aspartic acid (6.16 g/100 g protein), proline (5.76 g/100 g protein), glycine (4.88 g/100 g protein), serine (4.32 g/100 g protein), and alanine (2.72 g/100 g protein). Regarding essential amino acids, leucine (4.40 g/100 g protein) had the highest levels, followed by phenylalanine (3.52 g/100 g protein), valine (3.12 g/100 g protein), threonine (2.32 g/100 g protein), lysine (2.32 g/100 g protein), isoleucine (1.68 g/100 g protein), cysteine (1.52 g/100 g protein), tyrosine (1.52 g/100 g protein), histidine (1.28 g/100 g protein), and methionine (1.09 g/100 g protein). Overall, the amino acid content of DSOB was less than that of SOB but exhibited no significant differences. DSOB protein quality was further assessed by CS, EAAI, and BV. The CS of the essential amino acids in DSOB, based on egg protein, showed values between 31.11 and 58.18, and Iso was the first restricted amino acid; the EAAI of 44.84 and BV of 37.15 in DSOB indicated that overall levels of essential amino acids were relatively low.

As shown in Table 8, the amounts of four macroelements (Na, K, Ca, and Mg) and four microelements (Fe, Mn, Cu, and Zn) in DSOB and SOB were determined. Among the macroelements, Na (1121.67 mg/100 g dw) was present in the highest amount in DSOB, followed by Ca (1106.77 mg/100 g dw), Mg (320.88 mg/100 g dw), and K (259.63 mg/100 g dw). Regarding the microelements, Fe (50.35 mg/100 g dw) exhibited the highest level, followed by Mn (7.22 mg/100 g dw), Zn (3.6 mg/100 g dw), and Cu (0.40 mg/100 g dw). Overall, the DSOB mineral content was significantly higher than that of SOB, except for K and Cu. The daily requirement of macroelements is approximately 100 mg to maintain activity. However, only a few milligrams, even micrograms, of microelements are required per day, as their name suggests. Thus, DSOB was a good source of minerals for dietary supplementation, especially Ca and Fe.

Table 7. Amino acid composition (%) of proteins in defatted seeds of *Oenothera biennis* (DSOB) and seeds of *Oenothera biennis* (SOB).

Amino Acids	Amino Acid (g/100 g Protein)		p-Value [2]	CS [3]		p-Value
	DSOB	SOB		DSOB	SOB	
Essential amino acid						
Valine	3.12 ± 0.19 [1]	3.28 ± 0.20	0.493	47.27 ± 2.84	49.70 ± 2.98	0.487
Isoleucine	1.68 ± 0.10	1.72 ± 0.10	0.730	31.11 ± 1.87	31.85 ± 1.91	0.735
Leucine	4.40 ± 0.26	4.69 ± 0.28	0.384	51.16 ± 3.07	54.53 ± 3.27	0.388
Threonine	2.32 ± 0.14	2.50 ± 0.15	0.324	49.36 ± 2.96	53.19 ± 3.19	0.323
Lysine	2.32 ± 0.14	2.60 ± 0.16	0.175	33.14 ± 1.99	37.14 ± 2.23	0.170
Histidine	1.28 ± 0.08	1.48 ± 0.09	0.112	58.18 ± 3.49	67.27 ± 4.04	0.106
Tyrosine	1.52 ± 0.09	1.49 ± 0.09	0.773	54.19 ± 3.25	55.48 ± 3.33	0.736
Phenylalanine	3.52 ± 0.21	3.67 ± 0.22	0.555			
Cysteine	1.52 ± 0.09	1.47 ± 0.09	0.635	42.11 ± 2.53	44.91 ± 2.69	0.383
Methionine	0.88 ± 0.05	1.09 ± 0.07	0.047			
Non-essential amino acid						
Arginine	6.72 ± 0.40	7.27 ± 0.44	0.303			
Proline	5.76 ± 0.35	6.71 ± 0.40	0.096			
Aspartic	6.16 ± 0.37	6.72 ± 0.40	0.262			
Serine	4.32 ± 0.26	4.61 ± 0.28	0.384			
Glutamic acid	14.24 ± 0.85	15.63 ± 0.94	0.238			
Glycine	4.88 ± 0.29	5.23 ± 0.31	0.349			
Alanine	2.72 ± 0.16	2.81 ± 0.17	0.641			
EAAI [4]	44.84 ± 2.69	48.09 ± 2.89	0.350			
BV [5]	37.15 ± 2.93	40.68 ± 3.14	0.350			

[1] Results are presented as the mean ± standard deviation (n = 3). [2] p-value was calculated by Student's unpaired t-test. p < 0.05 was considered significant. [3] "CS" indicates chemical score. [4] "EAAI" indicates essential amino acid index. [5] "BV" indicates biological value.

Table 8. The contents of minerals in defatted seeds of *Oenothera biennis* (DSOB) and seeds of *Oenothera biennis* (SOB).

Sample	DSOB	SOB	p-Value [3]
Macroelements (mg/100 g dw [1])			
Na	1121.67 ± 3.45 [2]	999.44 ± 6.20	<0.001
K	259.63 ± 1.50	303.67 ± 0.87	<0.001
Ca	1106.77 ± 7.73	1001.69 ± 7.08	<0.001
Mg	320.88 ± 1.34	243.74 ± 1.01	<0.001
Microelements (mg/100 g dw)			
Fe	50.35 ± 0.63	42.86 ± 0.25	<0.001
Mn	7.22 ± 0.04	5.72 ± 0.12	<0.001
Cu	0.40 ± 0.003	0.43 ± 0.008	0.016
Zn	3.60 ± 0.001	2.02 ± 0.05	<0.001

[1] "dw" is short for dry weight. [2] Results are presented as the mean ± standard deviation (n = 3). [3] p-value was calculated by Student's unpaired t-test. p < 0.05 was considered significant.

Two organic acids (oxalic acid and citric acid) were identified in DSOB and SOB. As shown in Table 9, oxalic acid levels in DSOB are 255.30 mg/100 g dw, which is significantly lower than that in SOB (584.65 mg/100 g dw). In contrast, citric acid in DSOB (363.26 mg/100 g dw) was significantly greater than that in SOB (171.67 mg/100 g DW). Organic acids are a key indicator of food quality and serve as synergists to improve the activity of antioxidants. However, oxalic acid is an anti-nutrient whose high dietary intake results in the formation of calcium oxalate crystals, decreasing the absorption of Ca and increasing the risk of kidney stones. From this point of view, DSOB was more applicable for dietary supplementation than SOB.

Table 9. The contents of oxalic acid, citric acid, and tocopherols in defatted seeds of *Oenothera biennis* (DSOB) and seeds of *Oenothera biennis* (SOB).

Sample	DSOB	SOB	p-Value [3]
Oxalic acid (mg/100 g dw [1])	255.30 ± 30.00 [2]	584.65 ± 52.23	0.002
Citric acid (mg/100 g dw)	363.26 ± 18.05	171.67 ± 22.67	0.001
α-Tocopherol (mg/100 g dw)	ND [4]	ND	-[5]
β-Tocopherol (mg/100 g dw)	ND	ND	-
γ-Tocopherol (mg/100 g dw)	1.11 ± 0.001	11.61 ± 0.04	<0.001

[1] "dw" is short for dry weight. [2] Results are presented as the mean ± standard deviation ($n = 3$). [3] p-value was calculated by Student's unpaired t-test. $p < 0.05$ was considered significant. [4] "ND" indicates not detected. [5] "-" indicates not available.

The tocopherol contents (α-, β-, and γ-isoforms) have been determined, as shown in Table 9. Only one isoform of tocopherol (γ-) was identified and quantified in DSOB (1.11 mg/100 g dw) and SOB (11.61 mg/100 dw). Compared to SOB, the low tocopherol content in DSOB was likely related to its low fat content.

4. Conclusions

In summary, DSOB was found to be a reservoir of nutrients and polyphenols, and had remarkable inhibitory activity for α-glucosidase and aldose reductase, as well as antioxidant capacities. Moreover, 25 compounds in DSOB were temporarily identified using LC–ESI–QTOF/MS analysis, which were associated with the remarkable bioactivities of DSOB. DSOB also contained a high content of carbohydrates, proteins, and minerals. The fat content in DSOB decreased after EPO extraction from SOB, resulting in low levels of γ-tocopherol and energy. DSOB contained large amounts of fiber and low levels of sugars, providing a low glycemic index. Notably, DSOB was a good source of Ca and Fe, and the anti-nutrient and oxalic acid content was low. These findings imply that DSOB, that is, the waste material generated by SOB after EPO extraction, has potential as a functional food ingredient in diabetes.

Author Contributions: Conceptualization, Z.W. (Zhiqiang Wang) and H.Y.; methodology, Z.W. (Zhaoyang Wu) and G.Z.; investigation, Z.W. (Zhiqiang Wang), Z.W. (Zhaoyang Wu), and G.Z.; writing—original draft preparation, Z.W. (Zhiqiang Wang); writing—review and editing, S.S.L.; project administration, H.Y.; funding acquisition, Z.W. (Zhiqiang Wang) and H.Y. All authors have read and agreed to the published version of the manuscript.

Funding: This work was supported by the National Natural Science Foundation of China (81803401), Natural Science Foundation of Hebei Province (B2018201270, H2019201186), and the Education Department of Hebei Province (BJ2018035).

Institutional Review Board Statement: All animal experimental procedures were conducted in accordance with the guidelines and approval of the Institutional Animal Care and Use Committee (IACUC) of Hebei University (IACUC-20180051) at 6 December 2018.

Informed Consent Statement: Not applicable.

Data Availability Statement: The study did not report any data.

Conflicts of Interest: The authors declare no conflict of interest.

References

1. International Diabetes Federations. Diabetes Atlas. Available online: https://www.diabetesatlas.org/en/resources/ (accessed on 20 December 2019).
2. Veeresham, C.; Rama Rao, A.; Asres, K. Aldose reductase inhibitors of plant origin. *Phyther. Res.* **2014**, *28*, 317–333. [CrossRef] [PubMed]
3. Gleissner, C.A.; Galkina, E.; Nadler, J.L.; Ley, K. Mechanisms by which diabetes increases cardiovascular disease. *Drug Discov. Today. Dis. Mech.* **2007**, *4*, 131–140. [CrossRef]

4. UK Prospective Diabetes Study (UKPDS) Group. Intensive blood-glucose control with sulphonylureas or insulin compared with conventional treatment and risk of complications in patients with type 2 diabetes (UKPDS 33). *Lancet* **1998**, *352*, 837–853. [CrossRef]
5. Parkman, H.P.; Yates, K.P.; Hasler, W.L.; Nguyan, L.; Pasricha, P.J.; Snape, W.J.; Farrugia, G.; Calles, J.; Koch, K.L.; Abell, T.L.; et al. Dietary intake and nutritional deficiencies in patients with diabetic or idiopathic gastroparesis. *Gastroenterology* **2011**, *141*, 486–498. [CrossRef]
6. Wang, Z.; Hwang, S.H.; Lee, S.Y.; Lim, S.S. Fermentation of purple Jerusalem artichoke extract to improve the α-glucosidase inhibitory effect in vitro and ameliorate blood glucose in db/db mice. *Nutr. Res. Pract.* **2016**, *10*, 282–287. [CrossRef]
7. Kousaxidis, A.; Petrou, A.; Lavrentaki, V.; Fesatidou, M.; Nicolaou, I.; Geronikaki, A. Aldose reductase and protein tyrosine phosphatase 1B inhibitors as a promising therapeutic approach for diabetes mellitus. *Eur. J. Med. Chem.* **2020**, *207*, 112742. [CrossRef]
8. Yaribeygi, H.; Sathyapalan, T.; Atkin, S.L.; Sahebkar, A. Molecular mechanisms linking oxidative stress and diabetes mellitus. *Oxid. Med. Cell. Longev.* **2020**, *2020*, 8609213. [CrossRef]
9. Agrawal, A.A.; Hastings, A.P.; Johnson, M.T.J.; Maron, J.L.; Salminen, J.P. Insect herbivores drive real-time ecological and evolutionary change in plant populations. *Science* **2012**, *338*, 113–116. [CrossRef]
10. Pan, F.; Li, Y.; Luo, X.; Wang, X.; Wang, C.; Wen, B.; Guan, X.; Xu, Y.; Liu, B. Effect of the chemical refining process on composition and oxidative stability of evening primrose oil. *J. Food Process. Pres.* **2020**, *44*, e14800. [CrossRef]
11. Ghasemnezhad, A.; Honermeier, B. Yield, oil constituents, and protein content of evening primrose (*Oenothera biennis* L.) seeds depending on harvest time, harvest method and nitrogen application. *Ind. Crop Prod.* **2008**, *28*, 17–23. [CrossRef]
12. Kiss, A.K.; Naruszewicz, M. Polyphenolic compounds characterization and reactive nitrogen species scavenging capacity of Oenothera paradoxa defatted seed extracts. *Food Chem.* **2012**, *131*, 485–492. [CrossRef]
13. Kiss, A.K.; Derwińska, M.; Granica, S. Quantitative analysis of biologically active polyphenols in evening primrose (Oenothera paradoxa) seeds aqueous extracts. *Pol. J. Food Nutr. Sci.* **2011**, *61*, 109–113. [CrossRef]
14. Pająk, P.; Socha, R.; Broniek, J.; Królikowska, K.; Fortuna, T. Antioxidant properties, phenolic and mineral composition of germinated chia, golden flax, evening primrose, phacelia and fenugreek. *Food Chem.* **2019**, *275*, 69–76. [CrossRef]
15. Horwitz, W.; Latimer, G. *Official Methods of Analysis of AOAC International*; AOAC International: Gaithersburg, MD, USA, 2016.
16. Barros, L.; Pereira, E.; Calhelha, R.C.; Dueñas, M.; Carvalho, A.M.; Santos-Buelga, C.; Ferreira, I.C.F.R. Bioactivity and chemical characterization in hydrophilic and lipophilic compounds of *Chenopodium ambrosioides* L. *J. Funct. Foods* **2013**, *5*, 1732–1740. [CrossRef]
17. Wang, Z.; Hwang, S.H.; Guillen Quispe, Y.N.; Gonzales Arce, P.H.; Lim, S.S. Investigation of the antioxidant and aldose reductase inhibitory activities of extracts from Peruvian tea plant infusions. *Food Chem.* **2017**, *231*, 222–230. [CrossRef]
18. Kainama, H.; Fatmawati, S.; Santoso, M.; Papilaya, P.M.; Ersam, T. The relationship of free radical scavenging and total phenolic and flavonoid contents of *Garcinia lasoar* pam. *Pharm. Chem. J.* **2020**, *53*, 1151–1157. [CrossRef]
19. Kwon, S.H.; Wang, Z.; Hwang, S.H.; Kang, Y.H.; Lee, J.Y.; Lim, S.S. Comprehensive evaluation of the antioxidant capacity of Perilla frutescens leaves extract and isolation of free radical scavengers using step-wise HSCCC guided by DPPH-HPLC. *Int. J. Food Prop.* **2017**, *20*, S921–S934. [CrossRef]
20. Hazra, B.; Biswas, S.; Mandal, N. Antioxidant and free radical scavenging activity of *Spondias pinnata*. *BMC Complement. Altern. Med.* **2008**, *8*, 63. [CrossRef] [PubMed]
21. Wang, Z.; Guillen Quispe, Y.N.; Hwang, S.H.; Zuo, G.; Lim, S.S. Pistafolin B is the major aldose reductase inhibitor of the pods of tara [*Caesalpinia spinosa* (Molina) Kuntze]. *Ind. Crops Prod.* **2018**, *122*, 709–715. [CrossRef]
22. Yilmazer-Musa, M.; Griffith, A.M.; Michels, A.J.; Schneider, E.; Frei, B. Grape seed and tea extracts and catechin 3-gallates are potent inhibitors of α-amylase and α-glucosidase activity. *J. Agr. Food Chem.* **2012**, *60*, 8924–8929. [CrossRef]
23. Huang, G.J.; Hsieh, W.T.; Chang, H.Y.; Huang, S.S.; Lin, Y.C.; Kuo, Y.H. α-Glucosidase and aldose reductase inhibitory activities from the fruiting body of *Phellinus merrillii*. *J. Agr. Food Chem.* **2011**, *59*, 5702–5706. [CrossRef]
24. Wang, Z.; Shen, S.; Cui, Z.; Nie, H.; Han, D.; Yan, H. Screening and isolating major aldose reductase inhibitors from the seeds of evening primrose (*Oenothera biennis*). *Molecules* **2019**, *24*, 2709. [CrossRef] [PubMed]
25. Kalam, S.; Gul, M.Z.; Singh, R.; Ankati, S. Free radicals: Implications in etiology of chronic diseases and their amelioration through nutraceuticals. *Pharmacologia* **2015**, *6*, 11–20.
26. Johanses, J.S.; Harris, A.K.; Rychly, D.J.; Ergul, A. Oxidative stress and the use of antioxidants in diabetes: Linking basic science to clinical practice. *Cardiovasc. Diabetol.* **2005**, *4*, 5. [CrossRef]
27. Iglesias-Carres, L.; Mas-Capdevila, A.; Bravo, F.I.; Bladé, C.; Arola-Arnal, A.; Muguerza, B. Optimization of extraction methods for characterization of phenolic compounds in apricot fruit (*Prunus armeniaca*). *Food Funct.* **2019**, *10*, 6492–6502. [CrossRef] [PubMed]
28. Liu, R.H. Health benefits of fruit and vegetables are from additive and synergistic combinations of phytochemicals. *Am. J. Clin. Nutri.* **2003**, *78*, 517S–520S. [CrossRef]
29. Kim, D.H.; Kim, M.J.; Kim, D.W.; Kim, G.Y.; Kim, J.K.; Gebru, Y.A.; Choi, H.S.; Kim, Y.H.; Kim, M.K. Changes of phytochemical components (urushiols, polyphenols, gallotannins) and antioxidant capacity during fomitella fraxinea–mediated fermentation of toxicodendron vernicifluum bark. *Molecules* **2019**, *24*, 683. [CrossRef] [PubMed]

30. Muccilli, V.; Cardullo, N.; Spatafora, C.; Cunsolo, V.; Tringali, C. α-Glucosidase inhibition and antioxidant activity of an oenological commercial tannin. Extraction, fractionation and analysis by HPLC/ESI-MS/MS and 1H NMR. *Food Chem.* **2017**, *215*, 50–60. [CrossRef]
31. Xue, N.; Jia, Y.; Li, C.; He, B.; Yang, C.; Wang, J. Characterizations and assays of α-glucosidase inhibition activity on gallic acid cocrystals: Can the cocrystals be defined as a new chemical entity during binding with the α-glucosidase? *Molecules* **2020**, *25*, 1163. [CrossRef]
32. Mahindrakar, K.V.; Rathod, V.K. Ultrasonic assisted aqueous extraction of catechin and gallic acid from *Syzygium cumini* seed kernel and evaluation of total phenolic, flavonoid contents and antioxidant activity. *Chem. Eng. Process.* **2020**, *149*, 107841. [CrossRef]
33. Toshima, A.; Matsui, T.; Noguchi, M.; Qiu, J.; Tamaya, K.; Miyata, Y.; Tanaka, T.; Tanaka, K. Identification of alpha-glucosidase inhibitors from a new fermented tea obtained by tea-rolling processing of loquat (*Eriobotrya japonica*) and green tea leaves. *J. Sci. Food Agric.* **2010**, *90*, 1545–1550. [CrossRef] [PubMed]
34. Lee, J.; Jang, D.S.; Kim, N.H.; Lee, Y.M.; Kim, J.; Kim, J.S. Galloyl glucoses from the seeds of *Cornus officinalis* with inhibitory activity against protein glycation, aldose reductase, and cataractogenesis ex vivo. *Biol. Pharm. Bull.* **2011**, *34*, 443–446. [CrossRef] [PubMed]
35. Ononamadu, C.; Ihegboro, G.O.; Owolarafe, T.A.; Salawu, K.; Fadilu, M.; Ezeigwe, O.C.; Oshobu, M.L.; Nwachukwu, F.C. Identification of potential antioxidant and hypoglycemic compounds in aqueous-methanol fraction of methanolic extract of *Ocimum canum* leaves. *Anal. Bioanal. Chem. Res.* **2019**, *6*, 431–439.
36. Nguyen, M.T.T.; Nguyen, N.T.; Nguyen, H.X.; Huynh, T.N.N.; Min, B.S. Screening of α-glucosidase inhibitory activity of vietnamese medicinal plants: Isolation of active principles from *Oroxylum indicum*. *Nat. Prod. Sci.* **2012**, *18*, 47–51.
37. Zuo, G.L.; Kim, H.Y.; Guillen Quispe, Y.N.; Wang, Z.Q.; Hwang, S.H.; Shin, K.O.; Lim, S.S. Efficient separation of phytochemicals from *Muehlenbeckia volcanica* (Benth.) Endl. by polarity-stepwise elution counter-current chromatography and their antioxidant, antiglycation, and aldose reductase inhibition potentials. *Molecules* **2021**, *26*, 224. [CrossRef] [PubMed]
38. Al Olayan, E.M.; Aloufi, A.S.; Alamri, O.D.; El-Habit, O.H.; Abdel Moneim, A.E. Protocatechuic acid mitigates cadmium-induced neurotoxicity in rats: Role of oxidative stress, inflammation and apoptosis. *Sci. Total Environ.* **2020**, *723*, 137969. [CrossRef]
39. Zhao, L.; Wen, L.; Lu, Q.; Liu, R. Interaction mechanism between α-glucosidase and A-type trimer procyanidin revealed by integrated spectroscopic analysis techniques. *Int. J. Biol. Macromol.* **2020**, *143*, 173–180. [CrossRef]
40. Martins, G.R.; do Amaral, F.R.L.; Brum, F.L.; Mohana-Borges, R.; de Moura, S.S.T.; Ferreira, F.A.; Sangenito, L.S.; Santos, A.L.S.; Figueiredo, N.G.; da Silva, A.S.A. Chemical characterization, antioxidant and antimicrobial activities of açaí seed (*Euterpe oleracea* Mart.) extracts containing A- and B-type procyanidins. *LWT—Food Sci. Technol.* **2020**, *132*, 109830. [CrossRef]
41. Ado, M.A.; Abas, F.; Ismail, I.S.; Ghazali, H.M.; Shaari, K. Chemical profile and antiacetylcholinesterase, antityrosinase, antioxidant and α-glucosidase inhibitory activity of *Cynometra cauliflora* L. leaves. *J. Sci. Food Agric.* **2015**, *95*, 635–642. [CrossRef]
42. Wang, W.; Bostic, T.R.; Gu, L. Antioxidant capacities, procyanidins and pigments in avocados of different strains and cultivars. *Food Chem.* **2010**, *122*, 1193–1198. [CrossRef]
43. Kim, T.; Choi, H.J.; Eom, S.-H.; Lee, J.; Kim, T.H. Potential α-glucosidase inhibitors from thermal transformation of (+)-catechin. *Bioorg. Med. Chem. Lett.* **2014**, *24*, 1621–1624. [CrossRef]
44. Jung, H.A.; Jung, Y.J.; Yoon, N.Y.; Jeong, D.M.; Bae, H.J.; Kim, D.-W.; Na, D.H.; Choi, J.S. Inhibitory effects of *Nelumbo nucifera* leaves on rat lens aldose reductase, advanced glycation endproducts formation, and oxidative stress. *Food Chem. Toxicol.* **2008**, *46*, 3818–3826. [CrossRef] [PubMed]
45. Ahmadi, S.M.; Farhoosh, R.; Sharif, A.; Rezaie, M. Structure-antioxidant activity relationships of luteolin and catechin. *J. Food Sci.* **2020**, *85*, 298–305. [CrossRef] [PubMed]
46. Xu, R.; Bu, Y.G.; Zhao, M.L.; Tao, R.; Luo, J.; Li, Y. Studies on antioxidant and α-glucosidase inhibitory constituents of Chinese toon bud (*Toona sinensis*). *J. Funct. Foods* **2020**, *73*, 104108. [CrossRef]
47. Asnaashari, M.; Farhoosh, R.; Sharif, A. Antioxidant activity of gallic acid and methyl gallate in triacylglycerols of Kilka fish oil and its oil-in-water emulsion. *Food Chem.* **2014**, *159*, 439–444. [CrossRef] [PubMed]
48. Song, J.H.; Kim, S.; Yu, J.S.; Park, D.H.; Kim, S.Y.; Kang, K.S.; Lee, S.; Kim, K.H. Procyanidin B2 3″-O-gallate isolated from reynoutria elliptica prevents glutamate-induced HT22 cell death by blocking the accumulation of intracellular reactive oxygen species. *Biomolecules* **2019**, *9*, 412. [CrossRef]
49. Yan, X.-H.; Guo, Y.-W. Two new ellagic acid glycosides from leaves of *Diplopanax stachyanthus*. *J. Asian Nat. Prod. Res.* **2004**, *6*, 271–276. [CrossRef] [PubMed]
50. Lee, J.-H.; Talcott, S.T. Fruit maturity and juice extraction influences ellagic acid derivatives and other antioxidant polyphenolics in muscadine grapes. *J. Agric. Food Chem.* **2004**, *52*, 361–366. [CrossRef]
51. Grewal, A.S.; Thapa, K.; Kanojia, N.; Sharma, N.; Singh, S. Natural compounds as source of aldose reductase (AR) inhibitors for the treatment of diabetic complications: A Mini Review. *Curr. Drug Metab.* **2020**, *21*, 1091–1116. [CrossRef]
52. Xu, J.Z.; Yeung, S.Y.V.; Chang, Q.; Huang, Y.; Chen, Z.-Y. Comparison of antioxidant activity and bioavailability of tea epicatechins with their epimers. *Br. J. Nutr.* **2004**, *91*, 873–881.
53. Pallauf, K.; Rivas-Gonzalo, J.C.; del Castillo, M.D.; Cano, M.P.; de Pascual-Teresa, S. Characterization of the antioxidant composition of strawberry tree (*Arbutus unedo* L.) fruits. *J. Food Compos. Anal.* **2008**, *21*, 273–281. [CrossRef]

54. Benalla, W.; Bellahcen, S.; Bnouham, M. Antidiabetic medicinal plants as a source of alpha glucosidase inhibitors. *Curr. Diabetes Rev.* **2010**, *6*, 247–254. [CrossRef]
55. Wei, X.; Chen, D.; Yi, Y.; Qi, H.; Gao, X.; Fang, H.; Gu, Q.; Wang, L.; Gu, L. Syringic acid extracted from Herba dendrobii prevents diabetic cataract pathogenesis by inhibiting aldose reductase activity. *Evid. Based Complement. Alternat. Med.* **2012**, *2012*, 426537. [CrossRef]
56. Vo, Q.V.; Van Bay, M.; Nam, P.C.; Quang, D.T.; Flavel, M.; Hoa, N.T.; Mechler, A. Theoretical and experimental studies of the antioxidant and antinitrosant activity of syringic acid. *J. Org. Chem.* **2020**, *85*, 15514–15520. [CrossRef]
57. Miao, J.; Li, X.; Zhao, C.; Gao, X.; Wang, Y.; Gao, W. Active compounds, antioxidant activity and α-glucosidase inhibitory activity of different varieties of Chaenomeles fruits. *Food Chem.* **2018**, *248*, 330–339. [CrossRef] [PubMed]
58. Akileshwari, C.; Raghu, G.; Muthenna, P.; Mueller, N.H.; Suryanaryana, P.; Petrash, J.M.; Reddy, G.B. Bioflavonoid ellagic acid inhibits aldose reductase: Implications for prevention of diabetic complications. *J. Funct. Foods* **2014**, *6*, 374–383. [CrossRef]
59. Alfei, S.; Marengo, B.; Zuccari, G. Oxidative stress, antioxidant capabilities, and bioavailability: Ellagic acid or urolithins? *Antioxidants* **2020**, *9*, 707. [CrossRef]
60. Jiang, P.; Xiong, J.; Wang, F.; Grace, M.H.; Lila, M.A.; Xu, R. α-Amylase and α-glucosidase inhibitory activities of phenolic extracts from *Eucalyptus grandis* × *E. urophylla* Bark. *J. Chem.* **2017**, *2017*, 8516924. [CrossRef]
61. El-Zaeddi, H.; Calín-Sánchez, Á.; Nowicka, P.; Martínez-Tomé, J.; Noguera-Artiaga, L.; Burló, F.; Wojdyło, A.; Carbonell-Barrachina, Á.A. Preharvest treatments with malic, oxalic, and acetylsalicylic acids affect the phenolic composition and antioxidant capacity of coriander, dill and parsley. *Food Chem.* **2017**, *226*, 179–186. [CrossRef]
62. Rana, S.; Prakash, V.; Sagar, A. Medicinal and antioxidant properties of some medicinal plants. *J. Drug Deliv. Ther.* **2016**, *6*, 1–6. [CrossRef]
63. Harb, J.; Alseekh, S.; Tohge, T.; Fernie, A.R. Profiling of primary metabolites and flavonols in leaves of two table grape varieties collected from semiarid and temperate regions. *Phytochemistry* **2015**, *117*, 444–455. [CrossRef] [PubMed]
64. Tabopda, T.K.; Ngoupayo, J.; Liu, J.; Ali, M.S.; Khan, S.N.; Ngadjui, B.T.; Luu, B. Alpha-glucosidase inhibitors ellagic acid derivatives with immunoinhibitory properties from *Terminalia superba*. *Chem. Pharm. Bull.* **2008**, *56*, 847–850. [CrossRef]
65. Chen, J.; Xu, Y.; Ge, Z.; Zhu, W.; Xu, Z.; Li, C. Structural elucidation and antioxidant activity evaluation of key phenolic compounds isolated from longan (*Dimocarpus longan* Lour.) seeds. *J. Funct. Foods* **2015**, *17*, 872–880. [CrossRef]
66. Siebert, D.A.; Campos, J.S.; Alberton, M.D.; Vitali, L.; Micke, G.A. Dual electrophoretically-mediated microanalysis in multiple injection mode for the simultaneous determination of acetylcholinesterase and α-glucosidase activity applied to selected polyphenols. *Talanta* **2021**, *224*, 121773. [CrossRef]
67. Ulusoy, H.G.; Sanlier, N. A minireview of quercetin: From its metabolism to possible mechanisms of its biological activities. *Crit. Rev. Food Sci.* **2020**, *60*, 3290–3303. [CrossRef]
68. Song, X.; Wang, Y.; Gao, L. Mechanism of antioxidant properties of quercetin and quercetin-DNA complex. *J. Mol. Model.* **2020**, *26*, 133. [CrossRef]

Article

Olive Pomace-Derived Biomasses Fractionation through a Two-Step Extraction Based on the Use of Ultrasounds: Chemical Characteristics

María del Mar Contreras [1,2,*], Irene Gómez-Cruz [1,2], Inmaculada Romero [1,2] and Eulogio Castro [1,2]

1. Campus Las Lagunillas, Department of Chemical, Environmental and Materials Engineering, University of Jaén, 23071 Jaén, Spain; igcruz@ujaen.es (I.G.-C.); iromero@ujaen.es (I.R.); ecastro@ujaen.es (E.C.)
2. Center for Advanced Studies in Earth Sciences, Energy and Environment (CEACTEMA), University of Jaén, Campus Las Lagunillas, 23071 Jaén, Spain
* Correspondence: mcgamez@ujaen.es

Abstract: Olive-derived biomass is not only a renewable bioenergy resource but also it can be a source of bioproducts, including antioxidants. In this study, the antioxidant composition of extracted olive pomace (EOP) and a new byproduct, the residual fraction from olive pit cleaning (RFOPC or residual pulp) was characterized and compared to olive leafy biomass, which have been extensively studied as a source of antioxidants and other bioactive compounds with pharmacological properties. The chemical characterization showed that these byproducts contain a high amount of extractives; in the case of EOP, it was even higher (52.9%) than in olive leaves (OL) and olive mill leaves (OML) (35.8–45.1%). Then, ultrasound-assisted extraction (UAE) was applied to recover antioxidants from the extractive fraction of these biomasses. The solubilization of antioxidants was much higher for EOP, correlating well with the extractives content and the total extraction yield. Accordingly, this also affected the phenolic richness of the extracts and the differences between all biomasses were diminished. In any case, the phenolic profile and the hydroxytyrosol cluster were different. While OL, OML, and EOP contained mainly hydroxytyrosol derivatives and flavones, RFOPC presented novel trilignols. Other compounds were also characterized, including secoiridoids, hydroxylated fatty acids, triterpenoids, among others, depending on the bioresource. Moreover, after the UAE extraction step, alkaline extraction was applied recovering a liquid and a solid fraction. While the solid fraction could of interest for further valorization as a biofuel, the liquid fraction contained proteins, sugars, and soluble lignin, which conferred antioxidant properties to these extracts, and whose content depended on the biomass and conditions applied.

Keywords: antioxidants; biorefinery; olive-derived biomass; ultrasound-assisted extraction; valorization

1. Introduction

The healthy properties of olive leaves (OL) are recognized in the traditional medicine and also supported by several scientific reports. The potential of olive leaves extracts to formulate functional ingredients and to obtain antioxidant and antimicrobial preservatives is promising [1,2]. Currently, in the phytopharmacy sector, olive leaves and fruits extracts are key ingredients of dietary supplements and nutraceuticals (infusions, capsules, liquid solutions, etc.) due to their cardiovascular health promoting properties, among other effects. Moreover, the use of synthetic hydroxytyrosol has been approved as a novel ingredient to be added to oils and spreadable fats [3], which is a precedent for using natural extracts containing this compound. Furthermore, Rodrigues et al. [4] also suggested that the bioactive compounds present in olive by-products, including antioxidants, can become a source of anti-aging or hydration active ingredients for cosmetics.

Hydroxytyrosol and their derivatives are some of the active components both to improve health, as several clinical trials suggest [5–8], and the oxidative stability of oils [9,10].

Nonetheless, the hydroxytyrosol cluster composition depends on the olive biomass type and the extraction conditions [11,12]. For example, olive leaves and olive leafy byproducts are richer in oleuropein, while olive fruits and its derived byproduct, olive pomace, contain more hydroxtyrosol, among other derivatives [12–15].

In the olive pomace extracting industry, the extracted olive pomace (EOP) is obtained after the extraction of the residual oil contained in the olive pomace, generally, using hexane (Figure 1). This solid biomass is generated in high amounts; around 10–12% (w/w) of the olives processed in the mills. In Spain, the olive stones fragments are recovered from the olive pomace and sometimes cleaned by a pneumatic process to enhance their energetic potential (Figure 1). This residual fraction derived from the olive pits cleaning (RFOPC or residual pulp) consists of rests of olive pulp, mainly, skin crushed into fragments [11,16,17]. While crushed pits represent around 8–10% of olives weight, the average percentage of RFOPC in the latter fraction is up to 4% [17]. Interestingly, EOP and RFOPC can be produced in the same facility where the main products, olive oil and olive pomace oil, are produced, implying additional advantages for their valorization, i.e., reduced collection and transport costs. While olive pits and EOP are used as a relatively low-cost biofuel, the RFOPC has no current industrial application. Nonetheless, the application of the EOP as a biofuel has some constrains [18] and the removal of a part of the extractive fraction (non-structural components) could improve its energetic use [19]. In this regard, the extractive fraction contains valuable bioactive compounds, including phenolic compounds [11], and hence another alternative would be to obtain antioxidants from these cheap and abundant bioresources before applying other valorization strategies [19,20]. Therefore, their comprehensive characterization may give also clues about the phenolic composition, including the hydroxytyrosol cluster and the presence of other bioactive compounds.

In this context, to recover antioxidants like phenolic compounds from olive-derived byproducts, new trends included the use of ultrasound to assist the extraction process, favoring the mass transfer, shortening the extraction time and/or reducing the solvent necessities [14,21,22]. Nevertheless, the extraction of antioxidants generates a large residual fraction that is worthy of valorization since it can provide an extra income and move towards the circular bioeconomy. For this purpose, antioxidants can be obtained as a first step previous to a further fractionation of the rest of components present in the biomass [14,19,21]. Another alternative is to recover the antioxidants in the lateral streams obtained after the pretreatment of these biomasses, for example, for the conversion of the sugar fraction to biofuels. Nevertheless, it generally requires severe thermal treatments and more thermolabile bioactive compounds could be affected. For example, oleuropein seems to be resistant at least in part [23], but it depends on the conditions applied to the olive leafy biomass [21].

Therefore, in this work, an integrated scheme was applied to fractionate EOP and the new byproduct RFOPC and to characterize the fractions obtained for further valorization. This consisted of ultrasound-assisted extraction (UAE) as a first step to recover antioxidant extracts and an alkaline extraction as second step to fractionate the residual lignocelullosic fraction, according to Contreras et al. [14]. The phenolic composition of extracts obtained from EOP and RFOPC in the first step was characterized, including the hydroxytyrosol cluster, and the antioxidant activity measured, being compared to those extracts obtained from olive leafy biomasses. The second step enabled to recover a liquid fraction, whose composition was characterized in terms of protein, lignin, sugars, and antioxidant properties.

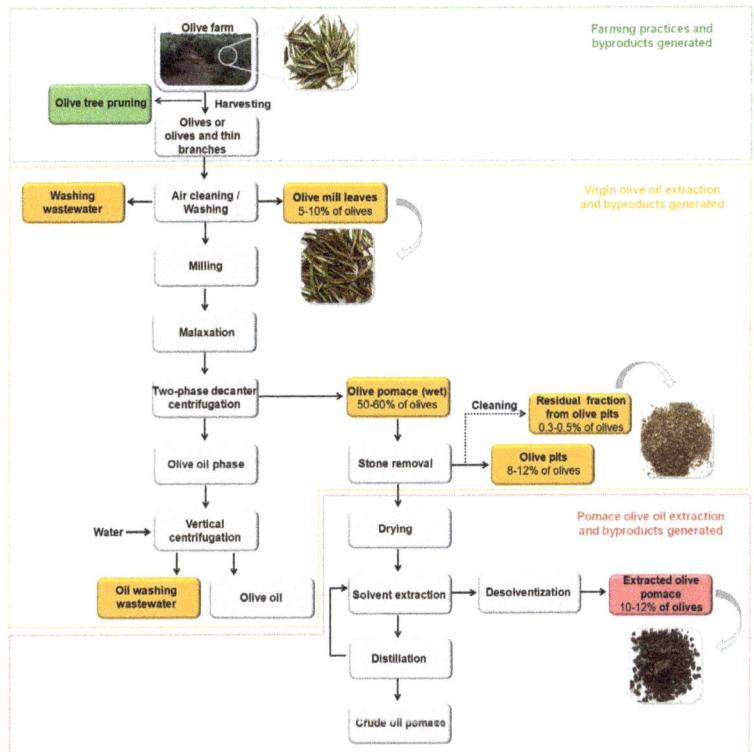

Figure 1. Simplified schemes of the extraction of virgin olive oil and pomace olive oil and the byproducts (squares in green, yellow, and pink) generated during the production steps.

2. Materials and Methods

2.1. Reagents, Standards, and Samples

2.1.1. Reagents and Standards

The following reagents were purchased from Sigma-Aldrich (St. Louis, MO, USA): Folin and Ciocalteu's phenol reagent, sodium carbonate, 2,2′-azobis (2-methylpropionamidine) dihydrochloride (AAPH), 2,4,6-tris (2-pyridyl)-s-triazine (TPTZ), 2,2′-azino-bis(3- ethylbenzothiazoline-6-sulfonic acid) (ABTS) diammonium salt, 6-hydroxy-2,5,7,8-tetramethylchroman-2-carboxylic acid (Trolox), fluorescein sodium salt, potassium persulfate, and ferric sulfate. The following reagents were bought from PanReac AppliChem (Barcelona, Spain): dehydrated sodium phosphate, sodium acetate, ferric chloride, hydrochloric acid, ethanol, formic acid, glacial acetic acid, acetonitrile, and acetone. Sodium hydroxide was purchased from VWR Chemicals (Radnor, PA, USA).

Phenolic standards (degree of purity ≥98%, w/w) were obtained from Extrasynthese (Genay, France) (hydroxytyrosol and oleuropein) and Sigma-Aldrich (St. Louis, MO, USA) (gallic acid, rutin, and caffeic acid).

2.1.2. Samples

EOP was obtained from the olive pomace extracting industry Oleocastellar S.A. (Castellar, Jaén, Spain) and RFOPC from Peláez Renovables (Jaén, Spain). The samples were ground using an Ultra Centrifugal Mill ZM 200 (1 mm sieve) (Retsch GmbH, Haan, Germany) before the determination of the chemical composition and extraction.

2.2. Determination of the Chemical Composition

The moisture and the content of extractives, lignin, carbohydrates, and ash of the samples were determined according to the standard National Renewable Energy Laboratory procedure [24]. Aqueous, ethanolic and hexane extractives were determined using Soxhlet extraction and gravimetric analysis. The characterization of carbohydrates and lignin was performed after acid hydrolysis. The liquid fraction was subjected to high-performance liquid chromatography (HPLC) analysis to quantify monomeric sugars and acid soluble lignin was determined spectrophotometrically at 205 nm. Acid insoluble lignin was determined by gravimetric analysis, taking into account the ash content in this fraction. Moreover, the crude protein content of the byproducts was determined by elemental analysis (TruSpec Micro, Leco, St. Joseph, MI, USA) using a conversion factor of 6.25. All analytical determinations were performed in triplicate.

2.3. Ultrasound-Assisted Extraction

The samples were mixed with ethanol/water (47:53, v/v) at a solid-to-liquid ratio of 6:100 (w/v, dry weight, d.w.) and sonicated in an ultrasonic bath (40 kHz) (Ultrasons, J.P. Selecta, Barcelona, Spain) for 50 min, according to our previous optimized method [14]. The mixture was centrifuged at 1717× g for 15 min, the supernatants were collected, and the recovered volume was measured. All the extractions were done in triplicate. Finally, a portion of the extracts was oven-dried (at 105 °C) till constant weight to estimate the total extraction yield, which was referred to the initial dry byproduct weight (%). Other portion was filtered with a syringe filter (nylon, 0.45 µm pore size) (SinerLab Group, Madrid, Spain) for further analysis.

2.4. Total Phenol Content and Antioxidant Capacity Assays

Total phenol content (TPC), Trolox equivalent antioxidant capacity (TEAC) and ferric ion reducing antioxidant power (FRAP) were determined using colorimetric assays in transparent microplates according to Medfai et al. [12]. Basically, these assays measure the ability to reduce the Folin and Ciocalteu's phenol reagent, $ABTS^{\bullet+}$, Fe^{3+}, respectively, which changes color when reduced and this change is correlated with the antioxidant concentration. For that, a Bio-Rad iMark™ microplate absorbance reader was used (Hercules, CA, USA) at 750 nm (Folin-Ciocalteu and TEAC assays) and 595 nm (FRAP assay). Using standard curves ($R^2 > 0.99$), the TPC results were expressed as gallic acid equivalents (GAE) (25 to 300 µg/mL) and the TEAC and FRAP results as Trolox equivalents (TE) (6 to 330 µM), respectively.

The oxygen radical absorbance capacity (ORAC) was performed using black microplates according to Medfai et al. [12]. Fluorescent measurements with excitation and emission wavelengths of 485 nm (±10) and 528 nm (±10), respectively, were obtained using a BioTek Synergy HT (Winooski, VT, USA) and acquired with Gen5 (BioTek) every min for 120 min. The data was normalized using the initial reading, the area under curve (AUC) for each well was estimated and the net AUC was calculated by subtracting the AUC corresponding to the blank. ORAC values were expressed as TE (standard curve from 0.5 to 20 µM; $R^2 > 0.99$).

The filtered aqueous-ethanolic extracts (Section 2.4) and the alkaline extracts obtained according to next Section 2.7 were measured. The latter samples were neutralized, centrifuged, and filtered (0.45 µm nylon syringe filters) before analysis.

Caffeic acid was used as positive control, obtaining the following values: TEAC = 1.23 ± 0.09 mmol equivalents of Trolox; FRAP = 1.15 ± 0.05 mmol equivalents of Trolox; ORAC = 4.29 ± 0.31 mmol equivalents of Trolox. These values agreed with those reported previously [25].

2.5. HPLC-Mass Spectrometry (MS) and Diode Array Analyses

Reversed phase (RP)-HPLC-MS and -MS^2 analyses were performed in an Agilent 1100 HPLC (Agilent Technologies, Waldbron, Germany) connected on-line to an ion trap (IT) via an electrospray interface (Esquire 6000; Bruker, Bremen, Germany), according to Medfai et al. [12]. Phenolic compounds were eluted at 0.35 mL/min using Milli-Q® water and formic acid (0.1%, v/v) as solvent A and acetonitrile and formic acid (0.1%, v/v) as solvent B. A Kinetex core-shell C18 column (2.1 × 50 mm, 2.7 μm) (Phenomenex, Barcelona, Spain) and a linear gradient of solvent B in A were used: 4%, 0 min; 7%, 1 min; 30%, 15 min; 40%, 4.5 min; 100%, 4.5 min; 100%, 2 min; 4%, 1.5 min; and 4%, 7 min. The injection volume was 10 μL.

MS spectra were recorded over the mass-to-charge (m/z) range of 100–1200 in the negative ionization mode and 4 spectra were averaged. Auto MS/MS analyses were performed at 0.6 V and acquired in the aforementioned range. About 2 spectra were averaged in the MS/MS analyses. The data were processed using DataAnalysis (version 4.0) from Bruker.

In addition, analyses by HPLC (Agilent 1200) coupled to quadrupole-time-of-flight (QTOF)-MS and MS/MS (Agilent 6530B Accurate Mass Q-TOF) were performed to obtain high resolution mass data. The interface was an electrospray ionization source. The column and the gradient use were the same as before. The MS parameters were applied according to Ammar et al. [26] in auto-MS mode, with some modifications. The spectra were acquired in the negative ionization mode, over the m/z range 60–1200 Da. For accurate m/z measurement reference mass correction was performed with a continuous infusion of trifluoroacetic acid ammonium salt (m/z 112.9856) and hexakis 1H,1H,3H–tetrafluoropropoxy) phosphazine (m/z 1033.9881) (Agilent Technologies, Waldbron, Germany). MassHunter Qualitative Analysis B.06.00 (Agilent Technologies) was applied for data treatment to generate molecular formula with a mass accuracy limit of 5 ppm.

Hydroxytyrosol and oleuropein were quantified using external standard calibration by RP-HPLC-diode array detection at 280 nm, according to Contreras et al. [21]. The curves ($R^2 > 0.99$) were $y = 20395x - 15047$ for hydroxytyrosol (1.25 to 500 mg/L) and $y = 5591x + 11,911$ for oleuropein (2.5 to 1000 mg/L).

2.6. Alkaline Extraction of the Residual Extraction Fraction and Determination of the Protein Content and Profile

Alkaline extraction was performed according to previous optimized conditions using a solid-to-liquid ratio of 1:10 (w/v) and sodium hydroxide 0.7 M and 0.4 M in a bath (JULABO GmbH, Seelbach, Germany) at 100 °C and 80 °C, respectively, for 240 min (at 150 rpm) [14]. After subsequent centrifugation, which was performed at 1717× g for 15 min (Herolab, Wiesloch, Germany), supernatants were collected for further analysis and the recovered volume measured. The solubilized protein in the alkaline extracts was determined using a Bradford kit assay from Bio-Rad (Irvine, CA, USA), with some modifications, and referred to a standard calibration curve of bovine serum albumin (BSA).

The protein extract was neutralized using HCl 2 M and proteins (100 μL) were precipitated with 400 μL of acetone at cold conditions for 20 min and followed by centrifugation at 10,000× g for 10 min. The protein pellets were dissolved in 50 μL of Laemmli sample buffer (with 2-mercaptoethanol at 5%, v/v). The separation was performed on Mini-PROTEAN® TGX ™ Precast Gels (Bio-Rad, Irvine, CA, USA) at 200 V in a Mini-PROTEAN® tetra cell (Bio-Rad, Irvine, CA, USA) and using Tris/Glycine/SDS buffer (Bio-Rad, Irvine, CA, USA) as running buffer. Gels were stained using Coomassie Brilliant Blue R-250 (Bio-Rad, Irvine, CA, USA) for 90 min and destained overnight.

2.7. Sugars Analysis and Lignin Determination in the Alkaline Extracts

Samples were acidified around 3.5 by adding HCl 2 M and centrifuged (10,000× g for 10 min) (MicroCen 16, Herolab GmbH Laborgeräte, Wiesloch, Germany). Then, the supernatants were collected and filtered (0.45 μm nylon filters). Additionally, oligomeric sugars were measured upon acid hydrolysis using sulfuric acid at 120 °C for 30 min. All samples were measured using HPLC with a refractive index detector and an ICSep ICE-COREGEL-87H3 column (Transgenomic, Inc., Omaha, NE, USA) [27].

Soluble lignin concentration was estimated in alkaline extracts after centrifugation (10,000× g for 10 min) according to Guerra [28] at 0.1 M NaOH as: $A/(\varepsilon \times l)$, where A is the absorbance at 280 nm and ε = 9.7 L/g cm. The measurements were performed by an UV-Vis spectrophotometer (UV-1800, Shimadzu Schweiz GmbH, Reinach BL, Switzerland) with a 1 cm quartz cuvette. The spectra were also recorded from 190 to 800 nm, which were processed by UVProbe 2.32 (Shimadzu Schweiz GmbH). The first derivative spectra were obtained using the latter software and applying a δ λ of 20.

2.8. Statistical Analysis

Data are expressed as mean ± standard deviation of three analyses. One-way analysis of variance (ANOVA) and LSD for multiple comparisons were performed using Statgraphics Centurion XVII (StatPoint Technologies, Inc., Warrenton, VA, USA). Pearson correlation was performed using Microsoft Excel 2007 (Redmond, WA, USA) and Statgraphics Centurion XVII.

3. Results and Discussion

3.1. Chemical Composition

The chemical composition and the elemental analysis results of the studied byproducts are shown in Table 1 and compared to olive leafy biomass. Although both byproducts, EOP and RFOPC, contain olive fruit rests, the chemical composition was quite different. Nevertheless, as in OL the major fractions were aqueous-ethanolic extractives (non-structural components), particularly, in EOP (52.85%), and lignin, particularly, in RFOPC (32.25%). The highest content of cellulose and hemicellulose was found in RFOPC (around 27.3%), while it contained the lowest protein amount (4.50%) compared to EOP (9.36%) and leaves (up to 9.34%). Moreover, EOP has a high content in ash (10.06%) as OML.

In general, these results agreed well with previous studies on EOP [20,30], but there is little information about RFOPC, as far as we know. It remarks that besides olive leafy biomasses, EOP and RFOPC contain a high amount of extractives, including free and oligomeric sugars and the sugar alcohol mannitol, which is a marketable sweetener and a drug [11]. In this regard, the content of mannitol was only remarkable in EOP (5.36% with respect to the extractives content), even higher than in leaves. Moreover, the extractive fraction is interesting since it presents olive bioactive components, as next sections highlight.

3.2. Extraction of Antioxidants by UAE as a First Valorization Step

3.2.1. Yield, Total Phenolic Content, and Antioxidant Characteristics of the Extracts

The aforementioned results suggest that the extractive fraction is worth of study because of its high content. Therefore, UAE was applied to recover antioxidant extracts from EOP and RFOPC using an aqueous ethanol solution as a first valorization step, according to Contreras et al. [14]. This method was selected to evaluate how the biomass type affects and thus olive leafy biomasses, EOP and RFOPC, were compared (Table 2).

EOP provided the highest total extraction yield by UAE and its capacity to retain solvent was low, indicating that the technical and theoretical yields for these parameters will be similar. Moreover, the solubilization of antioxidants, expressed in terms of biomass weight, from EOP was higher than that for OL and OML, while RFOPC showed the lowest values (Table 2). All these data showed good correlation values with the extractives content ($r > 0.958$), the total phenolic content and between each other ($r > 0.914$) (Table S1). This suggests that phenolic compounds are the main antioxidant compounds in the extracts.

Nonetheless, in terms of purity, the antioxidants extracts of RFOPC only showed a slightly lower potency than the former byproducts. It is explained by the fact that a lower amount of solids were released from RFOPC (Table 2), which can be correlated with its lower amount of extractives (Table 1).

As for olive leafy biomasses, EOP and RFOPC antioxidants showed a higher efficiency for scavenging peroxyl radicals by hydrogen atom transfer mechanisms (ORAC) than reduction properties by electron transfer mechanisms (TEAC and FRAP) under the conditions assayed [12]. Therefore, the antioxidants from these byproducts could have some similarities, being probably relevant to scavenge radicals in vivo and in food systems and further work is necessary. Moreover, the chemical characterization of the extracts is necessary to reveal information about the hydroxytyrosol cluster.

Table 1. Chemical composition of olive leafy and pomace-derived biomasses.

Component (%)	OL [4]	OML [4]	EOP	RFOPC
Chemical characterization				
Protein	9.34 ± 0.35	8.10 ± 0.38	9.36 ± 0.44	4.50 ± 0.33
Glucans [1]	6.98 ± 0.13	9.89 ± 0.57	6.96 ± 0.35	12.23 ± 0.96
Glucose	7.68 ± 0.14	10.88 ± 0.62	7.65 ± 0.38	13.45 ± 1.06
Hemicellulose [2]	5.69 ± 0.11	7.90 ± 0.18	8.17 ± 0.20	14.22 ± 0.99
Galactose	1.41 ± 0.07	1.58 ± 0.07	0.38 ± 0.02	0.48 ± 0.03
Mannose	0.60 ± 0.09	0.14 ± 0.01	ND	ND
Xylose	1.30 ± 0.10	4.57 ± 0.14	8.60 ± 0.23	15.20 ± 1.09
Arabinose	3.06 ± 0.06	2.59 ± 0.12	0.23 ± 0.02	0.37 ± 0.02
Acid soluble lignin	3.43 ± 0.06	2.58 ± 0.026	1.65 ± 0.011	1.46 ± 0.11
Acid insoluble lignin	12.66 ± 0.75	20.75 ± 0.43	19.22 ± 1.63	30.79 ± 2.51
Ash	5.07 ± 0.07	10.15 ± 0.10	10.06 ± 0.40	2.99 ± 0.02
Extractives	45.07 ± 1.49	35.77 ± 1.29	52.85 ± 0.72	26.29 ± 4.04
Aqueous	29.46 ± 0.34	23.08 ± 1.78	48.71 ± 0.54	8.39 ± 0.64
Ethanolic	15.61 ± 1.16	12.69 ± 0.51	4.14 ± 0.40	17.91 ± 3.48
Monomeric sugars [3]	8.79	2.86 ± 0.13	5.58 ± 0.04	1.54 ± 0.15
Oligomeric sugars [3]		8.46 ± 0.59	5.86 ± 0.68	10.27 ± 1.09
Glucose [3]	7.33 ± 0.17	7.61 ± 0.60	7.90 ± 0.36	1.31 ± 0.43
Galactose [3]	0.94 ± 0.03	1.36 ± 0.09	1.19 ± 0.07	1.93 ± 0.19
Mannose [3]	ND	0.03 ± 0.02	0.42 ± 0.03	0.46 ± 0.07
Xylose [3]	0.11 ± 0.01	0.31 ± 0.13	0.29 ± 0.26	4.97 ± 0.35
Arabinose [3]	0.41 ± 0.04	2.00 ± 0.44	1.64 ± 0.08	3.13 ± 0.34
Mannitol [3]	4.64 ± 0.23	2.63 ± 0.13	5.36 ± 0.07	0.64 ± 0.19
Fat	ND	9.85 ± 0.10	2.47 ± 0.34	8.72 ± 0.96
Acetyl groups	0.66 ± 0.01	0.77 ± 0.02	0.79 ± 0.59	1.97 ± 0.06
Elemental analysis				
N	1.49 ± 0.06	1.30 ± 0.06	1.50 ± 0.07	0.72 ± 0.05
C	48.38 ± 0.48	48.08 ± 0.66	49.65 ± 0.85	56.86 ± 1.24
H	6.48 ± 0.15	6.49 ± 0.08	6.23 ± 0.33	7.26 ± 0.40

[1] As glucose. [2] As hemicellulosic sugars. [3] With respect to aqueous extractives. [4] Results from previous studies [14,29]. ND, not determined; OL: olive leaves; OML: olive mill leaves; EOP: extracted olive pomace; RFOPC: residual fraction from olive pit cleaning.

3.2.2. Characterization of the Antioxidant Extracts by HPLC-MS Analyses
Phenolic Compounds

Two MS analyzers, an IT and a QTOF, were applied to get maximum information about the phenolic class and other phytochemicals present in the antioxidant extracts. Both have demonstrated multiclass potential to characterize phenolic compounds and other bioactives present in olive-derived biomasses [26,31]. The former enabled us to compare the RP-HPLC-MS profiles and some phenolic compounds with those characterized in our previous work on OL and OML [12,14]. The second one provided mass accurate measurements for structural confirmation and characterization of novel compounds [26,32]. Figure 2 depicts the MS profiles obtained by RP-HPLC-IT-MS of OL, OML, EOP, and RFOPC after UAE extraction. It shows that all samples show qualitative differences, especially those derived from olive pomace, which were also different between each other.

Table 2. Ratio of extracted volume (%), total extraction yield (%), total phenolic content (g of gallic acid equivalents/100 g), and antioxidant activity (mmol Trolox equivalents/100 g).

Biomass	Extracted Volume/Total Volume	Total Extraction Yield	Solubilization [1]				Extract Richness [2]			
			TPC [3]	TEAC [3]	FRAP [3]	ORAC [3]	TPC [3]	TEAC [3]	FRAP [3]	ORAC [3]
OL [4]	83.33 ± 1.44 c	24.00 ± 1.99 b	3.33 ± 0.42 b	11.03 ± 0.20 b	14.24 ± 0.78 b	72.69 ± 3.65 b	14.03 ± 2.99 a	46.18 ± 4.44 b	59.53 ± 4.48 a	272.77 ± 33.35 a
OML [4]	83.33 ± 0.00 c	15.77 ± 0.88 c	2.06 ± 0.06 c	11.94 ± 2.96 b	ND	ND	13.11 ± 0.68 a	74.72 ± 19.33 a	ND	ND
EOP	92.17 ± 0.00 a	37.44 ± 0.38 a	4.45 ± 0.09 a	25.25 ± 0.20 a	20.30 ± 1.28 a	95.30 ± 5.93 a	11.90 ± 0.30 a	67.43 ± 1.08 a	54.23 ± 3.62 a	254.44 ± 13.78 a
RFOPC	85.80 ± 0.25 b	6.70 ± 0.33 d	0.46 ± 0.02 d	2.61 ± 0.31 c	1.35 ± 0.16 c	11.50 ± 0.87 c	6.92 ± 0.33 b	39.14 ± 6.09 b	20.10 ± 1.83 b	171.78 ± 12.98 b

Different lowercase letters within a row indicate significant differences between the samples ($p < 0.05$). OL, olive leaves; OML, olive mill leaves; EOP, exhausted olive pomace; RFOPC, residual fraction from olive pits cleaning. ND, not determined. [1] Data expressed in terms of byproduct weight on dry basis. [2] Data expressed in terms of extract weight on dry basis. [3] EOP, extracted olive pomace; FRAP, ferric reducing antioxidant power assay; TEAC, Trolox equivalent antioxidant capacity; TPC, total phenolic content; OML, olive mill leaves; OL, olive leaves; ORAC, oxygen radical absorbance capacity; RFOPC, residual fraction from olive pit cleaning. [4] Results from previous studies [12,14].

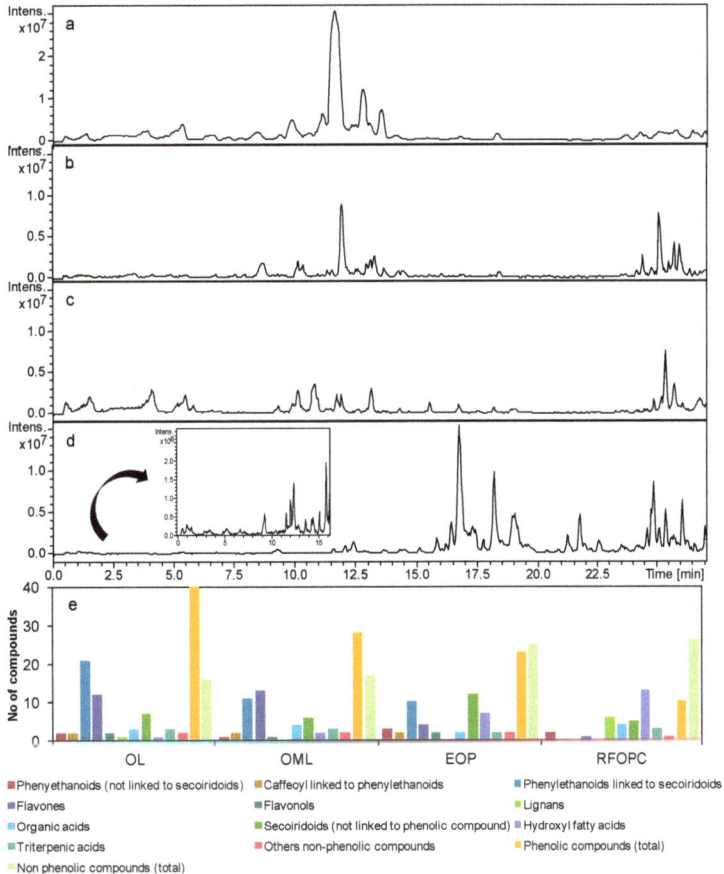

Figure 2. Base peak chromatograms of (**a**) olive leaves, (**b**) olive mill leaves, (**c**) exhausted olive pomace, and (**d**) residual fraction from olive pit cleaning obtained by RP-HPLC-IT-MS. (**e**) Number of compounds characterized per chemical class. OL: olive leaves; OML: olive mill leaves; EOP: extracted olive pomace; RFOPC: residual fraction from olive pit cleaning.

Then, the phenolic compounds were characterized by HPLC-MS and -MS/MS based on their accurate m/z value, molecular formula, and mass fragmentation patterns, which were compared to those in an in-source library and literature [12,26,33–35]. A total of 55 phenolic compounds were characterized in the four extract types and belonged to several phenolic classes, i.e., phenylethanoids (3), caffeoyl phenylethanoid derivatives (2), caffeoyl derivatives (2), phenyl ethanoids linked to secoiridoids (including oleuropein) (26), flavones (13), flavonols (2), and lignans derivatives (7) (Table 3). The number of compounds found in each extract type is shown in Figure 2e. OL was qualitatively richer in phenolic compounds than the rest of the samples, followed by OML. Not surprisingly, OML and OL shared most phenolic compounds since OML is composed of olive leaves and thin branches. Nonetheless, quercetin glucoside and some oleuropein derivatives were only present in OL; particularly, the novel derivatives 49, 50, 52–54. These compounds were characterized in our previous work [12], but here the QTOF analysis enabled us to establish their molecular formulae and confirm the presence of fragments related to oleuropein. On the basis of the MS information, it seems that they can be formed by the conjugation of oleuropein and hydroxycinnamic acids (compounds 46 and 49) and fatty acids (compounds 50, 52–54) (Table 3), but further nuclear magnetic resonance analysis is required for confirmation.

Table 3. Phenolic compounds characterized in olive leaves (OL), olive mill leaves (OML), exhausted olive pomace (EOP), and residual fraction from olive pit cleaning (RFOPC).

N°	RT [1] (Min)	Molecular Formula [2]	m/z [2,3]	Error (Ppm) [3]	Score [2]	MS/MS [1]	Compound	Class	OL	OML	EOP	RFOPC
1	1.2	$C_{14}H_{20}O_8$	315.11	−3.4	94	153, 135, 123	Hydroxytyrosol glucoside	Phenylethanoids (not linked to secoiridoids)	+	−	+	+
2	1.1	$C_8H_{10}O_3$	153.06	1.1	99	123	Hydroxytyrosol	Phenylethanoids (not linked to secoiridoids)	+	+	+	+
3	1.9	$C_{14}H_{20}O_7$	299.11	0.5	95	179, 161, 119, 101	Tyrosol glucoside	Phenylethanoids (not linked to secoiridoids)	−	−	+	−
4	7.3	$C_{27}H_{30}O_{15}$	593.15	−2.8	94	503, 473, 383, 353	Apigenin 6,8-di-C-glucoside	Flavones	+	+	−	−
5	8.1	$C_{27}H_{30}O_{16}$	609.15	−3.2	90	447, 285	Luteolin di-hexoside 1	Flavone	+	+	−	−
6	8.4	$C_{25}H_{32}O_{14}$	555.17	−3.5	91	537, 403, 323, 223	Hydroxyoleuropein 1	Phenylethanoids (linked to secoiridoids)	+	+	−	−
7	8.4	$C_{27}H_{30}O_{16}$	609.15	−2.7	94	447, 285	Luteolin di-hexoside 2	Flavone	+	+	−	−
8	8.5	$C_{24}H_{30}O_{13}$	525.16	−3.4	91	481, 389, 319, 195, 165	Demethyloleuropein	Phenylethanoids (linked to secoiridoids)	+	−	−	−
9	8.9	$C_{25}H_{32}O_{15}$	571.17	−3.5	92	523, 403, 359, 223, 179	Dihydroxyoleuropein	Phenylethanoids (linked to secoiridoids)	+	+	−	−
10	9.2	$C_{27}H_{30}O_{16}$	609.15	−3.5	91	447, 285	Luteolin di-hexoside 3	Flavones	+	+	−	−
11	9.3	$C_{25}H_{32}O_{14}$	555.17	−3.1	92	537, 403, 371, 323, 223	Hydroxyoleuropein 2	Phenylethanoids (linked to secoiridoids)	+	+	−	−
12	9.3	$C_{23}H_{32}O_{11}$	483.19	−3.5	93	347, 123	3,4-DHPEA-EDA derivative (+ hexose + H_2)	Phenylethanoids (linked to secoiridoids)	−	−	+	−
13	9.7	$C_{20}H_{32}O_{12}$	463.18	−3.0	95	347, 301	Quercetin glucoside	Flavonols	+	−	+	−
14	9.9	$C_{27}H_{30}O_{16}$	609.15	−3.2	92	447, 301, 179	Quercetin 3-O-rutinoside	Flavonols	+	+	+	−
15	9.9	$C_{29}H_{36}O_{15}$	623.20	−3.3	91	461, 315	Verbascoside	Caffeoyl phenylethanoid derivatives	+	+	+	−
16	10.0	$C_{21}H_{20}O_{11}$	447.10	−3.7	91	285	Luteolin 7-O-glucoside	Flavones	+	+	+	−
17	10.2	$C_{27}H_{30}O_{15}$	593.15	−3.6	90	285	Luteolin hexoside deoxyhexoside 1	Flavones	+	+	+	−
18	10.5	$C_{27}H_{30}O_{15}$	593.15	−3.6	92	447, 285	Luteolin hexoside deoxyhexoside 2	Flavones	+	+	+	−
19	10.6	$C_{31}H_{42}O_{18}$	701.23	−1.9	96	539, 377, 307, 275	Oleuropein hexoside 1	Phenylethanoids (linked to secoiridoids)	+	+	+	−
20	10.7	$C_{29}H_{36}O_{15}$	623.20	−3.7	90	461	Isoverbascoside	Caffeoyl phenylethanoid derivatives	+	+	+	−

Table 3. *Cont.*

N°	RT [1] (Min)	Molecular Formula [2]	m/z [2,3]	Error (Ppm) [3]	Score [2]	MS/MS [1]	Compound	Class	OL	OML	EOP	RFOPC
21	10.8	$C_{17}H_{20}O_7$	335.11	−4.0	93	317, 199, 153, 111	Hydroxyde (carboxymethyl) oleuropein aglycone	Phenylethanoids (linked to secoiridoids)	−	−	−	−
22	11.1	$C_{31}H_{42}O_{18}$	701.23	−2.4	93	539, 437, 377, 307, 275	Oleuropein hexoside 2	Phenylethanoids (linked to secoiridoids)	+	+	−	−
23	11.2	$C_{30}H_{34}O_{11}$	569.20	−2.6	95	551, 539, 393, 373, 177, 162	G(8-O-4)S(8-5)G [5] (−C)	Lignan derivatives	−	−	−	+
24	11.3	$C_{27}H_{30}O_{14}$	557.16	−3.7	90	269	Apigenin 7-O-rutinoside	Flavones	+	+	−	−
25	11.4	$C_{25}H_{32}O_{13}$	539.18	−4.6	85	403, 223	Oleouropein 1	Phenylethanoids (linked to secoiridoids)	−	+	+	−
26	11.5	$C_{21}H_{20}O_{11}$	447.09	−3.5	91	285	Luteolin 7-O-hexoside 1	Flavones	+	+	−	−
27	11.6	$C_{25}H_{32}O_{13}$	539.18	−3.1	92	403, 377, 307, 275, 223	Oleuropein	Phenylethanoids (linked to secoiridoids)	+	+	+	+
28	11.7	$C_{25}H_{28}O_{14}$	551.14	−3.6	92	507, 389, 341, 281, 251, 179, 161	Caffeoyl-6′-secologanoside	Hydroxycinnamics (linked to secoiridoids)	−	+	+	−
29	12.0	$C_{22}H_{22}O_{11}$	461.11	−4.2	90	446, 299, 284	Diosmetin 7-O-glucoside	Flavones	+	+	−	−
30	12.3	$C_{21}H_{20}O_{11}$	447.09	−3.5	92	285	Luteolin 7-O-hexoside isomer 2	Flavones	+	+	+	−
31	12.4	$C_{25}H_{32}O_{13}$	539.18	−2.9	92	377, 307, 275, 223	Oleouropein 2	Phenylethanoids (linked to secoiridoids)	+	+	+	−
32	12.7	$C_{25}H_{32}O_{13}$	539.18	−3.5	92	403, 377, 307, 275, 223	Oleouropein 3	Phenylethanoids (linked to secoiridoids)	+	+	+	−
33	13.0	$C_{27}H_{36}O_{14}$	583.20	−2.1	95	537, 403, 223, 179	Lucidumoside C	Phenylethanoids (linked to secoiridoids)	+	+	+	−
34	13.1	$C_{17}H_{20}O_6$	319.12	−3.6	94	183, 181, 153, 111	3,4-DHPEA-EDA	Phenylethanoids (linked to secoiridoids)	−	−	−	−
35	13.2	$C_{19}H_{22}O_8$	377.13	−4.3	91	307, 275	Oleuropein aglycone 1	Phenylethanoids (linked to secoiridoids)	+	−	+	−
36	13.2	$C_{25}H_{28}O_{13}$	535.15	−4.0	90	491, 389, 345, 265, 163	p-Coumaroyl-6′-secologanoside	Hydroxycinnamics (linked to secoiridoids)	−	−	+	+
37	13.5	$C_{25}H_{32}O_{12}$	523.18	−2.9	94	361, 291, 259, 223	Ligustroside	Phenylethanoids (linked to secoiridoids)	+	+	+	−
38	14.1	$C_{19}H_{22}O_8$	377.13	−4.0	91	307, 275	Oleuropein aglycone 2	Phenylethanoids (linked to secoiridoids)	+	−	−	−
39	14.4	$C_{21}H_{20}O_6$	367.12	−3.7	95	352, 337, 336, 322, 307, 177, 162	S(8-5)G [4] derivative 1 (−H_2O)	Lignan derivatives	−	−	−	+

Table 3. Cont.

N°	RT [1] (Min)	Molecular Formula [2]	m/z [2,3]	Error (Ppm) [3]	Score [2]	MS/MS [1]	Compound	Class	OL	OML	EOP	RFOPC
40	**14.6**	$C_{20}H_{18}O_5$	337.11	−3.9	93	322, 307, **291**, 177, 162	S(8-5)G [4] derivative 2 (−H_2O, −CH_2O)	Lignan derivatives	−	−	−	+
41	15.0	$C_{28}H_{34}O_{13}$	577.19	4.4	92	531, 415, 398, 285, 273, 239	(+)-1-Acetoxypinoresinol 4′-b-O-glucoside	Lignan derivatives	+	−	−	−
42	15.4	$C_{31}H_{28}O_{14}$	623.14	−2.2	94	323, 299, 285	Diosmetin di-hexoside	Flavones	+	+	−	−
43	15.4	$C_{42}H_{54}O_{23}$	925.30	0.9	97	539, 377, 307, 275	Jaspolyoside isomer 1	Phenylethanoids (linked to secoiridoids)	+	−	−	−
44	**15.8**	$C_{31}H_{36}O_{11}$	583.22	−3.2	93	565, 535, **413**, 387, 373, 357, 343, 195, 165	G(8-O-4)S(8-8)G [5]	Lignan derivatives	−	−	−	+
45	**16.2**	$C_{31}H_{34}O_{11}$	581.20	−2.9	93	563, 551, 533, 503, 385, 367, 355, 337, 336, 218, **195**, 177, 165	G(8-O-4)S(8-5)G [4] 1	Lignan derivatives	−	−	−	+
46	16.4	$C_{41}H_{50}O_{21}$	877.28	0.2	99	715, 701, 539, 377, 307, 275, 149	Oleuropein derivative 1 (oleuropein hexoside + $C_{10}H_8O_3$)	Phenylethanoids (linked to secoiridoids)	+	+	−	−
47	16.9	$C_{42}H_{54}O_{23}$	925.30	0.6	96	539, 377, 307, 275	Jaspolyoside isomer 2	Phenylethanoids (linked to secoiridoids)	+	−	−	−
48	**17.0**	$C_{31}H_{34}O_{11}$	581.20	−2.6	94	**551**, 367, 355, 337, 218, **195**, 165	G(8-O-4)S(8-5)G [4] 2	Lignan derivatives	−	−	−	+
49	17.0	$C_{34}H_{38}O_{15}$	685.21	−1.7	92	539, 377, 307, 275	Oleuropein derivative 2 (oleuropein + $C_9H_6O_2$)	Phenylethanoids (linked to secoiridoids)	+	−	−	−
50	18.3	$C_{41}H_{58}O_{20}$	869.34	0.6	95	829, 707, 539, 377, 325, 307, 275, 145	Oleuropein derivative 3 (oleuropein hexoside + $C_{10}H_{16}O_2$)	Phenylethanoids (linked to secoiridoids)	+	−	−	−
51	**18.7**	$C_{19}H_{22}O_7$	361.13	−1.6	98	329, 291, **225**, 193, 181	Hydroxytyrosol linked to desoxy elenolic acid	Phenylethanoids (linked to secoiridoids)	−	−	+	−
52	19.3	$C_{35}H_{48}O_{15}$	707.29	−0.5	97	539, 377, 307, 275	Oleuropein derivative 4 (oleuropein + $C_{10}H_{16}O_2$)	Phenylethanoids (linked to secoiridoids)	+	−	−	−
53	26.2	$C_{43}H_{60}O_{14}$	799.39	0.04	99	539, 377, 307, 277	Oleuropein derivative 6 (oleuropein + $C_{18}H_{28}O$)	Phenylethanoids (linked to secoiridoids)	+	−	−	−
54	26.8	$C_{41}H_{62}O_{14}$	777.41	0.23	98	539, 377, 307, 275	Oleuropein derivative 7 (oleuropein + $C_{16}H_{30}O$)	Phenylethanoids (linked to secoiridoids)	+	−	+	−
55	14.2	$C_{15}H_{10}O_6$	285.04	−2.12	98	175, 151	Luteolin	Flavones	−	−	+	−

+ Presence; − absence; bold letter indicates novel compounds. [1] By RP-HPLC-IT-MS. [2] By RP-HPLC-QTOF-MS. [3] [M−H]− ions. [4] G is coniferyl alcohol; S, sinapyl alcohol; G′ is coniferyl aldehyde. The first coniferyl alcohol is hydroxylated (+O). [5] G is coniferyl alcohol; S, sinapyl alcohol. The first coniferyl alcohol is hydroxylated (+O).

The EOP contained mainly hydroxytyrosol derivatives (Table 3), but the hydroxytyrosol cluster was different to that of OML and OL. It presented mainly free forms (hydroxytyrosol, hydroxytyrosol glucoside and tyrosol glucoside), hydroxytyrosol/tyrosol linked to secoiridoids, including oleuropein, ligustroside, and 3,4-DHPEA-EDA (dialdehydic form of elenolic acid linked to hydroxytyrosol), and to caffeic acid (verbascoside and isoverbascoside). Some of these compounds and the caffeoyl derivatives, caffeoyl- and p-coumaroyl-6'-secologanoside, have been found in olive pomace [35,36], and hence it remarks that they resist, at least in part, the processing of olive pomace to obtain pomace olive oil and EOP. Moreover, the hydroxytyrosol cluster included two novel hydroxytyrosol derivatives (compound 12 and 51). Compound 12 seems to be a glycosylated derivative of 3,4-DHPEA-EDA and compound 51 was formed by the linking between hydroxytyrosol and desoxy elenolic acid, whose molecular ion was found in the MS/MS spectrum and some product ions, e.g., m/z 181 derives from the loss of CO_2. This derivative of elenolic acid is present in olive oil [37].

RFOPC also contained some of the latter compounds, including hydroxytyrosol, but its phenolic profile was quite different. Interestingly, six novel phenolic compounds were found in RFOPC, i.e., trilignols (Figure 3). These compounds have previously been characterized in wild-type poplar (*Populus tremula* × *Populus tremuloides*) xylem as lignin oligomers [33] and in root exudates of *Arabidopsis thaliana* [38], but not in olive matrices. Basically, the phenolic compound at m/z 583 consists of coniferyl and sinapyl alcohol moieties linked by 8-8 linkage unit, where the syringyl moiety is linked to coniferyl alcohol via 8-O-4 linkage unit, which is hydroxylated in the position 7 (Figure 3a). Their fragmentation pattern is characterized by the presence of an ion at m/z 373, which is characteristic of X(8-8)X-containing trilignols [33]. Moreover, the fragmentation at the 8-O-4 linkage gave two main product ions, i.e., m/z 387 which correspond to the dilignol formed by coniferyl and sinapyl alcohols, and the counterpart m/z 195 (hydroxylated coniferyl alcohol). The other trilignols (m/z 581) is similar to the former, but it contains two fewer hydrogens in their structure. This indicated that the coniferyl alcohol was substituted by a coniferyl aldehyde moiety, which implies to connect via 8-5 linkage. This was characterized by the presence of the odd ion m/z 218 ($C_{12}H_{10}O_4^-$) that derive from the fragmentation of the phenylcoumaran (8-5) linkage. Furthermore, some potential degradation products could be detected at m/z 367 and m/z 337, sharing a similar structure to those fragments found at the same m/z values found in both IT- and QTOF-MS/MS spectra (Figure 3b). Both compounds showed the presence of an ion at m/z 177 with the molecular formula $C_{10}H_9O_3$, which correspond to coniferyl aldehyde, reaffirming their structures. Similarly, the compound at m/z 569 presented this fragment in MS/MS spectrum. Other fragments were at m/z 393 and 373, which could be hydroxylated coniferyl and sinapyl alcohols moieties (-C) and the latter linked to coniferyl aldehyde (-C). This indicates that the carbon loss occurs at the sinapyl moiety when compared to the compound at m/z 581.

Finally, the content of hydroxytyrosol and oleuropein was determined as a way to standardize and compare the extracts with those obtained in previous studies since both are reference olive phenolic compounds due to their biological properties, as commented in the introduction. The content of oleuropein was higher in OL followed by OML and EOP, both in terms of biomass and extract weight (Figure 4). Alternatively, EOP was the richest byproduct in hydroxytyrosol (0.8 g/100 g biomass weight and 2.1 g/100 g extract weight) followed by RFOPC. The values in EOP are higher than those reported in olive pomace [15,39] and other pomace byproducts [35,40] obtained by different technologies, including UAE and pressurized liquid extraction. Therefore, this work highlights that EOP is a source of hydroxytyrosol and its derivatives, which can be extracted before the use of EOP as a biofuel, for example.

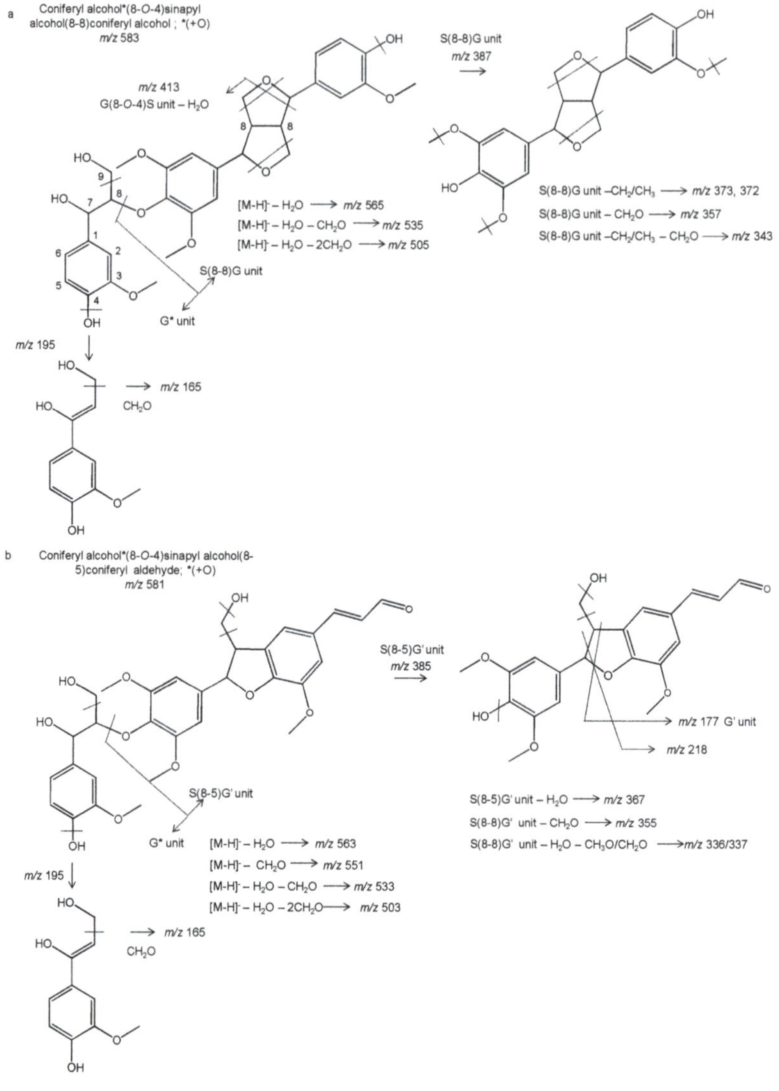

Figure 3. Tentative structure of novel free trilignols, (**a**) at m/z 583, (**b**) m/z 581 and derivatives characterized by mass spectrometry in olive byproducts.

Non-Phenolic Compounds

Table 4 shows the MS information of other compounds found in the extracts. In addition to the phenolic compounds, free secoiridoids, i.e., not linked to a phenolic moiety, were characterized including oleoside, secologanoside, elenolic acid hexosides and hydroxyelenolic acid, as well as six novel structures, which were proposed according to the MS information (Figure 5). Cyclic and acyclic forms are provided because it cannot be confirmed by MS what form could be or whether these forms coexist.

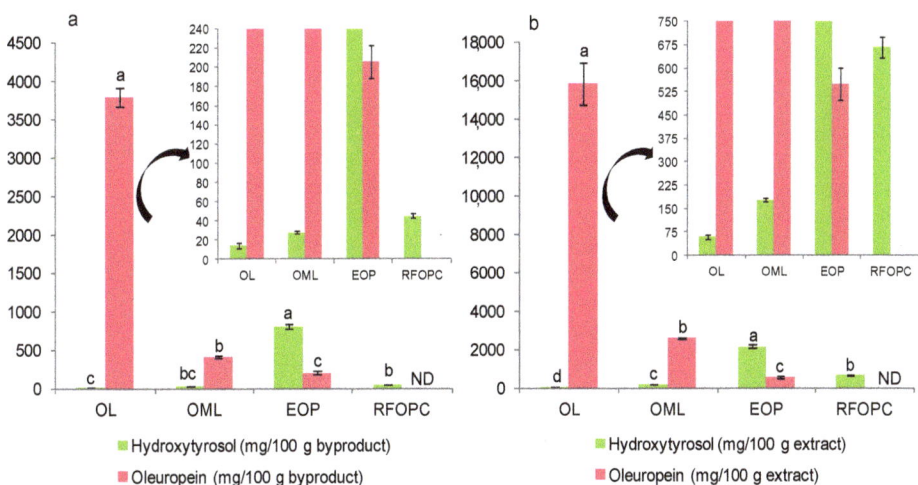

Figure 4. Content of hydroxytyrosol and oleuropein in olive byproducts in terms of: (**a**) Byproduct weight (d.w.) and (**b**) extract weight (d.w.). For each compound, different lowercase letters indicate significant differences between the samples ($p < 0.05$). ND, not detected.

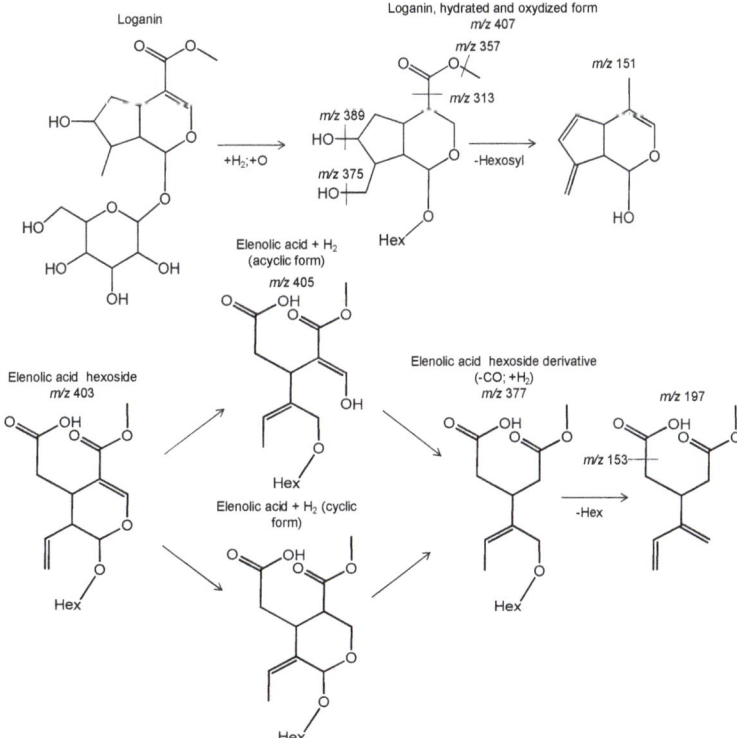

Figure 5. Tentative structure of novel free secoiridoids characterized by mass spectrometry in olive byproducts.

Table 4. Non-phenolic compounds characterized in the extracts from olive leaves (OL), olive mill leaves (OML), exhausted olive pomace (EOP), and residual fraction from olive pit cleaning (RFOPC) obtained by ultrasound-assisted extraction.

No	RT [1] (Min)	Molecular Formula [2]	m/z [2,3]	Error [3] (Ppm)	Score [3]	MS/MS Fragments [1]	Compound	OL	OML	EOP	RFOPC
1'	0.4	$C_6H_{14}O_6$	181.072	−0.2	99	163, 119	Mannitol	+	+	+	+
2'	0.5	$C_6H_8O_7$	191.020	−0.7	99	173, 111	Citric acid	+	+	+	+
3'	0.5	$C_7H_{12}O_6$	191.057	−2.1	99	—	Quinic acid	+	+	+	+
4'	0.5	$C_6H_{12}O_7$	195.052	−2.9	98	177, 129	Gluconic acid	+	+	−	+
5'	1.6	$C_{17}H_{28}O_{11}$	407.158	−4.5	90	389, 375, 357, 313, 151	Loganin derivative (+H_2; +O)	+	−	+	−
6'	3.7	$C_{12}H_{18}O_7S$	305.071	−3.9	94	225, 97	12−Hydroxyjasmonic acid sulfate	−	+	+	−
7'	2.5	$C_{16}H_{22}O_{11}$	389.110	−2.8	93	345, 302, 209, 165, 139, 121	Oleoside/Secologanoside	+	+	+	+
8'	3.3	$C_9H_{12}O_4$	183.066	0.8	99	139	Oleoside/secologanoside derivative (-glucosyl; −CO_2) or decarboxymethylelenolic acid	−	−	+	+
9'	3.9	$C_{16}H_{22}O_{11}$	389.111	−4	91	345, 302, 209, 187, 165, 139, 121	Oleoside/Secologanoside	−	+	+	−
10'	4.4	$C_{18}H_{28}O_{12}$	435.153	−4.1	90	389, 357, 313, 151	Loganin derivative (+H_2; +O; +CO)	−	+	+	−
11'	4.7	$C_{17}H_{24}O_{11}$	403.13	−4.3	90	371, 223, 179	Elenolic acid hexoside 1	−	+	−	−
12'	5.3	$C_{12}H_{20}O_7S$	307.09	−4.4	92	227, 165, 97	Dihydrohydroxyjasmonic acid sulfate derivative	−	+	+	+
13'	5.3	$C_{10}H_{14}O_4$	197.08	0.7	98	153	Elenolic acid hexoside derivative (+H_2; -CO; -Hexose)	−	−	−	+
14'	5.5	$C_{16}H_{26}O_{10}$	377.15	−4.4	90	197, 153	Elenolic acid hexoside derivative (+H_2; -CO)	+	+	+	+
15'	6.2	$C_{17}H_{26}O_{11}$	405.14	−4.7	90	373, 181	Elenolic acid hexoside derivative (+H_2)	−	+	+	−
16'	6.5	$C_{17}H_{24}O_{11}$	403.13	−4	92	371, 223, 179	Elenolic acid hexoside 2	−	+	−	−
17'	6.9	$C_{20}H_{34}O_{13}$	481.19	−3.9	90	371, 151	Unknown (elenolic acid hexoside derivative)	−	+	+	−
18'	7.5	$C_{11}H_{14}O_7$	257.07	−1	99	239, 225, 195, 137	Hydroxyelenolic acid	−	−	+	−
19'	7.8	$C_{17}H_{24}O_{11}$	403.13	−4.2	92	241, 223	Elenolic acid hexoside 3	+	−	+	−
20'	9.3	$C_9H_{16}O_4$	187.1	−1	99	125	Azelaic acid	−	+	−	+
21'	11.6	$C_{18}H_{34}O_6$	345.23	−2.5	96	201, 171	**Tetrahydroxyoctadecenoic acid isomer 1**	−	+	+	+
22'	12.1	$C_{18}H_{34}O_6$	345.23	−3.4	95	327, 309, 201, 171	**Tetrahydroxyoctadecenoic acid isomer 2**	−	−	+	+
23'	12.4	$C_{18}H_{34}O_6$	345.23	−4.2	91	327, 309, 201	**Tetrahydroxyoctadecenoic acid isomer 3**	−	−	+	+
24'	15.5	$C_{26}H_{38}O_{13}$	557.23	3.6	91	513, 345, 227, 185	6'-O-[(2E)-2,6-Dimethyl-8-hydroxy-2-octenoyloxy]-secologanoside	+	+	+	−
25'	15.8	$C_{18}H_{32}O_5$	327.22	−3.8	93	201, 171	Trihydroxyoctadecadienoic acid	−	−	−	+
26'	16.4	$C_{18}H_{32}O_6$	**343.21**	−3.8	93	325, 307, 245, 201	**Dihydroxyoctadecenedioic acid**	−	−	−	+
27'	16.7	$C_{18}H_{34}O_5$	329.23	−2.4	95	201	Trihydroxyoctadecenoic acid	−	−	+	+
28'	17.7	$C_{18}H_{34}O_6$	345.23	−4.2	91	327, 309, 265, 247	Dihydroxyoctadecanedioic acid	−	−	+	+
29'	18.1	$C_{18}H_{36}O_5$	331.25	−4	92	312.7	Trihydroxyoctadecanoic acid	−	+	+	+
30'	18.9	$C_{16}H_{32}O_4$	287.22	−3.3	95	269, 210	Dihydroxyhexadecanoic acid	+	+	+	+

Table 4. *Cont.*

No	RT [1] (Min)	Molecular Formula [2]	m/z [2,3]	Error [3] (Ppm)	Score [3]	MS/MS Fragments [1]	Compound	OL	OML	EOP	RFOPC
31'	21.7	$C_{30}H_{48}O_5$	487.34	−3.8	90	467	Hydroxylated derivative of maslinic acid	−	−	−	+
32'	21.2	$C_{20}H_{36}O_6$	371.25	−3.4	93	329, 311, 201	Trihydroxyoctadecenoic acid + C_2H_2O (acetyl)	−	−	+	+
33'	24.7	$C_{36}H_{64}O_9$	639.45	−2.9	93	329, 327, 201	Dimer formed by trihydroxyoctadecenoic acid and trihydroxyoctadecadienoic acid	−	−	+	+
34'	24.9	$C_{36}H_{66}O_9$	641.47	−2.3	93	329, 201	Di-trihydroxyoctadecenoic acid	−	−	+	+
35'	25.1	$C_{30}H_{48}O_4$	471.35	−3.4	94	453, 407	Pomolic acid	+	+	−	−
36'	25.3	$C_{34}H_{64}O_8$	599.45	−2.7	94	329, 287, 201	Dimer formed by trihydroxyoctadecenoic acid and dihydroxyhexadecanoic acid	−	−	+	+
37'	25.7	$C_{30}H_{48}O_4$	471.35	−3.3	93	423, 405	Maslinic acid	+	+	−	+
38'	26	$C_{18}H_{36}O_3$	299.26	−2.9	95	281, 253	Hydroxyoctadecanoic acid	−	−	−	+
39'	26.9	$C_{30}H_{48}O_3$	455.35	−3.4	93	407, 395	Oleanolic acid	+	+	+	+

+ Presence; − absence; Bold letter indicates novel compounds. [1] By RP-HPLC-IT-MS. [2] By RP-HPLC-QTOF-MS. [3] $[M-H]^-$ ions.

Other compounds detected were mannitol and some organic acids, a jasmonic acid derivative, hydroxyl fatty acids, and triterpenic acids (Table 4; Figure 2e). Their fragmentation patterns agreed well with previous studies [26,38,41–43].

Among them, hydroxy fatty acids, a type of oxylipins, are bioactive metabolites derived from the oxygenation of polyunsaturated fatty acids. They can be applied as starting materials for the synthesis of polymers and as additives for the manufacture of lubricants, emulsifiers, and stabilizers [44]. In plants, these compounds are formed after the release of free fatty acids from triglycerides due to the effect of lipolytic enzymes [42]. They forms part of cutin; a polymer of C16 and C18 fatty acids, with one or more hydroxy groups or epoxides, held together mainly by primary alcohol ester linkages [45]. Since cutin is present in fruit peels [46], it explains their presence in RFOPC, which is rich in olive skin [16,17]. In EOP, some of these compounds were also present. Moreover, although some of these compounds have been characterized in olive samples, the MS data revealed the presence of dimeric structures, which have not been reported before. Their m/z values correspond to the sum of the monomeric forms with a loss of water due to their linkage.

Regarding triterpenic acids, this class was composed of oleanolic, maslinic acid, pomolic acid, and hydroxylated maslinic acid. With the exception of the latter compound, oleanolic and maslinic acids have been reported before in OL, OML, and RFOPC [16,47]. Alternatively, hydroxylated maslinic acid has recently been characterized in olive oil and olive flour [37]. This is interesting since besides the presence of antioxidants, these compounds could give an extra value to the extracts due to their prominent bioactive properties, including cardioprotective properties [48].

3.3. Second Extraction Step Using Akaline Conditions: Evaluation of the Solubilization of Proteins, Sugars and Lignin

3.3.1. Solubilized Protein

Alkaline extraction can promote the solubilization of proteins, which can be purified for different applications in the food and other sectors [49,50]. Then, two conditions were tested to evaluate how the biomass type and alkaline extraction conditions affect the solubilization of proteins and the recovery values as a second valorization step: (i) 0.4 M NaOH, as solubilization agent, at 80 °C; and (ii) 0.7 M NaOH, as solubilization agent, at 100 °C [14,21]. The protein content varied from 3.7 g/100 g byproduct (RFOPC) to 6 g/100 g byproduct (EOP), i.e., a recovery between 49% (OML) and 100% (RFOPC) (Table 5). These recovery values for OL, OML and EOP are in the range of other agricultural and agro-industrial resources extracted using alkaline solutions [11,49], while RFOPC showed superior values as shown Zhang et al. [51] for tea leafy residue. Although EOP and RFOPC derive from the same source, olive pomace, the recovery values were different. Moreover, regardless of the applied conditions, the recovery was similar for EOP (around 60%), while for RFOPC the maximum value was achieved using strong alkaline-thermal conditions. The feedstock composition can affect the protein extractability as shown Sari et al. [52]. The latter authors suggested that cellulose and oil are the main constituents that hamper the extractability of proteins, but RFOPC has the highest value for both components. Hence, other factors can affect the separation of proteins, maybe, to a higher interaction with the lignocellulosic matrix, as commented on next.

Concerning the protein profile (Figure 6), all extracts contained protein bands around 100 kDa (band 3), 25 kDa (band 4) and 10 kDa (band 5), but the second one was very light in EOP. This band could be related to oleosins, which are alkaline proteins with molecular masses ranging between 15 and 26 kDa [53]. Moreover, lipoxygenase in olive fruit has a molecular weight around of 98 kDa [54]. Other band (band 6) with a molecular weight lower than bromophenol blue (0.67 kDa) was also observed, indicating the potential hydrolysis of proteins, as other studies on agri-food proteins suggested when using alkaline conditions for solubilization [11]. In fact, this and the 10 kDa band were also observed in OL after alkaline extraction [14,21]. They had a characteristic brown color. Among them, a protein/peptide around 10 kDa has also been detected in olive seeds [55]. Alternatively, two bands around 150 kDa and 250 kDa were detected in RFOPC when using strong alkaline conditions. Vioque et al. [56] suggested that olive pomace proteins can be highly denatured and/or associated with the fiber and other components and maybe the different processing that EOP and RFOCP are subjected to could affect these interactions or their solubilization in a different way, explaining their different solubility.

Table 5. Amount of protein solubilized (mg/100 g of byproduct, dry basis) and recovery (%, dry basis) obtained after the integration of antioxidants and alkaline extraction under mild (NaOH 0.4 M, 80 °C, 4 h) and strong alkaline conditions (NaOH 0.7 M, 100 °C, 4 h).

	Mild	Strong	Mild	Strong
Byproduct [1]	Solubilized Protein (g/L)	Solubilized Protein (g/L)	Recovery (%)	Recovery (%)
OML [2]	3.9 ± 0.1 [b]	5.1 ± 0.5 [a]	48.7 ± 1.5 [c]	63.1 ± 5.7 [b]
OL [2]	ND	5.2 ± 0.4	ND	55.5 ± 4.3 [b]
EOP	6.0 ± 0.3 [a]	5.4 ± 0.4 [a]	63.9 ± 2.3 [b]	58.2 ± 3.3 [b]
RFOPC	3.7 ± 0.3 [b]	4.5 ± 0.2 [a]	81.3 ± 5.8 [a]	100.1 ± 3.0 [a]

Different lowercase letters within a row indicate significant differences between the samples ($p < 0.05$). [1] EOP, extracted olive pomace; OL, olive leaves; OML, olive mill leaves; RFOP, residual fraction from olive pomace. [2] Data from [14,21]; ND, not determined.

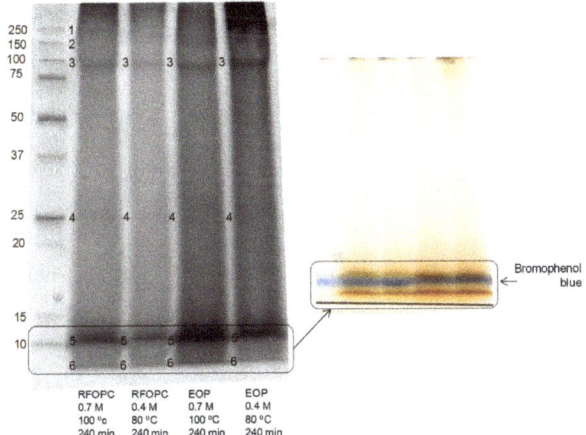

Figure 6. SDS-PAGE profiles of protein products obtained from the residual fraction from olive pit cleaning (RFOPC) (lanes 2 and 3) and extracted olive pomace (EOP) (lanes 4 and 5) by alkaline conditions after phenolic extraction. The protein markers kit was at lane 1.

3.3.2. Sugars, Sugar Alcohols, and Lignin

Alkaline treatments can also break the lignocellulosic matrix for further valorization of the polymeric sugars. It can promote the release of a part of the hemicellulosic sugars and lignin to obtain a solid fraction enriched in cellulose, which can be applied to obtain sugars and valuable sugars derivatives (bioethanol, organic acids, etc.) [11,57,58].

In this regard, glucose was absent in alkaline extracts from RFOCP, while its content was around 0.5 g/L in the EOP extracts. This suggests that the dissolution of the cellulose was negligible or poor under the conditions tested. Alternatively, the content of hemicellulosic sugars increased up to 2.7 g/L when stronger alkaline-thermal conditions were applied (Figure 7). These values were lower than those found in OML [21], indicating that the fractionation results depend on the biomass. Moreover, the released sugars were mainly in the form of dimers/oligomers (92–100%). Regarding sugar alcohols, a part of the mannitol in EOP was also extracted (around 0.7 g/L) (Figure 6), while other part is present in the antioxidant extracts, as commented before. The content was in the range of that found in OL after similar alkaline conditions [14], remarking again that EOP is a good source of this sugar alcohol.

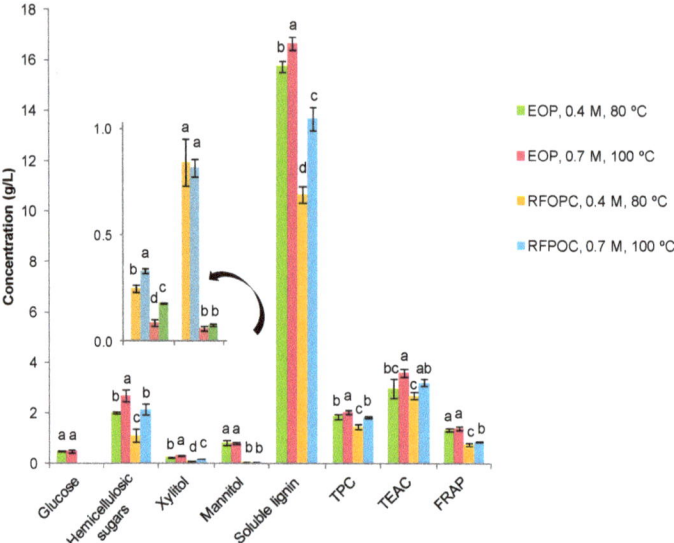

Figure 7. Content of sugar, sugar alcohols, and soluble lignin (g/L), total phenolic content (g gallic acid equivalents/L) and antioxidant activity (g Trolox equivalents/L) of the alkaline extracts obtained from the extracted olive pomace (EOP) and the residual fraction from olive pit cleaning (RFOPC). For each component, different lowercase letters indicate significant differences between the samples ($p < 0.05$).

The content of soluble lignin was also estimated since, as commented before, alkaline treatments can lead to a partial delignification [59]. Soluble lignin ranged from around 10.7 g/L to 16.7 g/L (Figure 7). In both cases, the major content was found in EOP extracts. Moreover, the UV spectra of the extracts were measured and show the main characteristic absorption bands for lignin, but the profiles presented some differences (Figure S1). This difference was more evidenced when applying the first derivative (Figure S1). In angiosperms, lignin is mainly composed of coniferyl alcohol and sinapyl alcohol-derived units in varying ratios that depend on the biomass type [60]. Thus, this is expected that the phenolic units are, at least in part, preserved and released after the alkaline treatment, being able to react with the Folin and Ciocalteu's phenol reagent. However, the biomass type and the treatment can modify the aromatic structures [61,62], explaining the differences found between the samples. In fact, the TPC values ranged from 1.47 g GAE/L to 2.06 g GAE/L, i.e., a solubilized amount ranging from 1.62 g GAE/100 g to 2.20 g GAE/100 g of byproduct (Figure 7), and it was correlated with the lignin content ($r = 0.960$). It also correlated with the antioxidant activities ($r > 0.784$) of these extracts. The highest values were observed for EOP. Other authors have shown that lignin has antioxidant properties and thereby it could serve as a natural source of antioxidants to replace synthetic ones [62]. Thus, this procedure is another way to obtain antioxidant extracts with solubilized lignin and particularly to valorize the EOP and the RFOPC residual fractions obtained after the first extraction step by UAE. The antioxidant activity was lower or in the range of the phenolic extracts obtained by UAE in the first valorization step, with an antioxidant activity ranging from 12.8 mmol TE/100 g to 17.2 mmol TE/100 g of byproduct in the TEAC assay and 3.6 mmol TE/100 g to 6.9 mmol TE/100 g of byproduct in the FRAP assay.

Overall, further studies are required to separate these components, mainly, protein and lignin, which can find application in different sectors, including the food and feed sectors.

Alternatively, the solid fraction with the rest of polymeric sugars, specially, glucose can be converted into biofuels or other building blocks derivatives within a biorefinery framework.

4. Conclusions

As a first step of valorization, UAE can be successfully applied to extract phenolic compounds from olive pomace-derived byproducts using aqueous-ethanol, particularly, EOP and RFOPC. The solubilization, extract richness and phenolic composition depended on the bioresource used. Among other phenolic compounds, EOP was a source of hydroxytyrosol and its derivatives: hydroxytyrosol glucoside, tyrosol glucoside, oleuropein, ligustroside, 3,4-DHPEA-EDA and its derivative, hydroxytyrosol linked to desoxy elenolic acid, verbascoside, and isoverbascoside. Alternatively, RFOPC resulted to be a source of novel trilignols. All of these extracts had good antioxidant properties compared to OL and OML extracts. Other compounds were present in the extracts, including free secoiridoids (i.e., not linked to phenolic compounds) and bioactive compounds such as triterpenic acids, while hydroxyl fatty acids were mainly present in RFOPC.

When the residual fraction obtained after UAE was subjected to an alkaline treatment for fractionation, the liquid fraction was rich in protein and it also contained soluble lignin, which conferred antioxidant properties to the extract. Due the selective release of protein and lignin in the liquid fraction, the recovery of a solid fraction rich in polymeric sugars is expected for further applications in biorefinery such as biofuel production. Overall, these results can be useful in a more sustainable olive sector promoting biorefinery approaches.

Supplementary Materials: The following are available online at https://www.mdpi.com/2304-8158/10/1/11/s1, Table S1: Pearson correlation values for the solubilization of phenolic compounds, Figure S1: UV-Vis spectra and their first derivative of alkaline extracts from the extracted olive pomace (EOP) (a and c, respectively) and the residual fraction from olive pit cleaning (RFOPC) (b and d, respectively).

Author Contributions: Conceptualization, M.d.M.C. and E.C.; methodology, M.d.M.C.; software, M.d.M.C.; formal analysis, M.d.M.C.; investigation, M.d.M.C.; writing—original draft preparation, M.d.M.C.; writing—review and editing, M.d.M.C., I.G.-C., and E.C.; supervision, M.d.M.C. and E.C.; project administration, M.d.M.C. and I.R.; funding acquisition, M.d.M.C., E.C., and I.R. All authors have read and agreed to the published version of the manuscript.

Funding: Authors thank Agencia Estatal de Investigación (MICINN, Spain) and Fondo Europeo de Desarrollo Regional, reference project ENE2017-85819-C2-1-R; FEDER UJA project 1260905 funded by "Programa Operativo FEDER 2014-2020" and "Consejería de Economía y Conocimiento de la Junta de Andalucía"; UJA grant R5/04/2017.

Institutional Review Board Statement: Not applicable.

Informed Consent Statement: Not applicable.

Data Availability Statement: The data presented in this study are available on request from the corresponding author. The data are not publicly available due to privacy concerns.

Acknowledgments: The technical and human support provided by CICT of the Universidad de Jaén is gratefully acknowledged.

Conflicts of Interest: The authors declare no conflict of interest.

Abbreviations

EOP	extracted (or exhausted) olive pomace
FRAP	ferric ion reducing antioxidant power
HPLC	high performance liquid chromatography
IT	ion trap
MS	mass spectrometry
OML	olive mill leaves
OL	olive leaves
ORAC	oxygen radical absorbance capacity
QTOF	quadrupole-time-of-flight
RFOPC	residual fraction from olive pit cleaning (or residual pulp)
RP	reversed phase
TEAC	Trolox equivalent antioxidant capacity
TPC	total phenol content
UAE	ultrasound-assisted extraction

References

1. Rahmanian, N.; Jafari, S.M.; Wani, T.A. Bioactive profile, dehydration, extraction and application of the bioactive components of olive leaves. *Trends Food Sci. Technol.* **2015**, *42*, 150–172. [CrossRef]
2. Talhaoui, N.; Taamalli, A.; Gómez-Caravaca, A.M.; Fernández-Gutiérrez, A.; Segura-Carretero, A. Phenolic compounds in olive leaves: Analytical determination, biotic and abiotic influence, and health benefits. *Food Res. Int.* **2015**, *77*, 92–108. [CrossRef]
3. EC Comission Implementing Regulation (EU) 2017/2470. *Off. J. Eur. Union* **2017**, *351*, 72–201.
4. Rodrigues, F.; Pimentel, F.B.; Oliveira, M.B.P.P. Olive by-products: Challenge application in cosmetic industry. *Ind. Crops Prod.* **2015**, *70*, 116–124. [CrossRef]
5. Carnevale, R.; Silvestri, R.; Loffredo, L.; Novo, M.; Cammisotto, V.; Castellani, V.; Bartimoccia, S.; Nocella, C.; Violi, F. Oleuropein, a component of extra virgin olive oil, lowers postprandial glycaemia in healthy subjects. *Br. J. Clin. Pharmacol.* **2018**, *84*, 1566–1574. [CrossRef] [PubMed]
6. Filip, R.; Possemiers, S.; Heyerick, A.; Pinheiro, I.; Raszewski, G.; Davicco, M.J.; Coxam, V. Twelve-month consumption of a polyphenol extract from olive (*Olea europaea*) in a double blind, randomized trial increases serum total osteocalcin levels and improves serum lipid profiles in postmenopausal women with osteopenia. *J. Nutr. Health Aging* **2015**, *19*, 77–86. [CrossRef]
7. Mosca, A.; Crudele, A.; Smeriglio, A.; Braghini, M.R.; Panera, N.; Comparcola, D.; Alterio, A.; Sartorelli, M.R.; Tozzi, G.; Raponi, M.; et al. Antioxidant activity of Hydroxytyrosol and Vitamin E reduces systemic inflammation in children with paediatric NAFLD. *Dig. Liver Dis.* **2020**, 1–5. [CrossRef] [PubMed]
8. Lopez-Huertas, E.; Fonolla, J. Hydroxytyrosol supplementation increases vitamin C levels in vivo. A human volunteer trial. *Redox Biol.* **2017**, *11*, 384–389. [CrossRef]
9. Bañares, C.; Martin, D.; Reglero, G.; Torres, C.F. Protective effect of hydroxytyrosol and rosemary extract in a comparative study of the oxidative stability of Echium oil. *Food Chem.* **2019**, *290*, 316–323. [CrossRef]
10. Hammouda, I.B.; Márquez-Ruiz, G.; Holgado, F.; Sonda, A.; Skalicka-Wozniak, K.; Bouaziz, M. RP-UHPLC-DAD-QTOF-MS as a powerful tool of oleuropein and ligstroside characterization in olive-leaf extract and their contribution to the improved performance of refined olive-pomace oil during heating. *J. Agric. Food Chem.* **2020**, *68*, 12039–12047. [CrossRef]
11. Contreras, M.d.M.; Romero, I.; Moya, M.; Castro, E. Olive-derived biomass as a renewable source of value-added products. *Process Biochem.* **2020**, *97*, 43–56. [CrossRef]
12. Medfai, W.; Contreras, M.d.M.; Lama-Muñoz, A.; Mhamdi, R.; Oueslati, I.; Castro, E. How cultivar and extraction conditions affect antioxidants type and extractability for olive leaves valorization. *ACS Sustain. Chem. Eng.* **2020**, *8*, 5107–5118. [CrossRef]
13. Talhaoui, N.; Gómez-Caravaca, A.M.; León, L.; De La Rosa, R.; Fernández-Gutiérrez, A.; Segura-Carretero, A. from olive fruits to olive oil: Phenolic compound transfer in six different olive cultivars grown under the same agronomical conditions. *Int. J. Mol. Sci.* **2016**, *17*, 337. [CrossRef] [PubMed]
14. de Contreras, M.M.; Lama-Muñoz, A.; Espínola, F.; Moya, M.; Romero, I.; Castro, E. Valorization of olive mill leaves through ultrasound-assisted extraction. *Food Chem.* **2020**, *314*, 126218. [CrossRef]
15. Pavez, I.; Lozano-Sánchez, J.; Borrás-Linares, I.; Nuñez, H.; Robert, P.; Segura-Carretero, A. Obtaining an extract rich in phenolic compounds from olive pomace by pressurized liquid extraction. *Molecules* **2019**, *24*, 3108. [CrossRef] [PubMed]
16. Romero, C.; Medina, E.; Mateo, M.A.; Brenes, M. New by-products rich in bioactive substances from the olive oil mill processing. *J. Sci. Food Agric.* **2018**, *98*, 225–230. [CrossRef]
17. Sánchez, J.M.; Pérez Jiménez, J.A.; Díaz Villanueva, M.J.; Serrano, A.; Núñez, N.; Giménez, J.L. New techniques developed to quantify the impurities of olive stone as solid biofuel. *Renew. Energy* **2015**, *78*, 566–572. [CrossRef]
18. Ruiz, E.; Romero-García, J.M.; Romero, I.; Manzanares, P.; Negro, M.J.; Castro, E. Olive-derived biomass as a source of energy and chemicals. *Biofuel Bioprod. Biorefin.* **2017**, *6*, 246–256. [CrossRef]
19. Martínez-Patiño, J.C.; Gómez-Cruz, I.; Romero, I.; Gullón, B.; Ruiz, E.; Brnčićc, M.; Castro, E. Ultrasound-assisted extraction as a first step in a biorefinery strategy for valorisation of extracted olive pomace. *Energies* **2019**, *12*, 2679. [CrossRef]

20. Gómez-Cruz, I.; Cara, C.; Romero, I.; Castro, E.; Gullón, B. Valorisation of exhausted olive pomace by an eco-friendly solvent extraction process of natural antioxidants. *Antioxidants* **2020**, *9*, 1010. [CrossRef]
21. Contreras, M.d.M.; Lama-Muñoz, A.; Gutiérrez-Pérez, J.M.; Espínola, F.; Moya, M.; Romero, I.; Castro, E. Integrated process for sequential extraction of bioactive phenolic compounds and proteins from mill and field olive leaves and effects on the lignocellulosic profile. *Foods* **2019**, *8*, 531. [CrossRef] [PubMed]
22. Lama-Muñoz, A.; Contreras, M.d.M.; Espínola, F.; Moya, M.; Romero, I.; Castro, E. Optimization of oleuropein and luteolin-7-O-glucoside extraction from olive leaves by ultrasound-assisted technology. *Energies* **2019**, *12*, 2486. [CrossRef]
23. Romero-García, J.M.; Lama-Muñoz, A.; Rodríguez-Gutiérrez, G.; Moya, M.; Ruiz, E.; Fernández-Bolaños, J.; Castro, E. Obtaining sugars and natural antioxidants from olive leaves by steam-explosion. *Food Chem.* **2016**, *210*, 457–465. [CrossRef]
24. Sluiter, A.; Hames, B.; Ruiz, R.; Scarlata, C.; Sluiter, J.; Templeton, D.; Nrel, D.C. *Determination of Structural Carbohydrates and Lignin in Biomass Determination of Structural Carbohydrates and Lignin in Biomass*; NREL/TP-510-42618; National Renewable Energy Laboratory: Golden, CO, USA, 2012.
25. Ammar, S.; Contreras, M.D.M.; Belguith-Hadrich, O.; Bouaziz, M.; Segura-Carretero, A. New insights into the qualitative phenolic profile of *Ficus carica L.* fruits and leaves from Tunisia using ultra-high-performance liquid chromatography coupled to quadrupole-time-of-flight mass spectrometry and their antioxidant activity. *RSC Adv.* **2015**, *5*, 20035–20050. [CrossRef]
26. Ammar, S.; Contreras, M.d.M.; Gargouri, B.; Segura-Carretero, A.; Bouaziz, M. RP-HPLC-DAD-ESI-QTOF-MS based metabolic profiling of the potential *Olea europaea* by-product "wood" and its comparison with leaf counterpart. *Phytochem. Anal.* **2017**, *28*, 217–229. [CrossRef]
27. Martínez-Patiño, J.C.; Ruiz, E.; Romero, I.; Cara, C.; López-Linares, J.C.; Castro, E. Combined acid/alkaline-peroxide pretreatment of olive tree biomass for bioethanol production. *Bioresour. Technol.* **2017**, *239*, 326–335. [CrossRef]
28. Guerra, A.; Ferraz, A.; Cotrim, A.R.; Da Silva, F.T. Polymerization of lignin fragments contained in a model effluent by polyphenoloxidases and horseradish peroxidase/hydrogen peroxide system. *Enzyme Microb. Technol.* **2000**, *26*, 315–323. [CrossRef]
29. Lama-Muñoz, A.; de Contreras, M.M.; Espínola, F.; Moya, M.; Romero, I.; Castro, E. Characterization of the lignocellulosic and sugars composition of different olive leaves cultivars. *Food Chem.* **2020**, *329*, 127153. [CrossRef]
30. Manzanares, P.; Ballesteros, I.; Negro, M.J.; González, A.; Oliva, J.M.; Ballesteros, M. Processing of extracted olive oil pomace residue by hydrothermal or dilute acid pretreatment and enzymatic hydrolysis in a biorefinery context. *Renew. Energy* **2020**, *145*, 1235–1245. [CrossRef]
31. Olmo-García, L.; Polari, J.J.; Li, X.; Bajoub, A.; Fernández-Gutiérrez, A.; Wang, S.C.; Carrasco-Pancorbo, A. Deep insight into the minor fractin of virgin olive oil by using LC-MS and GC-MS multi-class methodologies. *Food Chem.* **2018**, *261*, 184–193.
32. Contreras, M.d.M.; Algieri, F.; Rodriguez-Nogales, A.; Gálvez, J.; Segura-Carretero, A. Phytochemical profiling of anti-inflammatory Lavandula extracts via RP–HPLC–DAD–QTOF–MS and –MS/MS: Assessment of their qualitative and quantitative differences. *Electrophoresis* **2018**, *39*, 1284–1293. [CrossRef] [PubMed]
33. Morreel, K.; Dima, O.; Kim, H.; Lu, F.; Niculaes, C.; Vanholme, R.; Dauwe, R.; Goeminne, G.; Inzé, D.; Messens, E.; et al. Mass spectrometry-based sequencing of lignin oligomers. *Plant Physiol.* **2010**, *153*, 1464–1478. [CrossRef]
34. Olmo-García, L.; Kessler, N.; Neuweger, H.; Wendt, K.; Olmo-Peinado, J.M.; Fernández-Gutiérrez, A.; Baessmann, C.; Carrasco-Pancorbo, A. Unravelling the distribution of secondary metabolites in *Olea europaea L.*: Exhaustive characterization of eight olive-tree derived matrices by complementary platforms (LC-ESI/APCI-MS and GC-APCI-MS). *Molecules* **2018**, *23*, 2419. [CrossRef] [PubMed]
35. Ribeiro, T.B.; Oliveira, A.L.; Costa, C.; Nunes, J.; Vicente, A.A.; Pintado, M. Total and sustainable valorisation of olive pomace using a fractionation approach. *Appl. Sci.* **2020**, *10*, 6785. [CrossRef]
36. Peralbo-Molina, Á.; Priego-Capote, F.; Luque De Castro, M.D. Tentative identification of phenolic compounds in olive pomace extracts using liquid chromatography-tandem mass spectrometry with a quadrupole- quadrupole-time-of-flight mass detector. *J. Agric. Food Chem.* **2012**, *60*, 11542–11550. [CrossRef] [PubMed]
37. Olmo-García, L.; Monasterio, R.P.; Sánchez-Arévalo, C.M.; Fernández-Gutiérrez, A.; Olmo-Peinado, J.M.; Carrasco-Pancorbo, A. Characterization of New Olive Fruit Derived Products Obtained by Means of a Novel Processing Method Involving Stone Removal and Dehydration with Zero Waste Generation. *J. Agric. Food Chem.* **2019**, *67*, 9295–9306. [CrossRef]
38. Strehmel, N.; Böttcher, C.; Schmidt, S.; Scheel, D. Profiling of secondary metabolites in root exudates of Arabidopsis thaliana. *Phytochemistry* **2014**, *108*, 35–46. [CrossRef]
39. Nunes, M.A.; Costa, A.S.G.; Bessada, S.; Santos, J.; Puga, H.; Alves, R.C.; Freitas, V.; Oliveira, M.B.P.P. Olive pomace as a valuable source of bioactive compounds: A study regarding its lipid- and water-soluble components. *Sci. Total Environ.* **2018**, *644*, 229–236. [CrossRef]
40. Lozano-Sánchez, J.; Bendini, A.; Di Lecce, G.; Valli, E.; Gallina Toschi, T.; Segura-Carretero, A. Macro and micro functional components of a spreadable olive by-product (pâté) generated by new concept of two-phase decanter. *Eur. J. Lipid Sci. Technol.* **2017**, *119*, 1–9. [CrossRef]
41. Miersch, O.; Neumerkel, J.; Dippe, M.; Stenzel, I.; Wasternack, C. Hydroxylated jasmonates are commonly occurring metabolites of jasmonic acid and contribute to a partial switch-off in jasmonate signaling. *New Phytol.* **2008**, *177*, 114–127. [CrossRef]

42. Jiménez-Sánchez, C.; Lozano-Sánchez, J.; Rodríguez-Pérez, C.; Segura-Carretero, A.; Fernández-Gutiérrez, A. Comprehensive, untargeted, and qualitative RP-HPLC-ESI-QTOF/MS2 metabolite profiling of green asparagus (Asparagus officinalis). *J Food Compos. Anal.* **2016**, *46*, 78–87. [CrossRef]
43. Peragón, J.; Rufino-Palomares, E.E.; Muñoz-Espada, I.; Reyes-Zurita, F.J.; Lupiáñez, J.A. A new HPLC-MS method for measuring maslinic acid and oleanolic acid in HT29 and HepG2 human cancer cells. *Int. J. Mol. Sci.* **2015**, *16*, 21681–21694. [CrossRef]
44. Kyoung-Rok, K.; Deok-Kun, O. Production of hydroxy fatty acids by microbial fatty acid-hydroxylation enzymes. *Biotechnol. Adv.* **2013**, *31*, 1473–1485. [CrossRef]
45. Srivastava, L.M. Cell wall, cell division and cell growth. In *Molecular Cell Biology of the Growth and Differentiation of Plant Cells*; Srivastava, L.M., Ed.; Academic Press: San Diego, CA, USA, 2002; pp. 23–74.
46. Ladaniya, M.S. Fruit Biochemistry. In *Citrus Fruit: Biology, Technology and Evaluation*; Ladaniya, M., Ed.; Academic Press: San Diego, CA, USA, 2008; pp. 125–190.
47. Taamalli, A.; Lozano Sánchez, J.; Jebabli, H.; Trabelsi, N.; Abaza, L.; Segura Carretero, A.; Youl Cho, J.; Arráez Román, D. Monitoring the bioactive compounds status in *Olea europaea* according to collecting period and drying conditions. *Energies* **2019**, *12*, 947. [CrossRef]
48. de la Torre, R.; Carbó, M.; Pujadas, M.; Biel, S.; Mesa, M.D.; Covas, M.I.; Expósito, M.; Espejo, J.A.; Sanchez-Rodriguez, E.; Díaz-Pellicer, P.; et al. Pharmacokinetics of maslinic and oleanolic acids from olive oil–Effects on endothelial function in healthy adults. A randomized, controlled, dose–response study. *Food Chem.* **2020**, *322*, 126676. [CrossRef] [PubMed]
49. Sari, Y.W.; Mulder, W.J.; Sanders, J.P.M.; Bruins, M.E. Towards plant protein refinery: Review on protein extraction using alkali and potential enzymatic assistance. *J. Biotechnol.* **2015**, *10*, 1138–1157. [CrossRef]
50. Contreras, M.d.M.; Lama-Muñoz, A.; Manuel Gutiérrez-Pérez, J.; Espínola, F.; Moya, M.; Castro, E. Protein extraction from agri-food residues for integration in biorefinery: Potential techniques and current status. *Bioresour. Technol.* **2019**, *280*, 459–477. [CrossRef] [PubMed]
51. Zhang, C.; Sanders, J.P.M.; Bruins, M.E. Critical parameters in cost-effective alkaline extraction for high protein yield from leaves. *Biomass Bioenergy* **2014**, *67*, 466–472. [CrossRef]
52. Sari, Y.W.; Syafitri, U.; Sanders, J.P.M.; Bruins, M.E. How biomass composition determines protein extractability. *Ind. Crops Prod.* **2015**, *70*, 125–133. [CrossRef]
53. Montealegre, C.; Esteve, C.; García, M.C.; García-Ruiz, C.; Marina, M.L. Proteins in olive fruit and oil. *Crit. Rev. Food Sci. Nutr.* **2014**, *54*, 611–624. [CrossRef]
54. Lorenzi, V.; Maury, J.; Casanova, J.; Berti, L. Purification, product characterization and kinetic properties of lipoxygenase from olive fruit (*Olea europaea* L.). *Plant Physiol. Biochem.* **2006**, *44*, 450–454. [CrossRef] [PubMed]
55. Esteve, C.; D'Amato, A.; Marina, M.L.; García, M.C.; Citterio, A.; Righetti, P.G. Identification of olive (*Olea europaea*) seed and pulp proteins by nLC-MS/MS via combinatorial peptide ligand libraries. *J. Proteom.* **2012**, *75*, 2396–2403. [CrossRef] [PubMed]
56. Vioque, J.; Clemente, A.; Sáchez-Vioque, R.; Pedroche, J.; Millán, F. Effect of alcalase on olive pomace protein extraction. *J Am. Oil Chem. Soc.* **2000**, *77*, 181–185. [CrossRef]
57. McIntosh, S.; Vancov, T. Optimisation of dilute alkaline pretreatment for enzymatic saccharification of wheat straw. *Biomass Bioenergy* **2011**, *35*, 3094–3103. [CrossRef]
58. Rabetafika, H.N.; Bchir, B.; Blecker, C.; Paquot, M.; Wathelet, B. Comparative study of alkaline extraction process of hemicelluloses from pear pomace. *Biomass Bioenergy* **2014**, *61*, 254–264. [CrossRef]
59. Soares, M.L.; Gouveia, E.R. Influence of the alkaline delignification on the simultaneous saccharification and fermentation (SSF) of sugar cane bagasse. *Bioresour. Technol.* **2013**, *147*, 645–648. [CrossRef]
60. Erdocia, X.; Prado, R.; Corcuera, M.Á.; Labidi, J. Effect of different organosolv treatments on the structure and properties of olive tree pruning lignin. *J. Ind. Eng. Chem.* **2014**, *20*, 1103–1108. [CrossRef]
61. Lin, S.Y. Ultraviolet Spectrophotometry. In *Methods in Lignin Chemistry*; Lin, S.Y., Dence, C.W., Eds.; Springer: Berlin/Heidelberg, Germany, 2012; pp. 217–232.
62. García, A.; Spigno, G.; Labidi, J. Antioxidant and biocide behaviour of lignin fractions from apple tree pruning residues. *Ind. Crops Prod.* **2017**, *104*, 242–252. [CrossRef]

Article

Corn Bioethanol Side Streams: A Potential Sustainable Source of Fat-Soluble Bioactive Molecules for High-Value Applications

Gabriella Di Lena *, Jose Sanchez del Pulgar, Ginevra Lombardi Boccia, Irene Casini and Stefano Ferrari Nicoli

CREA Research Centre for Food and Nutrition, Via Ardeatina 546, 00178 Rome, Italy; jose.sanchezdelpulgar@crea.gov.it (J.S.d.P.); g.lombardiboccia@crea.gov.it (G.L.B.); irene.casini@crea.gov.it (I.C.); stefano.nicoli@crea.gov.it (S.F.N.)
* Correspondence: gabriella.dilena@crea.gov.it; Tel.: +39-06-51494445

Received: 19 October 2020; Accepted: 27 November 2020; Published: 2 December 2020

Abstract: This paper reports data from a characterization study conducted on the unsaponifiable lipid fraction of dry-grind corn bioethanol side streams. Phytosterols, squalene, tocopherols, tocotrienols, and carotenoids were quantified by High Performance Liquid Chromatography with Diode-Array Detector (HPLC-DAD) and Liquid Chromatography-tandem Mass Spectrometry (LC-MS/MS) in different lots of post-fermentation corn oil and thin stillage collected from a bioethanol plant over a time-span of one year. Fat-soluble bioactives were present at high levels in corn oil, with a prevalence of plant sterols over tocols and squalene. Beta-sitosterol and sitostanol accounted altogether for more than 60% of total sterols. The carotenoid profile was that typical of corn, with lutein and zeaxanthin as the prevalent molecules. The unsaponifiable lipid fraction profile of thin stillage was qualitatively similar to that of post-fermentation corn oil but, in quantitative terms, the amounts of valuable biomolecules were much lower because of the very high dilution of this side stream. Results indicate that post-fermentation corn oil is a promising and sustainable source of health-promoting bioactive molecules. The concomitant presence of a variegate complex of bioactive molecules with high antioxidant potentialities and their potential multifaceted market applications as functional ingredients for food, nutraceutical, and cosmeceutical formulations, make the perspective of their recovery a promising strategy to create new bio-based value chains and maximize the sustainability of corn dry-grind bioethanol biorefineries.

Keywords: bioethanol co-products; post-fermentation corn oil; distiller's corn oil; thin stillage; by-products; valorization; bioactive molecules; phytosterols; squalene; tocopherols; tocotrienols; tocols; carotenoids

1. Introduction

Ensuring the access to affordable, reliable, and sustainable energy is one of the Sustainable Development Goals of the 2030 Agenda for Sustainable Development adopted by the United Nations General Assembly in 2015. Biofuels are sustainable and renewable alternatives to fossil fuels, with the advantage of lower carbon and greenhouse gas (GHG) emissions. Global fuel ethanol production reached 115 billion L in 2019, with United States and Brazil, accounting altogether for over 80% of the world production, as the top producers [1]. In Europe, bioethanol, after biodiesel, is the second contributor of renewable energy sources to the transport sector, with a production capacity that in 2019 reached 9.9 billion L, corresponding to over 72% GHG savings [2].

To increase the competitiveness and sustainability of biofuels with respect to fossil fuels, a significant reduction of their production costs is necessary. A promising approach is the valorization of co-products and side streams arising from biofuels production and the maximization of the biomass-to-products value chains by creating biofuel-driven biorefineries.

Ethanol biorefineries are essential drivers of the energy transition, as they convert biomass into low-carbon fuels and also into a range of other valuable low-carbon co-products that may be conveyed to new value chains. Bioethanol production is mainly provided by first-generation biorefineries, with corn as the major feedstock [3]. The recovery of valuable compounds from corn bio-ethanol co-products and their application as ingredients for high-value market products, may open opportunities for the creation of new bio-based value chains. This approach is in line with the principles of circular economy, aiming at ensuring high quality, functional, and safe products to all, while reducing carbon and environmental footprints [4,5]. Although the present work has been conducted in the framework of the European Union's Horizon 2020-BBI-JU Program (Project EXConsEED, GA n. 792054), the approach proposed has a global interest, with a high potential impact for the big ethanol producers outside Europe, especially United States, a key player in corn bioethanol production.

In the food and feed industry sectors the development of bio-based products is driven by increasing business and consumers' demand for healthy, green, and sustainable products. The global demand of proteins and health-promoting bioactive substances for the food, feed, and nutraceutical market is steadily increasing. With the actual prospects of a growing world population and the increasing needs of proteins and highly nutrient food within the next decades, largely surpassing the current production capabilities and natural resources, the individuation of new natural and sustainable sources of nutrients and functional ingredients for the food and feed sectors represents a challenge.

Agrifood residues and wastes are potential sustainable sources of proteins and bioactive molecules. Research conducted in the past decade has shown that plant-derived wastes and by-products may be exploited for the development of functional food products [6–9].

First-generation biofuel biorefineries may be a starting point for new bio-based value chains. Their by-products and side streams still retain a series of valuable bioactive compounds that, if properly recovered and valorized, may open new perspectives for integrated biorefinery systems with the concomitant result of increasing the competitiveness and maximizing the efficiency of the biofuel production process.

Corn, the most common feedstock for bioethanol production in Europe, is inherently rich of bioactive phytochemicals, i.e., plant sterols, tocopherols, tocotrienols, phenolic compounds, and carotenoids [10]. These bioactives withstand the industrial processes of fuel ethanol production and may be found in the side streams together with yeast residues and metabolites resulting from the fermentation process. The dry-grind corn bioethanol production process gives two side streams, post-fermentation corn oil and thin stillage, currently not fully valorized. Post-fermentation corn oil, obtained by centrifugation of corn syrup, may be utilized for bio-diesel production. However, its full potentialities are actually unexploited since molecules other than fatty acids, reported to hinder the efficiency of biodiesel production, remain unused, while could first be recovered and conveyed to high-end applications.

Thin stillage, obtained after centrifugation of thick/whole stillage, is a highly diluted stream containing suspended particles and dissolved nutrients that originate from spent corn grains and yeast cells in bioethanol biorefineries. Currently, its main use in the feed sector mixed to distiller's dried grains with solubles (DDGS) requires high energy-demanding evaporation steps that reduce the economic sustainability of the whole biotech process.

With a circular economy approach the bio-molecules present in corn ethanol side streams, phytosterols, phenolics, tocols, squalene, and carotenoids, retaining health protective properties (i.e., antioxidant, anti-inflammatory and anti-aging) could be recovered and valorized by their reintroduction in productive processes as functional ingredients of high-value products (food, nutraceutics, cosmetics). The first step to succeed in this objective is to gather detailed information of the chemical composition of the side streams. In a previous paper we have described the proximate composition, mineral content, and fatty acid profile of bioethanol post-fermentation corn oil and thin stillage [11].

In this study we report the results of a characterization study focused on the unsaponifiable lipid fraction of the two side-streams. Plant sterols, tocopherols, tocotrienols, squalene, and carotenoids have been analyzed in different lots of post-fermentation corn oil and thin stillage, collected over a 1-year period at a dry-grind corn bioethanol plant.

2. Materials and Methods

2.1. Collection of Side Streams

Post-fermentation corn oil and thin stillage were obtained from the industrial dry-grind corn bioethanol plant ENVIRAL a.s. (Leopoldov, Slovack Republic). The original feedstock was a yellow non-genetically modified corn (*Zea mays*) grown in the Central East Europe region. Post-fermentation corn oil and thin stillage were sampled from the dry-grind corn ethanol facility as described previously [11].

In the period June 2018–September 2019, approximately at monthly intervals, 1 L corn oil and 3 L thin stillage were delivered (24–48 h from collection) to CREA-Research Centre for Food and Nutrition (Rome, Italy) for analyses. The first samples (lot 1) were obtained from the 2017 corn harvest. The new harvest season started in September 2018 and by October 2018 (lot 2) the newly harvested corn was utilized for the bioethanol production process. Up to eleven lots of corn oil and seven lots of thin stillage have been analyzed. The sampling date at ENVIRAL's plants of the different side stream lots analyzed in the study are the following: Lot 1 = 29 June 2018, Lot 2 = 22 October 2018, Lot 3 = 26 November 2018, Lot 4 = 9 December 2018, Lot 5 = 18 January 2019, Lot 6 = 25 February 2019, Lot 7 = 22 April 2019, Lot 8 = 23 May 2019, Lot 9 = 21 June 2019, Lot 10 = 29 July 2019, Lot 11 = 26 September 2019.

2.2. Chemicals

Pure standards of tocopherols, ergosterol, stigmasterol, campesterol, β-sitosterol, squalene, β–carotene, lutein, zeaxanthin, and β–cryptoxanthin were purchased from Sigma-Aldrich Inc. (St. Louis, MO, USA). Brassicasterol and sitostanol were from AVANTI Polar Lipids Inc. (Alabaster, AL, USA). Tocotrienols were purchased from Cayman Chemical Company (Ann Arbor, MI, USA). All solvents were of analytical or high performance liquid chromatography (HPLC) grade as required. Potassium hydroxide, and butylated hydroxytoluene (BHT) were from Carlo Erba. Tert-butyl-hydroquinone (TBHQ) was from Fluka Chemie AG (Buchs, Switzerland). Deionized water was provided by an Arium® pro UV Water Purification System (Sartorius Stedim Biotech GmbH, Goettingen, Germany).

2.3. Sample Treatment

Upon arrival at CREA laboratories, corn oil and thin stillage samples were immediately refrigerated (+4 °C). Corn oil was preserved from light and heat and analyzed without any pre-treatment. Before sampling for analyses, the oil was brought to room temperature and gently shaken in order to re-suspend any solid material sedimented at the bottom of the bottle. Thin stillage was subject to lyophilization before analyses. Total lipids were extracted from the freeze-dried thin stillage (about 5 g) with methanol, chloroform, and water according to the method of Bligh and Dyer [12]. Lipid extracts were analyzed for tocopherols, tocotrienols, plant sterols, and squalene. A specific protocol for extraction of carotenoids from thin stillage was applied as follows. Freeze-dried thin stillage (about 500 mg) was extracted with acetone: methanol (70:30 v/v containing BHT (500 mg L^{-1}) in screw-capped test tubes and allowed to stand 30 min at +3 °C. The mixture was vortexed for 10 s at a high speed (Reax 2000 Vortex mixer, Heidolph, Schwabach, Germany) and centrifuged at 1400× g, 5 °C for 15 min and the supernatant was collected. The procedure was repeated until the supernatant and residue were colorless. Extracts were combined, evaporated under vacuum (R-210 Rotavapor™, Bűchi, Switzerland) at 30 °C and brought to a known volume of acetone: methanol (70:30 v/v).

2.4. Chemical Analyses

2.4.1. Analytical Procedures

Analytes in corn oil and extracts of thin stillage were separated and quantified by HPLC before and after saponification to account for the presence of free compounds and of the total amounts released after ester hydrolysis. For analyses of free compounds, direct analyses of diluted corn oil and lipid extracts of thin stillage were accomplished. An aliquot of corn oil or lipid extract was diluted in a known volume of methyl tert-butyl ether (MTBE)/methanol (1:1, v/v) to be immediately filtered through 0.2-µm syringe filters (Minisart RC4, Sartorius Stedim Biotech) and injected (20 µL) into the HPLC. The total amount of each component, including also the bound fractions released after ester hydrolysis, was determined after saponification. For tocopherols, tocotrienols, plant sterols, and squalene evaluation, saponification in ethanolic potassium hydroxide (10% w/v) in the presence of TBHQ dissolved in methanol (1% w/v) occurred at 70 °C for a total of 15 min. For the determination of carotenoid contents, saponification was conducted overnight at room temperature [13]. Saponification occurred in screw-capped ambered test tubes under a nitrogen atmosphere. Unsaponifiables were recovered with n-hexane/ethyl acetate (9:1 v/v), evaporated to dryness with the aid of a nitrogen stream and dissolved in a suitable amount of MTBE/methanol (1:1, v/v), to be filtered through 0.2-µm syringe filters before chromatographic injection. All steps were conducted avoiding any direct exposure to light.

2.4.2. High-Performance-Liquid-Chromatography (HPLC)

Chromatographic analyses were carried out on a 1100 Series Agilent HPLC System (Agilent Technologies, Santa Clara, CA, USA) equipped with a quaternary pump, solvent degasser, column thermostat, and photodiode-array (DAD) detector. Tocopherols, tocotrienols, plant sterols, and squalene were determined simultaneously on a reversed-phase Ultrasphere C-18 column (25 cm × 4.6 mm inner diameter, 5 µm, Beckman, Palo Alto, CA., USA) coupled with a C18 guard column (15 cm × 4.6 mm, 5 µm). The mobile phase consisted of acetonitrile/methanol (50:50, v/v) in isocratic conditions at a flow rate of 1.5 mL min^{-1}. Runs were monitored at 215 nm and 282 nm and thermostated at 25 °C. Baseline separation of analytes was accomplished except the pairs β-tocopherol/γ-tocopherol, β-tocotrienol/γ-tocotrienol, campesterol/stigmasterol that coeluted.

Separation of carotenoids occurred on a reversed-phase C-30 column (25 cm × 4.6 mm inner diameter, 5 µm) coupled with a C30 guard cartridge (10 mm, 4 mm, particle size 5 µm), both from YMC Co., Ltd. (Basel, Switzerland). The mobile phase consisted of methanol (eluent A), MTBE (eluent B), and water (eluent C). The gradient program was as follows: time 0: 81% A—15% B—4% C, time 90 min: 7% A—90% B—3% C. Flow rate was 0.7 mL min^{-1} and the column temperature was kept constant at 25 °C. Injection volume was 20 µL. Carotenoids were integrated at 450 nm. Chromatograms were also registered at 325 nm to monitor the elution of steryl ferulate esters in direct extracts. Ultraviolet-visible spectra were recorded over the range 250–680 nm in steps of 2 nm.

Analytes were identified by comparing retention times and UV–Vis absorption spectra to those of authentic standards. Peak areas were used to determine the analyte concentrations in the samples by reference to standard curves obtained by chromatographing pure substances under identical conditions. Data were analyzed with the Agilent ChemStation Software. The analyte contents are expressed as mg per kg of product.

2.4.3. Liquid-Chromatography-Tandem Mass Spectrometry (LC-MS/MS)

Untreated corn oil and thin stillage extracts were also analyzed in the LC-MS/MS system in order to improve the identification of the detected compounds. Analyses were performed on an Agilent

1200 quaternary pump coupled to a 6410 series triple quadrupole. The ion source was an APCI operated in positive mode. Chromatographic separation was conducted on an ACME C18-120A, 100 mm × 2.1 mm column, with 3 μm particle size. Mobile phase was acetonitrile/methanol (50:50, v/v) in isocratic conditions, at a flow rate of 300 μL min^{-1}. The APCI ionization parameters were as follows: Gas temperature 350 °C, vaporizer 375 °C, gas flow 6 L min^{-1}, nebulizer 60 psi, capillary voltage 3000 V and corona current 8 μA. Because of a the higher sensibility of the MS/MS detector regarding to the DAD, the concentration of sitostanol in the samples was measured in this system. For doing it, the triple quadrupole analyzer was operated in the MRM mode, with precursor ion 399.7 (corresponding to the molecular weight with the loss of a water molecule and an H$^+$ gain) and the following product ions: 95.1, used for quantification, 135.1, 109.1, and 81.1. In all cases the collision energy was 29 V, with the exception of transition 399.1 → 81.1, where it was set at 45 V. Quantification was performed by interpolation on a calibration curve constructed with pure sitostanol standard analyzed in the same conditions, in the range of 0.04 to 1.0 μg mL^{-1}.

2.5. Quality Assurance

The concentrations of stock solutions of pure standards were determined spectrophotometrically using their specific absorption coefficients. External linear calibration curves of analytical standards, with a minimum of five concentration levels, were constructed for each analyte. The DAD response for each analyte was linear within the calibration ranges with correlation coefficients exceeding 0.998. Repeatability was estimated by calculating the coefficient of variation (CV) after repeated runs of a standard solution containing each compound at the level found in samples. After HPLC runs, the purity of analytes was checked by matching the UV/Vis spectra of each peak with those of the standards. The standard reference material NIST 3278 (Tocopherols in edible oils) was analyzed for validation of the method and quality control of tocopherol data.

2.6. Data Treatment

For each parameter, analyses of single lots of post-fermentation corn oil and thin stillage were performed at least in duplicate. Data for single lots, mean, standard deviation, coefficient of variation (CV), and range of values detected during the experimental period were calculated with Microsoft Excel software, 2013 version.

3. Results and Discussion

3.1. Phytosterols and Squalene

The sterol profile of post-fermentation corn oil sampled monthly from July 2018 to September 2019 and the mean and standard deviation of all values detected are reported in Table 1.

Beta-sitosterol, sitostanol, campesterol + stigmasterol and ergosterol were the sterols identified, based on the comparison with pure standards for retention time and UV spectra characteristics. Peak identity was confirmed in the LC-MS/MS system by comparing the retention time and ionization and fragmentation pattern with those of pure standards. A sterol eluting in correspondence of the retention time of brassicasterol was tentatively identified as δ-5-avenasterol based on LC-MS/MS data, UV spectrum, and literature indications [14,15]. Other minor peaks with UV spectrum and MS/MS ionization/fragmentation characteristics compatible with phytosterols were detected.

Analyses carried out before and after oil saponification allowed the quantification of free and total sterols released after hydrolysis of esters with fatty acids or phenolic acids. The sum of the identified free sterols in direct analysis of corn oil ranged from about 5700 to 8383 mg kg^{-1} in the lots examined. These values more than doubled after saponification (15,832–17,912 mg kg^{-1}), indicating that at least 50% of the sterols was in bound form.

Table 1. Free and total amounts of phytosterols and squalene in post-fermentation corn oil from a dry-grind corn ethanol plant. Data refer to individual lots collected at monthly intervals from July 2018 to September 2019 and overall mean, standard deviation (sd), and range of values observed ($n = 11$) *.

	Post-Fermentation Corn Oil													
Lot	ERG		AVN [a]		STG + CAMP		β-SITO		STN		Σ STEROLS		SQUA	
	mg kg^{-1} Corn Oil													
	Free	Total	Free	Total	Free	Total	Free	Total	Free	Total	Free	Total	Free	Total
1	456	461	2081	2796	1178	2193	3073	7253	598	-	7387	-	867	832
2	413	433	1578	2636	1171	2616	3277	7652	529	4574	6968	17,911	875	843
3	408	411	1662	2248	1052	2186	3100	6742	540	-	6762	-	874	837
4	412	408	1618	2708	1113	2418	3289	7418	540	4277	6972	17,229	899	817
5	253	259	986	2131	1156	2358	4172	7238	569	-	7135	-	947	940
6	288	282	1412	2358	1149	2444	3528	7137	695	4638	7071	16,860	882	831
7	381	402	1428	2362	1034	2330	4180	7460	677	4829	7700	17,382	744	745
8	275	297	982	2066	983	2278	2834	6973	625	4379	5700	15,993	817	800
9	320	336	1353	2202	1081	2430	2803	7027	739	4472	6296	16,467	788	764
10	340	371	1107	2382	1022	2436	5250	7509	663	3134	8383	15,832	767	764
11	314	322	1151	2343	1017	2041	3449	6931	-	-	-	-	803	776
Mean	351	362	1396	2385	1087	2339	3541	7213	617	4329	7037	16,811	842	814
sd	66.7	66.6	333	236	69.6	158	730	280	73.8	556	731	759	62.4	54.2
min	253	259	982	2066	983	2041	2803	6742	529	3134	5700	15,832	744	745
max	456	461	2081	2796	1178	2616	5250	7652	739	4829	8383	17,912	947	940

* Details on the origin of corn feedstock and on lot timings are provided in Materials and Methods. Data for each lot represent mean of duplicate measurements. [a] tentative identification (AVN), quantified as brassicasterol-equivalent; - not available. ERG ergosterol, AVN Δ5-avenasterol, STG + CAMP stigmasterol + campesterol, β-SITO β-sitosterol, STN sitostanol, SQUA squalene.

Beta-sitosterol was the most abundant sterol in corn oil, with values as high as 6742–7652 mg kg^{-1} after saponification (corresponding to 42.3–47.4% of total sterols). Sitostanol, the saturated equivalent of sitosterol, was detected at very low amounts in free form (529–739 mg kg^{-1}) while after saponification resulted to be the second most abundant sterol (3134–4829 mg kg^{-1}, correspondent to 19.8–27.8% of total sterols). This is an indication that sitostanol is mostly present in post-fermentation corn oil in esterified form, in line with the literature on the sterol profile of corn, indicating stanols as highly present in endosperm and bran mainly as ferulate esters [14,16–18]. Campesterol + stigmasterol (2041–2616 mg kg^{-1}, 13.4–15.4% of total sterols) and the sterol tentatively identified as δ-5-avenasterol (2066–2796 mg kg^{-1}, 12.9–15.7% of total sterols) were present in significant amounts in the saponified extract, about half of which in free form. Ergosterol, not inherently present in corn, but essential component of yeast cells membrane, was presumably found in corn oil as a result of the fermentation process. Its levels, ranging from 259 to 461 mg kg^{-1} (corresponding to 1.7–2.4% of total sterols) were not affected by saponification, meaning that this sterol is present mostly in free form. Squalene was present in corn oil at concentrations corresponding to 745–940 mg kg^{-1}.

Results obtained on post-fermentation corn oil show that plant sterols are present at considerable levels in this side stream. Compared to a commercial corn oil, one of the richest sources of phytosterols among vegetable oils, with as high as 0.7–0.8% w/w content of phytosterols [19–22], post-fermentation corn oil showed much higher sterol levels. This is due to the fact that while a commercial corn oil originates from the germ fraction, post-fermentation corn oil derives from the whole kernel and therefore retains the whole set of phytosterols, phytostanols, and their ferulate esters highly present in the aleurone, pericarp, and endosperm fractions. Moreover, the ethanol produced during fermentation acts as an extractant of sterols and other fat-soluble compounds from the whole fermenting mass, including yeast cells, as evident from the presence of ergosterol, the prevalent sterol in the cell membranes of yeasts, virtually absent in corn. Phytosterol and squalene contents in the different lots examined showed a low variability (CV < 15%), indicating a stable quality of the feedstock and standardized process conditions in the bioethanol plant over the period of study.

Thin stillage is a liquid stream generated in large amounts by the corn dry grind ethanol industry after centrifugation of heavy stillage. Although the majority of undissolved solids are removed with centrifugation, thin stillage still contains, along with a large part of water (90–93%), a residual lipid fraction quantified in the range 1.5–2.3% [11]. The HPLC profile of the unsaponifiable lipid fraction components of thin stillage extracts has shown a qualitative profile comparable to that of corn oil. Chromatographic analyses were performed after saponification. The sterol profile of thin stillage extracts is reported in Table 2, where the values are reported both on a wet mass and on a dry mass basis.

Table 2. Total amounts of phytosterols and squalene in thin stillage from a dry-grind corn ethanol plant. Data refer to individual lots collected at monthly intervals from July 2018 to April 2019 and overall mean, standard deviation (sd), and range of values observed (n = 7) *.

	Thin Stillage						
Lot	ERG	AVN [a]	STG + CAMP	β-SITO	STN	Σ STEROLS	SQUA
	mg kg^{-1} Thin Stillage (Wet Mass Basis)						
1	11.3	61.3	54.0	170	-	-	19.2
2	5.16	31.4	29.5	88.3	50.6	205	10.1
3	8.33	38.0	40.4	135	-	-	16.2
4	7.62	45.6	41.3	141	86.6	322	15.3
5	5.20	35.8	34.8	103	-	-	12.9
6	5.12	32.2	30.1	110	63.1	241	12.0
7	4.87	18.0	35.0	101	47.4	207	11.4
Mean	6.80	37.5	37.9	121	61.9	244	13.9
sd	2.41	13.4	8.43	28.6	17.8	54.8	3.17
min	4.87	18.0	29.5	88.3	47.4	205	10.1
max	11.3	61.3	54.0	170	86.6	322	19.2
	mg kg^{-1} Thin Stillage (Dry Mass Basis)						
1	141	767	676	2134	-	-	240
2	63.7	388	364	1090	625	2532	125
3	106	483	514	1718	-	-	206
4	88.2	528	478	1632	1002	3728	178
5	60.9	420	409	1210	-	-	151
6	62.5	394	368	1349	771	2944	146
7	54.6	202	393	1138	532	2320	128
Mean	82.4	455	457	1467	732	2881	168
sd	31.7	172	112	379	205	621	42.7
min	54.6	202	364	1090	532	2320	125
max	141	767	676	2134	1002	3728	240

* Details on the origin of corn feedstock and on lot timings are provided in Materials and Methods. Data for each lot represent the mean of duplicate measurements. [a] tentative identification (AVN), quantified as brassicasterol-equivalent; - not available. ERG ergosterol, AVN Δ5-avenasterol, STG + CAMP stigmasterol + campesterol, β-SITO β-sitosterol, STN sitostanol, SQUA squalene.

As observed for corn oil, thin stillage contained β-sitosterol as the prevalent sterol (88.3–170 mg kg^{-1} on a wet mass basis), followed by sitostanol (47.4–86.6 mg kg^{-1}), campesterol + stigmasterol (29.5–54.0 mg kg^{-1}), a sterol tentatively identified as δ-5-avenasterol (18.0–61.3 mg kg^{-1}) and trace amounts of ergosterol (4.87–11.3 mg kg^{-1}). The percent distribution of single sterols in post-fermentation corn oil and thin stillage were similar as can be seen in Figure S1. In absolute terms, the sterol content of thin stillage is very low on a wet mass basis (205–322 mg kg^{-1} wet mass) compared to corn oil (15832–17912 mg kg^{-1}). Squalene, at levels comprised between 10.1 and 19.2 mg kg^{-1} wet mass, was also detected in thin stillage.

The perspective to recover plant sterols and squalene from corn bioethanol co-products for further application in food and nutraceutical products adds value and sustainability to the whole fuel ethanol process. Widely known as cholesterol-lowering compounds, plant sterols are currently approved by regulatory agencies (FDA, EFSA) as food ingredients. Plant sterols, stanols, and their esters are nutritionally relevant nutrients because of their abilities to reduce blood cholesterol levels via partial inhibition of intestinal cholesterol absorption, to inhibit the growth of cancer cells, enhance the immune response, and act as anti-inflammatory and anti-oxidant factors [23–25]. Free sterols are the physiologically active form, known for their cholesterol-lowering properties made possible by inhibition of cholesterol absorption in the small intestine. The stanols and sterols esterified to phenolic acids present in corn are mostly hydrolyzed in the intestine. Steryl ferulates and hydroxycinnamate esters, are chain-breaking antioxidants and have proven cholesterol-lowering properties [26–29]. The development of functional food products enriched with plant sterols is a feasible way to provide consumers with novel healthy food products able to lower serum cholesterol levels [30,31].

Squalene, a polyunsaturated triterpene containing six isoprene units, is naturally present in animal and plant organisms, and in yeast, as an intermediate metabolite in the synthesis of sterols. As a minor constituent of food typical of the Mediterranean diet, squalene has been indicated as a key component in the prevention of cardiovascular heart disease, protection from cancer, and aging. Because of its unique properties, (i.e., drug carrier, adjuvant for vaccines, protective against cancer and other disease, skin repairing properties, UV-protecting properties, antibacterial properties, anti-wrinkle properties) squalene is also indicated in several pharmaceutical and cosmetic applications [32,33]. Although shark liver oil is a major source of squalene in nature, the growing concern for the protection of aquatic animals and the accumulation of persistent chemical pollutants at the high levels of the marine food chain, make plant sources of squalene a sustainable and highly attractive alternative.

3.2. Tocopherols and Tocotrienols

The tocol profile of post-fermentation corn oil sampled from July 2018 to September 2019 and the average and standard deviation of all values detected are reported in Table 3. Analyses carried out before and after oil saponification allowed the quantification of free and total tocols released after ester hydrolysis. The coelution of β- and γ-homologues of tocopherols and tocotrienols, common in reversed-phase LC systems, is of no relevance in the case of corn, where β-homologues of tocopherols and tocotrienols are known to be absent or negligible [14,34,35].

Table 3. Free and total amounts of tocopherols (T) and tocotrienols (T3) in post-fermentation corn oil from a dry-grind corn ethanol plant. Data refer to individual lots collected at monthly intervals from July 2018 to September 2019 and overall mean, standard deviation (sd), and range of values observed (n = 11) *.

Lot	α-T		γ-T [a]		δ-T		Σ T		α-T3		γ-T3 [a]		δ-T3		Σ T3		Σ (T + T3)	
	\multicolumn{18}{c}{Post-Fermentation Corn Oil}																
	\multicolumn{18}{c}{mg kg^{-1} Corn Oil}																
	Free	Total	Free	Total	Free	Total	Free	Total	Free	Total	Free	Total	Free	Total	Free	Total	Free	Total
1	179	175	700	706	25.2	18.2	904	899	216	133	—	175	0.00	8.13	-	316	-	1215
2	195	204	753	773	24.8	25.6	973	1004	220	154	—	196	0.00	8.45	-	358	-	1362
3	224	222	748	733	22.4	17.2	994	972	259	185	—	214	0.00	6.08	-	405	-	1377
4	210	216	789	775	24.0	24.4	1023	1016	226	189	—	241	0.00	7.72	-	438	-	1454
5	205	206	733	711	23.5	20.9	962	938	220	183	—	214	0.00	8.31	-	406	-	1344
6	224	212	737	701	21.6	26.1	982	939	220	178	—	202	0.00	4.10	-	384	-	1323
7	224	238	725	766	23.3	21.6	972	1026	208	174	—	203	0.00	9.56	-	386	-	1412
8	201	224	727	731	28.3	30.1	956	985	196	175	—	210	0.00	8.52	-	394	-	1379
9	188	235	694	716	22.3	17.1	904	969	181	168	—	189	0.00	8.80	-	365	-	1334
10	219	224	761	761	21.1	22.0	1002	1006	227	156	—	193	0.00	5.23	-	354	-	1360
11	194	220	761	793	20.6	22.9	975	1036	169	162	—	186	0.00	9.63	-	357	-	1393
Mean	206	216	739	742	23.4	22.4	968	981	213	169	—	202	0.00	7.68	-	379	-	1359
sd	15.7	17.0	27.8	32.3	2.18	4.04	36.6	42.3	24.3	16.6	—	17.9	0.00	1.78	-	33.0	-	60.5
min	179	175	694	701	20.6	17.1	904	899	169	133	—	175	0.00	4.10	-	316	-	1215
max	224	238	789	793	28.3	30.1	1023	1036	259	189	—	241	0.00	9.63	-	438	-	1454

* Details on the origin of corn feedstock and on lot timings are provided in Materials and Methods. Data for each lot represent the mean of duplicate measurements. [a] may contain low or trace amounts of β-homologue; — data not available for the presence of co-eluting compound(s); - not calculated because free γ-T3 is missing.

Tocols in post-fermentation corn oil were found to be present mostly in their free form, as evidenced by the comparison of levels obtained before and after saponification. Gamma-tocotrienol in direct analysis of corn oil coeluted with one or more unidentified compounds not present in the saponified extract, probably one or more different sterol esters, as evidenced by the analysis of peak spectra characteristics and purity. This coelution did not allow to quantify the amount of free γ-tocotrienol and hence the total amounts of free tocotrienols and free tocols.

Post-fermentation corn oil was characterized by the prevalence of tocopherols (981 ± 42.3 mg kg^{-1}, corresponding to 72% of total tocols) over tocotrienols (379 ± 33 mg kg^{-1}, corresponding to 28% of total tocols), with γ-tocopherol prevailing over α- and δ- homologues, as typical for corn [31,35]. The levels of γ-tocopherol after saponification accounted for an average value of 742 mg kg^{-1}, followed by α-tocopherol (216 mg kg^{-1}) and very minor amounts of δ-tocopherol (22.4 mg kg^{-1}).

Tocols are inherently present in corn, where they play an antioxidant role, protecting the unsaturated fatty acids from oxidation. In particular, tocopherols are concentrated in corn germ, while tocotrienols are preferentially located in the endosperm and in the outer portions of the kernel.

This explains why in post-fermentation corn oil, which derives from the whole kernel, tocopherol levels are comparable to those of an unrefined corn germ oil, while tocotrienol levels are quite higher [21,35].

Values detected in the different lots examined showed quite stable tocol contents and a low variability (CV < 15%), indicating standardized process conditions in the bioethanol plant over the time and a resistance of tocols to the fuel ethanol production conditions. The prevalence of γ-homologues of tocopherols and tocotrienols over α- and δ-homologues here observed is a characteristic feature of corn that may be of interest for final applications [36,37]. In fact, γ-tocopherol is reported to retain higher antioxidant properties compared to α-tocopherol and to act in synergy with it in biological systems, also protecting from inflammation [38].

The data here reported are in accordance with those reported in literature for ethanol-extracted corn kernel oil and co-products of corn bioethanol production [14,39,40].

The tocol profile of thin stillage and the average and standard deviation of values detected in the different lots after saponification of lipid extracts are reported in Table 4. The amounts are expressed both on a wet mass and dry matter basis. Because of its high dilution, in absolute terms thin stillage has a very low concentration of tocols (average value 20.9 mg kg^{-1} thin stillage). As observed for post-fermentation corn oil, the tocol profile of thin stillage was characterized by the presence of tocopherols (73%) dominating over tocotrienols (27%), with γ-homologues prevailing over α- and δ- homologues a. The relative proportions of each tocopherol and tocotrienol homologue identified in post-fermentation corn oil and thin stillage were quite similar, as can be seen in Figure S2. As observed for phytosterols, in absolute terms, the tocol amounts in thin tillage on a wet mass basis were very low compared to post-fermentation corn oil.

Besides retaining vitamin E activity and playing as potent antioxidants, tocols cover multiple functions in biological systems such as gene expression regulation, signal transduction, and modulation of cell functions through modulation of protein–membrane interactions [41]. All tocopherols possess a high antioxidant activity and are important tools in the prevention of cardiovascular disease and cancer. While most of the studies on vitamin E have been first focused on α-tocopherol, the primary form in most living organisms, further evidences have shown that its homologues have superior biological properties that may be useful for prevention and therapy against chronic diseases [42]. Most recently, tocotrienols have raised increasing interest because of their hypocholesterolemic, neuroprotective, anti-thrombotic, and anti-tumor effects, suggesting that they may serve as effective agents in the prevention and/or treatment of cancer and cardiovascular and neurodegenerative diseases [43–46]. The enrichment of food products with natural extracts rich of tocols and other natural antioxidants and bioactives is the best strategy to ensure that daily requirements are met and at the same time to improve the healthiness and oxidative stability of processed food [47,48].

Table 4. Total amounts of tocopherols (T) and tocotrienols (T3) in thin stillage from a dry-grind corn ethanol plant. Data refer to individual lots collected at monthly intervals from July 2018 to April 2019 and overall mean, standard deviation (sd), and range of values observed (n = 7) *.

Lot	α-T	γ-T [a]	δ-T	Σ T	α-T3	γ-T3 [a]	δ-T3	Σ T3	Σ T + T3
				mg kg^{-1} Thin Stillage Wet Mass					
1	3.95	16.5	0.57	21.0	2.99	3.96	0.05	6.99	28.0
2	1.89	8.21	0.22	10.3	1.32	2.00	0.00	3.32	13.6
3	3.63	14.0	0.29	17.9	3.14	4.12	0.00	7.26	25.2
4	3.77	13.6	0.39	17.8	3.11	3.90	0.00	7.02	24.8
5	2.70	9.80	0.40	12.9	2.30	2.90	0.00	5.20	18.1
6	2.87	10.3	0.26	13.4	2.31	2.83	0.00	5.14	18.5
7	2.85	10.5	0.46	13.8	2.00	2.55	0.00	4.55	18.4
Mean	3.09	11.8	0.37	15.3	2.45	3.18	0.01	5.64	20.9
sd	0.73	2.92	0.12	3.70	0.68	0.82	0.02	1.49	5.11
min	1.89	8.21	0.22	10.3	1.32	2.00	0.00	3.32	13.6
max	3.95	16.5	0.57	21.0	3.14	4.12	0.05	7.26	28.0
				mg kg^{-1} Thin Stillage Dry Mass					
1	49.4	206	7.15	263	37.4	49.6	0.61	87.5	350
2	23.3	101	2.73	127	16.3	24.7	0.00	41.0	168
3	46.2	178	3.72	228	40.0	52.4	0.00	92.4	320
4	43.6	158	4.51	206	36.0	45.2	0.00	81.2	287
5	31.5	116	4.80	152	26.7	34.2	0.00	60.9	213
6	35.1	125	3.12	163	28.2	34.6	0.00	62.8	226
7	31.9	118	5.16	155	22.4	28.6	0.00	51.0	206
Mean	37.3	143	4.46	185	29.6	38.5	0.09	68.1	253
sd	9.39	38.5	1.48	48.6	8.65	10.7	0.23	19.4	67.0
min	23.3	101	2.73	127	16.3	24.7	0.00	41.0	168
max	49.4	206	7.15	263	40.0	52.4	0.61	92.4	351

* Details on the origin of corn feedstock and on lot timings are provided in Materials and Methods. Data for each lot represent the mean of duplicate measurements. [a] may contain low or trace amounts of β-homologue.

3.3. Carotenoids

Levels of single carotenoids and average and standard deviation of total amounts observed in post-fermentation corn oil are reported in Table 5. The carotenoid profile of post-fermentation corn oil was that typical of corn, with lutein and zeaxanthin, accounting altogether for over 60% of the total carotenoids, as the prevalent molecules. Minor amounts of β-cryptoxanthin and traces of β-carotene were also present. Cis-isomers of lutein and zeaxanthin, most probably resulting from the high temperatures reached during the biotech process, were also observed and quantified as lutein-equivalents. One of these compounds (N.I.C. 6) was the third most concentrated carotenoid, followed by β-cryptoxanthin.

Values detected in the different lots examined showed higher variability than that found for other molecules, which could be due to differences in the feedstock more than to process conditions in the bioethanol plant.

The average amounts of single carotenoids identified in the different lots of thin stillage analyzed and the average and standard deviation of all values, expressed both on a wet mass and on a dry mass basis, are reported in Table 6. The carotenoid profile of thin stillage was qualitatively very similar to that of post-fermentation corn oil while the concentration, as expected, was much lower. As for corn oil, the amount of carotenoids in the different lots of thin stillage analyzed showed high variability.

Table 5. Free and total amounts of carotenoids in post-fermentation corn oil from a dry-grind corn ethanol plant. Data refer to individual lots collected at monthly intervals from July 2018 to April 2019 and overall mean, standard deviation (sd), and range of values observed (n = 6) *.

Lot	Lutein		Zeaxanthin		β-cryptoxanthin		N.I.C. 1		N.I.C. 2		N.I.C. 3		N.I.C. 4		N.I.C. 5		N.I.C. 6		Total	
	Free	Total	Free	Total	Free	Total	Free	Total	Free	Total	Free	Total	Free	Total	Free	Total	Free	Total	Σ Free	Σ Total
									mg kg^{-1} Corn Oil											
1	68.1	84.4	53.7	69.8	14.3	18.0	9.60	12.1	7.17	8.31	10.5	13.2	6.66	8.96	6.40	8.26	19.6	24.2	196	240
2	64.4	87.0	47.2	75.4	11.8	19.6	9.14	13.9	5.73	8.18	10.5	15.2	6.01	9.05	5.87	8.28	14.1	24.4	175	254
3	88.7	93.8	66.9	72.4	16.9	20.4	12.6	13.1	8.79	9.46	13.7	13.7	7.54	8.85	7.08	7.66	21.2	25.4	243	257
4	80.5	79.4	64.3	61.9	18.3	18.9	10.8	13.2	7.58	7.79	12.5	12.3	7.18	7.79	6.64	7.01	21.1	21.6	229	223
6	84.0	84.4	61.7	62.8	17.9	19.8	11.6	11.7	7.82	8.29	11.9	12.0	7.31	8.55	6.90	7.44	19.1	21.7	228	229
7	88.3	-	58.7	-	16.2	-	10.1	-	6.71	-	11.4	-	8.36	-	8.20	-	21.0	-	229	-
Mean	79.0	85.8	58.7	68.5	15.9	19.3	10.6	12.8	7.3	8.41	11.8	13.3	7.17	8.64	6.85	7.73	19.3	23.4	217	240
sd	10.4	5.28	7.27	5.92	2.46	0.95	1.28	0.89	1.04	0.63	1.23	1.29	0.80	0.51	0.79	0.55	2.73	1.71	25.6	14.9
min	64.4	79.4	47.2	61.9	11.8	18.0	9.14	11.7	5.73	7.79	10.5	12.0	6.01	7.79	5.87	7.01	14.0	21.6	175	223
max	88.7	93.8	66.9	75.4	18.3	20.4	12.6	13.9	8.79	9.46	13.7	15.2	8.36	9.05	8.20	8.28	21.2	25.4	243	257

* Details on the origin of corn feedstock and on lot timings are provided in Materials and Methods. Data for each lot represent the mean of triplicate measurements. N.I.C. 1–6, not identified cis-isomers of carotenoids quantified as lutein-equivalent (in order of elution); - not available.

Table 6. Free and total amounts of carotenoids in thin stillage from a dry-grind corn ethanol plant. Data refer to individual lots collected at monthly intervals from July 2018 to April 2019 and overall mean, standard deviation (*sd*), and range of values observed ($n = 7$) *.

Lot	Lutein		Zeaxanthin		β-cryptoxanthin		N.I.C. 1		N.I.C. 2		N.I.C. 3		N.I.C. 4		N.I.C. 5		N.I.C. 6		Total	
	Free	Total	Free	Total	Free	Total	Free	Total	Free	Total	Free	Total	Free	Total	Free	Total	Free	Total	Σ Free	Σ Total
mg kg^{-1} Thin Stillage Wet Mass																				
1	2.62	2.98	2.61	3.23	0.54	0.75	0.30	0.40	0.20	0.23	0.36	0.42	0.24	0.27	0.23	0.27	0.73	0.88	7.84	9.44
2	1.68	2.11	1.39	2.24	0.27	0.62	0.19	0.23	0.07	0.17	0.20	0.40	0.15	0.19	0.14	0.18	0.41	0.61	4.50	6.76
3	2.15	2.78	1.91	3.16	0.39	0.83	0.3	0.37	0.1	0.24	0.19	0.44	0.19	0.23	0.17	0.23	0.58	0.87	5.96	9.14
4	2.28	2.76	2.04	3.04	0.41	0.81	0.29	0.33	0.1	0.23	0.18	0.41	0.19	0.23	0.17	0.24	0.58	0.85	6.24	8.91
5	2.19	2.45	2.09	2.42	0.42	0.63	0.24	0.29	0.19	0.21	0.32	0.38	0.17	0.20	0.16	0.20	0.50	0.66	6.28	7.45
6	2.35	2.6	2.15	2.48	0.47	0.66	0.29	0.35	0.21	0.24	0.33	0.38	0.21	0.25	0.2	0.25	0.60	0.75	6.81	7.95
7	2.07	2.31	2.03	2.25	0.46	0.59	0.19	0.25	0.16	0.17	0.27	0.33	0.18	0.21	0.18	0.22	0.54	0.66	6.09	6.98
Mean	**2.19**	**2.57**	**2.03**	**2.69**	**0.42**	**0.70**	**0.26**	**0.32**	**0.15**	**0.21**	**0.26**	**0.39**	**0.19**	**0.23**	**0.18**	**0.23**	**0.56**	**0.75**	**6.25**	**8.09**
sd	0.28	0.30	0.36	0.44	0.08	0.09	0.05	0.06	0.06	0.03	0.08	0.04	0.03	0.03	0.03	0.03	0.1	0.12	1.00	1.08
min	1.68	2.11	1.39	2.24	0.27	0.59	0.19	0.23	0.07	0.17	0.18	0.33	0.15	0.19	0.14	0.18	0.41	0.61	4.50	6.76
max	2.62	2.98	2.61	3.23	0.54	0.83	0.30	0.40	0.21	0.24	0.36	0.44	0.24	0.27	0.23	0.27	0.73	0.88	7.84	9.44
mg kg^{-1} Thin Stillage Dry Mass																				
1	32.7	37.4	32.7	40.4	6.81	9.37	3.82	4.99	2.52	2.91	4.53	5.31	2.98	3.44	2.88	3.34	9.11	10.99	98	118
2	20.8	26.1	17.1	27.7	3.34	7.69	2.38	2.84	0.87	2.15	2.46	4.93	1.83	2.30	1.69	2.25	5.00	7.49	56	83
3	27.4	35.4	24.2	40.2	4.95	10.5	3.77	4.70	1.22	3.00	2.36	5.61	2.36	2.95	2.11	2.90	7.40	11.1	76	116
4	26.4	31.9	23.6	35.2	4.72	9.33	3.35	3.88	1.15	2.64	2.04	4.75	2.18	2.71	2.02	2.78	6.75	9.89	72	103
5	25.8	28.7	24.5	28.4	4.94	7.43	2.80	3.45	2.21	2.49	3.77	4.44	2.02	2.37	1.91	2.35	5.82	7.70	74	87
6	28.7	31.8	26.2	30.3	5.68	8.08	3.59	4.22	2.62	2.89	4.05	4.65	2.58	3.03	2.42	3.04	7.28	9.16	83	97
7	23.3	26.0	22.7	25.3	5.20	6.62	2.16	2.78	1.78	1.86	3.06	3.69	2.05	2.39	1.97	2.42	6.07	7.40	68	78
Media	**26.4**	**31.0**	**24.5**	**32.5**	**5.09**	**8.43**	**3.12**	**3.83**	**1.77**	**2.56**	**3.18**	**4.77**	**2.29**	**2.74**	**2.14**	**2.73**	**6.77**	**9.10**	**75**	**98**
sd	3.83	4.39	4.63	6.13	1.05	1.35	0.68	0.86	0.70	0.43	0.95	0.62	0.39	0.42	0.39	0.40	1.33	1.61	13	16
min	20.8	26.0	17.1	25.3	3.34	6.62	2.16	2.78	0.87	1.86	2.04	3.69	1.83	2.30	1.69	2.25	5.00	7.40	56	78
max	32.7	37.4	32.7	40.4	6.81	10.5	3.82	4.99	2.62	3.00	4.53	5.61	2.98	3.44	2.88	3.34	9.11	11.1	98.1	118

* Details on the origin of corn feedstock and on lot timings are provided in Materials and Methods. Data for each lot represent the mean of triplicate measurements. N.I.C. 1–6, not identified cis-isomers of carotenoids quantified as lutein-equivalent (in order of elution).

In cereals, carotenoids are important phytonutrients responsible for the yellow color of the endosperm, where they occur either in free or esterified forms, mostly with palmitic and linoleic acid. In corn the major carotenoids are the xanthophylls zeaxanthin and lutein, isomers differing by the position of a double bond in the β-ionone ring, with minor amounts of β-cryptoxanthin and β-carotene.

In terms of health benefits, carotenoids are powerful antioxidants protecting cells against reactive oxygen species and free radicals, and playing an important role in the maintenance of good health and disease prevention. Several studies have indicated their protective effect against chronic degenerative, inflammatory, metabolic, and age-related diseases and their immunomodulatory properties [49–52]. In particular lutein and zeaxanthin, the prevalent xanthophylls in corn, are interesting molecules for the food, pharmaceutical, and nutraceutical sectors because of their strong antioxidant properties and their important role in the maintenance of the normal visual function in humans. As essential components of the eye macula, they protect the retina from the oxidative damages responsible of age-related macular degeneration and cataract [53]. Although fresh vegetables and fruits are a rich source of carotenoids in our diet, the formulation of functional food enriched with carotenoids is a suitable and successful strategy to compensate for nutritional losses occurring during the technological processes or to integrate them in food matrixes not inherently rich of these molecules [54,55]

4. Conclusions

This study highlights the potentialities of dry-grind corn bioethanol side streams as sustainable sources of bioactive compounds for high-value applications. Chromatographic analyses on post-fermentation corn oil indicated that this fuel ethanol co-product is particularly rich of plant sterols and stanols and also retains the whole set of tocopherols, tocotrienols, and carotenoids originating from the corn kernel.

The huge volumes of thin stillage produced during the dry-grind corn bioethanol process make the perspective to recover valuable molecules therein a very attractive one. Yet, it still represents a challenge as the high dilution of this stream strongly reduces the affordability and sustainability of any recovery and separation process.

The low variability observed during the year of the chemical profile of the two side streams, in spite of the different origin and seasonality of corn feedstock lots and of the complex biotechnological processes, represents an important element for their industrial utilization.

With a circular economy approach the bioactive molecules present in corn bioethanol co-products, valuable for their antioxidant, anti-inflammatory, hypocholesterolemic, anti-aging, and several other beneficial properties, could be recovered through appropriate technologies and re-introduced in productive processes as ingredients of a wide range of high-value functional products in the food, nutraceutical, and cosmeceutical sectors.

Even though post-fermentation corn oil is currently utilized by biofuel biorefineries as a feedstock for biodiesel production, its full potentials are actually unexploited since molecules other than fatty acids remain unused. A preliminary separation and fractionation of sterols and other unsaponifiables from corn bioethanol oil would therefore not only maximize the efficiency of biodiesel production, actually hindered by the presence of these molecules, but would also add value to the whole biotech process, opening perspectives for the creation of integrated biorefineries and new value chains.

Supplementary Materials: The following are available online at http://www.mdpi.com/2304-8158/9/12/1788/s1. Figure S1: Relative distribution of phytosterols in post-fermentation corn oil and thin stillage from a dry-grind corn ethanol plant. Figure S2: Relative distribution of tocopherols (T) and tocotrienols (T3) in post-fermentation corn oil and thin stillage from a dry-grind corn ethanol plant.

Author Contributions: Conceptualization, G.D.L.; validation, G.D.L. and J.S.d.P.; investigation, J.S.d.P., I.C., and S.F.N.; data curation and visualization, G.D.L. and J.S.d.P.; writing—original draft preparation, G.D.L.; writing—review and editing, G.D.L., J.S.d.P., and G.L.B.; supervision, G.D.L.; funding acquisition, G.D.L. All authors have read and agreed to the published version of the manuscript.

Funding: This study was carried out in the frame of the EXCornsEED project. This project received funding from the Bio Based Industries Joint Undertaking under the European Union's Horizon 2020 research and innovation program under grant agreement n° 792054.

Conflicts of Interest: The authors declare no conflict of interest.

References

1. International Energy Agency. *Renewables 2020—Analysis and Forecast to 2025*; IEA: Paris, France, 2020.
2. ePURE. European Renewable Ethanol—Key Figures. 2019. Available online: https://www.epure.org/media/2044/200813-def-pr-epure-infographic-european-renewable-ethanol-key-figures-2019_web.pdf (accessed on 20 November 2020).
3. European Commission. *Renewable Energy Progress Report. Report from the Commission to the European Parliament, the Council, the European Economic and Social Committee and the Committee of the Regions*; European Commission: Brussels, Belgium, 2017.
4. European Commission. *A sustainable Bioeconomy for Europe. Strengthening the Connection between Economy, Society and the Environment. Updated Bioeconomy Strategy 2018. Directorate-General for Research and Innovation*; European Commission: Brussels, Belgium, 2018.
5. European Commission. *Communication from the Commission to the European Parliament, the Council, the European Economic and Social Committee and the Committee of the Regions. A New Circular Economy Action Plan for a Cleaner and More Competitive Europe*; European Commission: Brussels, Belgium, 2020.
6. Oreopoulou, V.; Tzia, C. Utilization of plant by-products for the recovery of proteins, dietary fibers, antioxidants, and colorants. In *Utilization of By-Products and Treatment of Waste in the Food Industry*; Oreopoulou, V., Russ, W., Eds.; Springer: New York, NY, USA, 2007; Volume 3, pp. 209–232. [CrossRef]
7. Teixeira, A.; Baenas, N.; Dominguez-Perles, R.; Barros, L.; Rosa, E.; Moreno, D.A.; Garcia-Viguera, C. Natural bioactive compounds from winery by-products as health promoters: A review. *Int. J. Mol. Sci.* **2014**, *15*, 15638–15678. [CrossRef] [PubMed]
8. Mirabella, N.; Castellani, V.; Sala, S. Current options for the valorization of food manufacturing waste: A review. *J. Clean. Prod.* **2014**, *65*, 28–41. [CrossRef]
9. Ben-Othman, S.; Jõudu, I.; Bhat, R. Bioactives from agri-food wastes: Present insights and future challenges. *Molecules* **2020**, *25*, 510. [CrossRef]
10. Sheng, S.; Li, T.; Liu, R. Corn phytochemicals and their health benefits. *Food Sci. Hum. Wellness* **2018**, *7*, 185–195. [CrossRef]
11. Di Lena, G.; Ondrejíčková, P.; Pulgar, J.S.D.; Cyprichová, V.; Ježovič, T.; Lucarini, M.; Lombardi Boccia, G.; Ferrari Nicoli, S.; Gabrielli, P.; Aguzzi, A.; et al. Towards a Valorization of Corn Bioethanol Side Streams: Chemical Characterization of Post Fermentation Corn Oil and Thin Stillage. *Molecules* **2020**, *25*, 3549. [CrossRef]
12. Bligh, E.G.; Dyer, W.J. A rapid method of total lipid extraction and purification. *Can. J. Biochem. Physiol.* **1959**, *37*, 911–917. [CrossRef]
13. Di Lena, G.; Casini, I.; Lucarini, M.; Lombardi-Boccia, G. Carotenoid profiling of five microalgae species from large-scale production. *Food Res. Int.* **2019**, *120*, 810–818. [CrossRef]
14. Winkler, J.K.; Rennick, K.A.; Eller, F.J.; Vaughn, S.F. Phytosterol and tocopherol components in extracts of corn distiller's dried grain. *J. Agric. Food Chem.* **2007**, *55*, 6482–6486. [CrossRef]
15. Majoni, S.; Wang, T. Characterization of oil precipitate and oil extracted from condensed corn distillers solubles. *JAOCS J. Am. Oil Chem. Soc.* **2010**, *87*, 205–213. [CrossRef]
16. Seitz, L.M. Stanol and sterol esters of ferulic and p-coumaric acids in wheat, corn, rye, and triticale. *J. Agric. Food Chem.* **1989**, *37*, 662–667. [CrossRef]

17. Harrabi, S.; St-Amand, A.; Sakouhi, F.; Sebei, K.; Kallel, H.; Mayer, P.M.; Boukhchina, S. Phytostanols and phytosterols distributions in corn kernel. *Food Chem.* **2008**, *111*, 115–120. [CrossRef]
18. Esche, R.; Scholz, B.; Engel, K.H. Analysis of free phytosterols/stanols and their intact fatty acid and phenolic acid esters in various corn cultivars. *J. Cereal Sci.* **2013**, *58*, 333–340. [CrossRef]
19. Ostlund, R.E., Jr.; Racette, S.B.; Okeke, A.; Stenson, W.F. Phytosterols that are naturally present in commercial corn oil significantly reduce cholesterol absorption in humans. *Am. J. Clin. Nutr.* **2002**, *75*, 1000–1004. [CrossRef]
20. Phillips, K.M.; Ruggio, D.M.; Toivo, J.I.; Swank, M.A.; Simpkins, A.H. Free and esterified sterol composition of edible oils and fats. *J. Food Compos. Anal.* **2002**, *15*, 123–142. [CrossRef]
21. Schwartz, H.; Ollilainen, V.; Piironen, V.; Lampi, A.-M. Tocopherol, tocotrienol and plant sterol contents of vegetable oils and industrial fats. *J. Food Compos. Anal.* **2008**, *21*, 152–161. [CrossRef]
22. Normén, L.; Ellegård, L.; Brants, H.; Dutta, P.; Andersson, H. A phytosterol database: Fatty foods consumed in Sweden and the Netherlands. *J. Food Compos. Anal.* **2007**, *20*, 193–201. [CrossRef]
23. De Jong, A.; Plat, J.; Bast, A.; Godschalk, R.W.L.; Basu, S.; Mensink, R.P. Effects of plant sterol and stanol ester consumption on lipid metabolism antioxidant status, and markers of oxidative stress, endothelial function, and low-grade inflammation in patients on current statin treatment. *Eur. J. Clin. Nutr.* **2008**, *62*, 263–273. [CrossRef]
24. Woyengo, T.A.; Ramprasath, V.R.; Jones, P.J.H. Anticancer effects of phytosterols. *Eur. J. Clin. Nutr.* **2009**, *63*, 813–820. [CrossRef]
25. Talati, R.; Sobieraj, D.M.; Makanji, S.S.; Phung, O.J.; Coleman, C.I. The comparative efficacy of plant sterols and stanols on serum lipids: A systematic review and meta-analysis. *J. Am. Diet. Assoc.* **2010**, *110*, 719–726. [CrossRef]
26. Ghatak, S.B.; Panchal, S.J. Gamma-oryzanol—A multi-purpose steryl ferulate. *Curr. Nutr. Food Sci.* **2011**, *7*, 10–20. [CrossRef]
27. Zhu, D.; Sánchez-Ferrer, A.; Nyström, L. Antioxidant Activity of Individual Steryl Ferulates from Various Cereal Grain Sources. *J. Nat. Prod.* **2016**, *79*, 308–316. [CrossRef]
28. Jones, P.J.H.; Shamloo, M.; MacKay, D.S.; Rideout, T.C.; Myrie, S.B.; Plat, J.; Roullet, J.-B.; Baer, D.J.; Calkins, K.L.; Davis, H.R.; et al. Progress and perspectives in plant sterol and plant stanol research. *Nutr. Rev.* **2018**, *76*, 725–746. [CrossRef]
29. Moreau, R.A.; Nyström, L.; Whitaker, B.D.; Winkler-Moser, J.K.; Baer, D.J.; Gebauer, S.K.; Hicks, K.B. Phytosterols and their derivatives: Structural diversity, distribution, metabolism, analysis, and health-promoting uses. *Prog. Lipid Res.* **2018**, *70*, 35–61. [CrossRef]
30. Garcia-Llatas, G.; Cilla, A.; Alegría, A.; Lagarda, M.J. Bioavailability of plant sterol-enriched milk-based fruit beverages: In vivo and in vitro studies. *J. Funct. Foods* **2015**, *14*, 44–50. [CrossRef]
31. Gies, M.; Servent, A.; Borel, P.; Dhuique-Mayer, C. Phytosterol vehicles used in a functional product modify carotenoid/cholesterol bioaccessibility and uptake by Caco-2 cells. *J. Funct. Foods* **2020**, *68*, 103920. [CrossRef]
32. Reddy, L.; Couvreur, P. Squalene: A natural triterpene for use in disease management and therapy. *Adv. Drug Deliv. Rev.* **2009**, *61*, 1412–1426. [CrossRef]
33. Popa, I.; Băbeanu, N.; Niță, S.; Popa, O. Squalene-Natural resources and applications. *Farmacia* **2014**, *62*, 840–862.
34. Hossain, A.; Jayadeep, A. Determination of tocopherol and tocotrienol contents in maize by in vitro digestion and chemical methods. *J. Cereal Sci.* **2018**, *83*, 90–95. [CrossRef]
35. Grilo, E.C.; Costa, P.N.; Gurgel, C.S.S.; Beserra, A.F.L.; Almeida, F.N.S.; Dimenstein, R. Alpha-tocopherol and gamma-tocopherol concentration in vegetable oils. *Food Sci. Technol.* **2014**, *34*, 379–385. [CrossRef]
36. Gibreel, A.; Sandercock, J.R.; Lan, J.; Goonewardene, L.A.; Scott, A.C.; Zijlstra, R.T.; Curtis, J.M.; Bressler, D.C. Evaluation of value-added components of dried distiller's grain with and phytosterols distributions in corn kernel. *Food Chem.* **2011**, *111*, 115–120. [CrossRef]
37. Winkler-Moser, J.K.; Breyer, L. Composition and oxidative stability of crude oil extracts of corn germ and distillers grains. *Ind. Crops Prod.* **2011**, *33*, 572–578. [CrossRef]
38. Saldeen, K.; Saldeen, T. Importance of tocopherols beyond α-tocopherol: Evidence from animal and human studies. *Nutr. Res.* **2005**, *25*, 877–889. [CrossRef]
39. Moreau, R.A.; Hicks, K.B. The composition of corn oil obtained by the alcohol extraction of ground corn. *JAOCS J. Am. Oil Chem. Soc.* **2005**, *82*, 809–815. [CrossRef]

40. Shin, E.-C.; Shurson, G.C.; Gallaher, D.D. Antioxidant capacity and phytochemical content of 16 sources of corn distillers dried grains with solubles (DDGS). *Anim. Nutr.* **2018**, *4*, 435–441. [CrossRef]
41. Zingg, J.M. Modulation of signal transduction by vitamin E. *Mol. Asp. Med.* **2007**, *28*, 481–506. [CrossRef]
42. Ahsan, H.; Ahad, A.; Siddiqui, W.A. A review of characterization of tocotrienols from plant oils and foods. *J. Chem. Biol.* **2015**, *8*, 45–59. [CrossRef]
43. Sen, C.K.; Khanna, S.; Roy, S. Tocotrienols in health and disease: The other half of the natural vitamin E family. *Mol. Asp. Med.* **2007**, *28*, 692–728. [CrossRef]
44. Khanna, S.; Parinandi, N.L.; Kotha, S.R.; Roy, S.; Rink, C.; Bibus, D.; Sen, C.K. Nanomolar vitamin E α-tocotrienol inhibits glutamate induced activation of phospholipase A2 and causes neuroprotection. *J. Neurochem.* **2010**, *112*, 1249–1260. [CrossRef]
45. Peh, H.Y.; Tan, W.S.D.; Liao, W.; Wong, W.S.F. Vitamin E therapy beyond cancer: Tocopherol versus tocotrienol. *Pharmacol. Ther.* **2016**, *162*, 152–169. [CrossRef]
46. Ramanathan, N.; Tan, E.; Loh, L.J.; Soh, B.S.; Yap, W.N. Tocotrienol is a cardioprotective agent against ageing-associated cardiovascular disease and its associated morbidities. *Nutr. Metab.* **2018**, *15*, 6. [CrossRef]
47. Prasanth Kumar, P.K.; Jeyarani, T.; Gopala Krishna, A.G. Physicochemical characteristics of phytonutrient retained red palm olein and butter-fat blends and its utilization for formulating chocolate spread. *J. Food Sci. Technol.* **2016**, *53*, 3060–3072. [CrossRef]
48. Bakota, E.L.; Winkler-Moser, J.K.; Hwang, H.-S.; Bowman, M.J.; Palmquist, D.E.; Liu, S.X. Solvent fractionation of rice bran oil to produce a spreadable rice bran product. *Eur. J. Lipid Sci. Technol.* **2013**, *115*, 847–857. [CrossRef]
49. Obulesu, M.; Dowlathabad, M.R.; Bramhachari, P.V. Carotenoids and Alzheimer's Disease: An insight into therapeutic role of retinoids in animal models. *Neurochem. Int.* **2011**, *59*, 535–541. [CrossRef]
50. Wang, Y.; Cui, R.; Xiao, Y.; Fang, J.; Xu, Q. Effect of carotene and lycopene on the risk of prostate cancer: A systematic review and dose-response meta-analysis of observational studies. *PLoS ONE* **2015**, *10*, e0137427. [CrossRef]
51. Murillo, A.G.; Fernandez, M.L. Potential of dietary non-provitamin a carotenoids in the prevention and treatment of diabetic microvascular complications. *Adv. Nutr.* **2016**, *7*, 14–24. [CrossRef]
52. Cheng, H.M.; Koutsidis, G.; Lodge, J.K.; Ashor, A.; Siervo, M.; Lara, J. Tomato and lycopene supplementation and cardiovascular risk factors: A systematic review and meta-analysis. *Atherosclerosis* **2017**, *257*, 100–108. [CrossRef]
53. Carpentier, S.; Knaus, M.; Suh, M. Associations between lutein, zeaxanthin, and age-related macular degeneration: An overview. *Crit. Rev. Food Sci. Nutr.* **2009**, *49*, 313–326. [CrossRef]
54. Xavier, A.A.O.; Carvajal-Lérida, I.; Garrido-Fernández, J.; Pérez-Gálvez, A. In vitro bioaccessibility of lutein from cupcakes fortified with a water-soluble lutein esters formulation. *J. Food Compos. Anal.* **2018**, *68*, 60–64. [CrossRef]
55. Ursache, F.M.; Andronoiu, D.G.; Ghinea, I.O.; Barbu, V.; Ioniţă, E.; Cotârleţ, M.; Dumitraşcu, L.; Botez, E.; Râpeanu, G.; Stănciuc, N. Valorizations of carotenoids from sea buckthorn extract by microencapsulation and formulation of value-added food products. *J. Food Eng.* **2018**, *219*, 16–24. [CrossRef]

Publisher's Note: MDPI stays neutral with regard to jurisdictional claims in published maps and institutional affiliations.

© 2020 by the authors. Licensee MDPI, Basel, Switzerland. This article is an open access article distributed under the terms and conditions of the Creative Commons Attribution (CC BY) license (http://creativecommons.org/licenses/by/4.0/).

Article

Aspergillus oryzae Fermented Rice Bran: A Byproduct with Enhanced Bioactive Compounds and Antioxidant Potential

Sneh Punia [1,*], Kawaljit Singh Sandhu [2,*], Simona Grasso [3], Sukhvinder Singh Purewal [2], Maninder Kaur [4], Anil Kumar Siroha [1], Krishan Kumar [1], Vikas Kumar [1] and Manoj Kumar [5,*]

1 Department of Food Science & Technology, Chaudhary Devi Lal University, Sirsa 125055, India; siroha01@gmail.com (A.K.S.); k.kumar4032@gmail.com (K.K.); vk.pandit415@gmail.com (V.K.)
2 Department of Food Science & Technology, Maharaja Ranjit Singh Punjab Technical University, Bathinda 151001, India; purewal.0029@gmail.com
3 Institute of Food, Nutrition and Health, University of Reading, Reading RG6 6UR, UK; simona.grasso@ucdconnect.ie
4 Department of Food Science & Technology, Guru Nanak Dev University, Amritsar 143005, India; mandyvirk@rediffmail.com
5 Chemical and Biochemical Processing Division, ICAR—Central Institute for Research on Cotton Technology, Mumbai 400019, India
* Correspondence: snehpunia69@gmail.com or dimplepoonia@gmail.com (S.P.); kawsandhu@rediffmail.com (K.S.S.); manoj.kumar13@icar.gov.in (M.K.)

Abstract: Rice bran (RB) is a byproduct of the rice industry (milling). For the fermentation process and to add value to it, RB was sprayed with fungal spores (*Aspergillus oryzae* MTCC 3107). The impact of fermentation duration on antioxidant properties was studied. Total phenolic content (TPC) determined using the Folin–Ciocalteu method, increased during fermentation until the 4th day. The antioxidant activity analyzed using the 2,2 Diphenyl–1′ picrylhydrazyl (DPPH) assay, total antioxidant activity (TAC), 2,2′-azinobis 3-ethylbenzothiazoline-6-sulfonic acid ($ABTS^+$) assay, reducing power assay (RPA) and hydroxyl free radical scavenging activity (HFRSA) for fermented rice bran (FRB) were determined and compared to unfermented rice bran (URB). TAC, DPPH, $ABTS^+$ and RPA of FRB increased till 4th day of fermentation, and then decreased. The specific bioactive constituents in extracts (Ethanol 50%) from FRB and URB were identified using high performance liquid chromatography (HPLC). HPLC confirmed a significant ($p < 0.05$) increase in gallic acid and ascorbic acid. On the 4th day of fermentation, the concentrations of gallic acid and ascorbic acid were 23.3 and 12.7 μg/g, respectively. The outcome of present investigation confirms that antioxidant potential and TPC of rice bran may be augmented using SSF.

Keywords: rice bran; solid state fermentation; antioxidant activity; bioactive compounds; *Aspergillus oryzae*; HPLC; total phenolic content; reducing power assay

Citation: Punia, S.; Sandhu, K.S.; Grasso, S.; Purewal, S.S.; Kaur, M.; Siroha, A.K.; Kumar, K.; Kumar, V.; Kumar, M. *Aspergillus oryzae* Fermented Rice Bran: A Byproduct with Enhanced Bioactive Compounds and Antioxidant Potential. *Foods* **2021**, *10*, 70. https://doi.org/10.3390/foods10010070

Received: 9 December 2020
Accepted: 26 December 2020
Published: 31 December 2020

Publisher's Note: MDPI stays neutral with regard to jurisdictional claims in published maps and institutional affiliations.

Copyright: © 2020 by the authors. Licensee MDPI, Basel, Switzerland. This article is an open access article distributed under the terms and conditions of the Creative Commons Attribution (CC BY) license (https://creativecommons.org/licenses/by/4.0/).

1. Introduction

Rice (*Oryza sativa*) belongs to the grass family and is the most widely consumed grass by a significant proportion of human population, especially in Asian regions. It is an agricultural commodity with the third highest worldwide production [1]. The total worldwide production of rice was about 769,657,791 tonnes in an area of 167,249,103 ha, of which India produced 168,500,000 tonnes [1]. Rice bran (RB) is the major byproduct of milling industry, especially processing rice, and ultimately represents 5–10% of the total grain. RB constitutes crude protein (11–13%), oil (20%) and dietary fibers (22.9%), including hemicelluloses, arabinogalactan, arabinoxylan, xyloglycan, and raffinose with good sources of bioactive Υ-oryzanol, Vitamin-E and minerals [2–4].

In routine practice, RB is used as feed for animals or in the production of edible cooking oils [5]. In the context of making our economies more circular and our diets more sustainable, there is a growing need and interest to valorize byproducts into new sustainable food ingredients with high nutritional value. Fungal fermentation is a promising method to process agricultural byproducts and to produce value added products [6]. SSF usually starts with the growth of fungal strains on substrate with little or no free water, with several advantages, including low costs, low environmental impact and high reproducibility [7].

Scientific reports supporting effect of fermentation on the antioxidant levels of various substrates, including barley [8], pearl millet [9,10], wheat [11], and rice bran [12,13], and reported their enhancement after SSF. This is a commonly used approach by the scientific community for the improvement of bioactive content of agro-industrial residues and assisted in reducing the environmental pollution caused by these residues. It is also evident from the findings that SSF may be used to improve product functional properties and as a tool to develop cereals with beneficial nutritional properties. SSF using *Rhizopus oligosporus* and *Monascus purpureus* enhanced the quality of fermented RB in terms of antioxidant property and total phenolic compounds [12]. Authors achieved maximum antioxidant capacity (more than 5-fold compared to untreated RB), and total phenolic content (more than 8-fold compared to untreated RB), when RB was fermented with the mixed cultures of *R. oligosporus* and *M. purpureus*. The other strain, *Rhizopus oryzae*, was also investigated by another group of researchers who established that SSF using *R. oryzae* improves the overall nutritional profile of the RB with excellent antioxidant activities [13].

The RB fraction has not been as much of a focus of research compared to polished rice. RB has well known for health beneficial properties due to its bioactive compounds, and during the present experimental work, an attempt was made to further enhance the antioxidant content of RB using *Aspergillus oryzae* as a starter culture. *Aspergillus oryzae* in particular was used because fungal strains, especially those belonging to the *Aspergillus* group, are well known for their potential to produce hydrolytic enzymes which resulted in enhanced production of bioactive compounds, especially cinnamic acids in fermented substrates during SSF [14]. SSF could be an important process to prepare antioxidant rich products with industrial applications. This process is comparatively cheaper than any other method of modulating nutrients. Furthermore, the efficacy of fungal strains towards the improvement of nutrients may vary with the substrate nutritional profile. Using SSF, those substrates could also be processed in the form of antioxidant rich food/feed which initially considered as waste. *Aspergillus oryzae* is widely used for fermentation of different natural resources such as rice [15,16]; brown rice and rice bran [17]. *Aspergillus oryzae* is a famous fungus commonly used for the preparation of local foods and beverages in Japan for the preparation of sweet potato, sake, shōchū, soy sauce and miso. Hence, it is evident that use of *Aspergillus oryzae* is common practice aiming to produce foods with high nutraceutical values. This is the first study investigating the use of SSF on RB using *Aspergillus oryzae* to evaluate the effect on antioxidant properties and bioactive compounds.

2. Experimental Details

2.1. Chemicals

Organic solvents and chemical reagents such as catechin, gallic acid, 2,2-diphenyl-1-picrylhydrazyl (DPPH), 2,2′-azinobis 3-ethylbenzothiazoline-6-sulfonic acid (ABTS$^+$), ascorbic acid used were of analytical grade and procured from HiMedia and Sigma-Aldrich. HPLC grade standards were procured (HiMedia, India) and used during HPLC analysis for estimating specific compounds in RB extracts. The glassware used in the present experimental part was of Borosilicate. Before using, glassware was washed with Labolene detergent and rinsed with tap water and sterilized in an oven at 100 °C for 1.5 h.

2.2. Isolation of Rice Bran (RB)

The experimental sample (Paddy cultivar PB-1121) was obtained from a local market in Sirsa, India. Sample grains were washed, dried and stored in airtight containers. A paddy dehusker (Khera, Delhi, India) was used for de-husking the paddy and a rice polisher

(Khera, Delhi, India) was used to separate the bran. RB was converted to a powdered form using mixer-grinder (Bajaj, India) and stored in deep freezer ($-20\ °C$; Vestfrost, India).

2.3. Starter Culture for Solid State Fermentation (SSF)

Fungal culture (*Aspergillus oryzae* MTCC 3107) for fermentation of RB was procured. Starter culture was grown on CYEA media (czapek yeast extract agar) and CYEB (czapek yeast extract broth) at $25 \pm 2\ °C$. Steam sterilized RB was inoculated by spraying spores as a suspension (2 mL, 1×10^5). RB sample which was not sprayed with fungal spores assigned name as URB.

2.4. SSF of RB

The experimental sample (50 g powdered RB) was used as substrate in Erlenmeyer flasks (250 mL). The sample was soaked in CYB (czapek yeast broth, 1:1 w/v) at ambient conditions for 10–12 h. Substrate was sprayed with the spores suspension and incubated (7 days, $25 \pm 2\ °C$). Choice and types of media used for starter culture growth during SSF merely vary with the fungal strain or starter culture type. Fermented rice bran (FRB) was removed from flasks after a predetermined interval of time (24 h) and dried in an oven (Narang Scientific Works, New Delhi, India) at $45\ °C$ ($24\ °C$ 48 h). Fermented and unfermented substrate was converted to flour (Sujata, India). URB and FRB flour was defatted with hexane (1:5 w/v, 3 times, 5 min), dried in an oven (NSW, India) and extracted with organic phase ethanol (50%) at $45\ °C$ for 30 min. Before performing the extraction process, flour samples were sieved to attain uniform sized particles for extracting bioactive compounds from them.

2.5. Evaluation of Phytochemical Composition

The detection of specific phytochemicals in the URB and FRB extracts was carried out using various qualitative tests as per standard methods described by [18]. Different tests were performed to assess the chemical composition of URB and FRB, such as saponin, steroids, flavonoids, coumarins and alkaloids.

2.5.1. Total Phenolic Content (TPC)

TPC in FRB and URB extracts was estimated by following FCR (Folin–Ciocalteu reagent) method [19]. An aliquot (100 µL) of extract was allowed to react with FC reagent (500 µL) and after an incubation period of 5 min, sodium carbonate (1500 µL) was added to the reaction mixture and total volume (10 mL) was prepared with distilled water. Absorbance was recorded at 765 nm against a blank. The standard used during the TPC assay was gallic acid.

2.5.2. Determination of Saponins

An amount of 5 mL of water was kept in storage vial following the addition of 1 mL of URB and FRB extract and the tube was shaken vigorously. The formation of lather confirms saponin presence.

2.5.3. Determination of Steroids

To detect steroids, 2 mL of URB and FRB extracts were taken, and 2 mL of chloroform was added followed by 2 mL of conc. H_2SO_4, red color in the chloroform layer showed the presence of steroids.

2.5.4. Determination of Flavonoid

To detect flavonoids in URB and FRB, 1 mL of 10% Lead Acetate was added to 1 mL of URB and FRB extracts; the formation of yellow-colored precipitates is an indication for the presence of flavonoids.

2.5.5. Condensed Tannin Content (CTC)

Quantification of CTC in URB and FRB extracts was calculated by vanillin: HCl protocol [20]. Aliquot of URB and FRB extracts (100 µL) was taken separately in storage vial (5 mL) followed by Vanillin–HCl (1:0.5) addition. The mixture was incubated (ambient temp. for 10 min) and absorbance was recorded.

2.5.6. Determination of Coumarins

To determine the presence of coumarins 1 mL of 10% NaOH was added to 1 mL of URB and FRB extracts. The formation of yellowish color showed the presence of coumarins.

2.5.7. Alkaloids

Alkaloids in URB and FRB extracts were estimated using three different tests.

i. Wagner's test: URB and FRB extracts (2 mL) were treated with Wagner's reagent (2 mL), the formation of precipitate (reddish-brown) confirmed alkaloids in sample extract.
ii. Mayer's test: To URB and FRB extract (1 mL), Mayer's reagent (2 mL) was added, and precipitate (dull white) confirmed alkaloids presence in sample extract.
iii. Hager's test: To URB and FRB extract (1 mL), Hager's reagent (3 mL) was added, and the formation of precipitate (yellow) confirmed alkaloids in sample extract.

2.5.8. Qualitative and Quantitative High Performance Liquid Chromatography (HPLC) Analysis

HPLC analysis of extracts (URB and FRB) was performed as per an already published report on antioxidants [21]. Extracts were prepared using the already published report [22]. HPLC analysis of URB and FRB extract for the estimation of bioactive compounds was performed (Shimadzu 10 AVP HPLC system). HPLC is an important step during analysis of extracts at different levels as it helps in validating the specific effects and finalizing the concept related to specific process. For the HPLC analysis of extracts and determination of specific bioactive compounds in them, a Shimadzu 10 AVP HPLC system was used which comprises SCL10 AVP system controller and two pumps (LC-10 AVP) CTO-10 AVP column oven with injection (Rheodyne 7120) value (20 µL sample loop) and photodiode-array detector (SPD-M10 AVP). Analytical HPLC column (Gemini-NX C18) (250 × 4.6 mm, 3 µm) with a guard column (40 × 3 mm, 3 µm) both from Phenomenex (Torrance, CA, USA) was used. Experimental performance was conducted at a rate (0.5 mL/min) using acetic acid (1.5% v/v solvent A) and aqueous ethanol:acetonitrile (40:50 v/v) mixture (solvent B) under the following gradient program: 0–8 min. 70% acetic acid, 8–19 min. 60% acetic acid, and 19–30 min. 30% acetic acid. Injection volume was 10 µL. The analysis was completed with different chromatograms formation at 280 nm.

2.6. Assessment of Antioxidant Properties in URB and FRB

2.6.1. DPPH (2,2-Diphenyl–1' picrylhydrazyl) Assay

Detection of DPPH scavenging activity in URB and FRB extracts was estimated [23]. First, 100 µL of URB and FRB extract was added (test tubes) followed by addition of 3 mL of DPPH (100 µM). Absorbance of URB and FRB extracts treated with DPPH solution was taken at 517 nm after 30 min of reaction process. Formula for calculation of percent (%) DPPH inhibition is mentioned below

$$\text{Percent (\%) DPPH inhibition} = (A_C - A_E/A_C) \times 100 \quad (1)$$

A_C (absorbance of control); A_E (absorbance of extracts).

2.6.2. ABTS Assay

Scavenging of the ABTS solution by URB and FRB extracts was calculated [24,25]. During the ABTS assay, URB and FRB extracts were allowed to react with potassium persulfate treated ABTS solution (16 h incubation). After 10 min of reaction time, the absorbance was taken (732 nm). The assay helps to study how extracts behaves under the oxidative stress conditions formed during normal biological processes. Percent (%) ABTS inhibition was calculated as mentioned below:

$$\text{Percent (\%) ABTS inhibition} = (A_C - A_E/A_C) \times 100 \quad (2)$$

A_C (absorbance of control); A_E (absorbance of extracts).

2.6.3. HFRSA Assay

The radical scavenging potential of URB and FRB extracts was estimated [26].

$$\text{Scavenged OH \%} = [(A_C - A_E)/A_C \times 100] \quad (3)$$

A_C (absorbance of control); A_E (absorbance of extracts).

2.6.4. Total Antioxidant Capacity (TAC)

The antioxidant capacity of extracts during TAC assay was determined by following the method as described by Prieto et al. [27]. URB and FRB extracts were analyzed for antioxidant capacity using sodium hydrogen orthophosphate (28 mM); conc. H_2SO_4 (0.6 M) and ammonium molybdate (4 mM) at 95 °C for 90 min.

2.6.5. Reducing Power Assay (RPA)

RPA of extracts (URB and FRB) was measured as per standardized method [28]. URB and FRB extract (100 µL) was allowed to react with aqueous potassium ferricyanide solution (1%; 100 µL) in water bath (50 °C for 30 min) and after reaction in water bath trichloroacetic acid (1% TCA; 100 µL) was added following incubation under dark (15 min). Dilution with double distilled water was done to achieve final volume (10 mL). Absorbance of colored complex was recorded at 700 nm. Activity was measured against quercetin (standard).

2.7. Statistical Analysis

Triplicate observations were processed through ANOVA using Minitab software (Version 16, Minitab Inc., State College, PA, USA).

3. Results and Discussion

3.1. Effect of SSF on Phytochemicals and TPC

Preliminary screening was carried out for URB and FRB to detect the presence of phytochemicals and the results of phytochemical analysis are shown in Table 1.

Table 1. Chemical composition of unfermented rice bran (URB) and fermented rice bran (FRB) (4th day) extracts.

Phytochemical	URB	FRB (4th Day)
Coumarins	+	+
Flavonoids	-	-
Saponin	-	-
Steroid	-	-
Alkaloids	-	-

(+) present whereas (-) sign showed absence.

Result shows the presence of coumarins and sugars in the extracts of samples. Shahidi [29] stated that phenolic and poly-phenolic compounds include a main class of secondary metabolites that act as free radical scavengers, reducer of low-density lipoprotein (LDL) and oxidation of cholesterol. TPC of URB was found to be 1.08 mg gallic acid equivalents (GAE)/g; significant ($p < 0.05$) difference among rice bran fermented for different durations was observed (Table 2).

Table 2. Impact of time of fermentation on total phenolic content (TPC) and condensed tannin content (CTC) of rice bran (RB) extracts.

Fermentation Time (Days)	TPC (g GAE/g dwb)	Percent (%) Change in TPC after SSF	CTC (mg CE/g dwb)	Percent (%) Change in CTC after SSF
URB	1.08 ± 0.13 [a]	–	34.6 ± 0.06 [a]	–
1	4.15 ± 0.09 [b]	↑284%	261 ± 0.05 [d]	↑652%
2	5.11 ± 0.10 [c]	↑373%	227 ± 0.07 [c]	↑555%
3	6.52 ± 0.11 [e]	↑503%	365 ± 0.13 [g]	↑952%
4	8.83 ± 0.21 [g]	↑717%	295 ± 0.11 [f]	↑750%
5	6.96 ± 0.34 [f]	↑544%	269 ± 0.02 [e]	↑675%
6	6.37 ± 0.19 [d,e]	↑489%	268 ± 0.07 [e]	↑674%
7	5.86 ± 0.08 [d]	↑442%	159 ± 0.09 [b]	↑358%

GAE: Gallic acid equivalent; SSF: Solid substrate fermentation; TPC: Total phenolic content; CTC: Condensed tannin content. Means followed by same superscript within a column do not differ significantly ($p < 0.05$). Subscripts show the % increase (↑) from unfermented sample for corresponding properties.

Increase in TPC was observed till the 4th day of fermentation, and further increase in time of fermentation TPC was decreased. On 4th day of fermentation, TPC was 8.83 mg GAE/g and the percentage increase in TPC content in FRB as compared to URB was 717%. Significant enhancement in amount of TPC during fermentation is a strong indication of the positive effect of SSF on substrate as later on antioxidant properties are solely dependent on the amount and type of phenolics. Fungal fermentation is considered as important phenomenon as desirable changes in nutritional profile could be achieved within short span of time. Schmidt et al. [13] observed two-fold increments of TPC in RB after SSF using *R. oryzae*. The increase in TPC till the 4th day may be due to hydrolytic enzymes produced during SSF [10]; however, the degradation of gallic acid to aliphatic compounds might be responsible for the decrease afterwards [30].

3.2. Effect of SSF on CTC

CTC is important as other secondary metabolites in the food. They have high antioxidant activity in vitro compared to monomeric phenolic compounds [31]. Significant ($p < 0.05$) differences were observed in CTC between URB and FRB and the values ranged between 34.6 to 365 mg CE/100 g extract (Table 2). An increase in CTC of FRB was observed till the 3rd day of fermentation, and thereafter the reverse was observed. Up to 10-fold increases with a percentage increase of 952% were observed on 3rd day of fermentation. Releases of enzymes take place after fermentation, which results in the production of plant chemicals such as tannin, alkaloids and phenylpropanoids.

3.3. Effect of SSF on Specific Bioactive Constituents

Phenolic compounds increase their antioxidant activity by various methods [32] and the effectiveness of these phenolic compounds as antioxidants mainly depends on their chemical structures, relative orientation, and number of OH groups attached to the aromatic ring [33]. Qualitative and quantitative measurements for the detection of specific phenolic compounds in URB and FRB were determined using HPLC. Standards *viz.* gallic acid, ascorbic acid, catechin and vanillin were used to evaluate the bioactive compounds (phenolic acids) in both URB and FRB. Bioactive compounds present in URB were significantly modulated during SSF which was also confirmed during HPLC analysis. The changes in the amount and type of bioactive compounds depend on enzymatic activity during the fermentation process. The presence of specific compounds in fermented products also makes them important substrate for pharmaceutical industries as it could also be used in preparation of various health benefiting formulations. The results of bioactive compounds in URB and FRB extracts are shown in Table 3 and Figures 1 and 2.

Table 3. Phenolic acid composition of un-fermented and fermented rice bran (4th day).

Compounds	URB	FRB (4th Day)
Ascorbic acid (µg/g)	11.1 [a]	12.7 [b]
Gallic acid (µg/g)	14.8 [a]	23.3 [b]
Catechin (µg/g)	9.6 [b]	2.8 [b]
Vanillin (µg/g)	5.8 [a]	1.2 [a]

Means followed by similar superscript within a column do not differ significantly ($p < 0.05$).

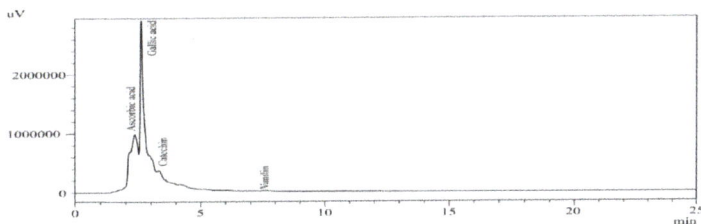

Figure 1. Phenolic acid composition of un-fermented rice bran at 280 nm.

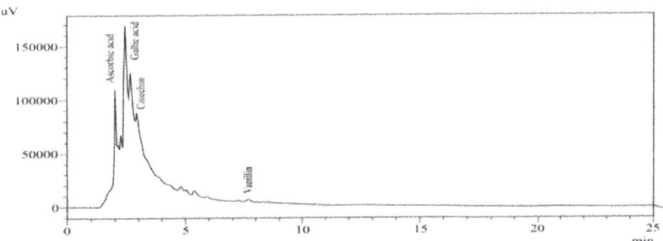

Figure 2. Phenolic acid composition of fermented rice bran (4th day) at 280 nm.

Results from the quantitative analysis of sample showed that the quantity of identified bioactive compounds varied from 1.2 to 23.3 µg/g, respectively. The composition of phenolic acid was significantly ($p < 0.05$) influenced by the duration of SSF. The effect of SSF on bioactive profiles depends on the type of substrate, starter culture and the extraction conditions after SSF [9,34]. After SSF, an increase in gallic acid (23.3 µg/g) was observed, followed by ascorbic acid (12.7 µg/g). Catechin (5.8 µg/g) and vanillin (1.2 µg/g) decreased after 4 days of fermentation. The reduction in catechin and vanillin may be due to the degradation of these compounds by microflora. The results are in line with the other researchers, where a significant ($p < 0.05$) enhancement in ferulic, sinapic, vanillic, caffeic, syringic, and 4-hydroxybenzoic acids of RB fermented with *Rhizopus oligosporus* and *Monascus purpureus* [12]. The authors also reported that the amount of ferulic acid was 8-fold higher compared to the untreated RB samples, suggesting the novelty of SSF for enhancing the phytochemical content. The increased phenolic acid content in RB is mainly caused due to the cleavage of compounds in conjunction with lignin [35]. In another study, it was reported that SSF using *Rhizopus oryzae* resulted in a 110% improvement in the phenolic compounds content [13]. The content of vanillin, chlorogenic acid, and p-hydroxybenzoic acid were increased throughout the fermentation process. The highest increment was detected in the ferulic acid (764.7 mg/g on dry weight basis) after SSF of 120 h. Authors concluded that *R. oryzae* produces certain enzymes which degrade the rigid cell wall of the RB and resulted in the release of the ferulic acid [13]. The increment in total phenolic content, and antioxidant properties may also attribute to the hydrolytic enzyme present in the fungal strain. These enzymes act upon the substrate (RB) and increase the access of hydroxy functional groups on the phenolic compounds. This improves the number of phenolic groups and, as a result, improves the antioxidant properties of the treated sample [12]. It is evident from the results that fermented Rb can be an important functional ingredient in the development of innovative food items with high phenolic content and antioxidant properties.

3.4. Effect of SSF on Antioxidant Activity

Evaluating the scavenging activity is a critical process, and a number of analytical methods may be used to evaluate the antioxidant activity of sample prepared from natural resources [36,37]. The DPPH radical scavenging assay reflects the capacity of the extract to transfer electrons or hydrogen atoms whereas $ABTS^+$ radical scavenging activity shows the hydrogen donating and the chain-breaking ability of the extract [38]. DPPH is popularly used as choice assay for analyzing the antioxidant properties of plant samples in a short period of time compared to other antioxidant assays. For URB, the free radical scavenging capacity of the sample using DPPH and $ABTS^+$ was observed to be 75.4 and 35.3%, respectively (Table 4).

Table 4. Effect of duration of fermentation on DPPH, ABTS inhibition, total antioxidant capacity (TAC), hydroxyl free radical scavenging activity (HFRSA) and reducing power assay (RPA).

Fermentation Time (Days)	DPPH (% Inhibition)	ABTS (% Inhibition)	TAC (mg AAE/g dwb)	HFRSA (% Inhibition)	RPA (mg QE/g dwb)
URB	75.4 ± 0.11 [a]	35.3 ± 0.48 [a]	7.3 ± 0.46 [a]	13.3 ± 0.90 [a]	0.7 ± 0.18 [a]
1	77.8 ± 0.33 ↑3.08 [e]	75.8 ± 0.89 ↑116.5 [d]	9.7 ± 0.32 ↑31 [b]	28.8 ± 0.42 ↑116 [e]	2.7 ± 0.39 ↑260 [b]
2	78.5 ± 0.24 ↑4.12 [d,e]	78.5 ± 0.77 ↑123.5 [e]	13.5 ± 0.38 ↑87 [c]	25.3 ± 0.66 ↑90 [c]	3.5 ± 0.16 ↑368 [c]
3	83.1 ± 0.20 ↑10.12 [f]	79.8 ± 0.54 ↑127.9 [f]	14.7 ± 0.19 ↑99 [d,e]	25.9 ± 0.48 ↑94 [c]	8.5 ± 0.22 ↑1040 [e]
4	85.4 ± 0.23 ↑12.66 [g]	82.7 ± 0.71 ↑136.5 [g]	15.4 ± 0.24 ↑103 [f]	28.5 ± 0.24 ↑114 [e]	16.5 ± 0.24 ↑2102 [g]
5	77 ± 0.19 ↑2.09 [d]	75.2 ± 0.85 ↑114.6 [d]	14.8 ± 0.61 ↑101 [e]	26.1 ± 0.56 ↑96 [d]	12.6 ± 0.21 ↑1589 [f]
6	76.2 ± 0.56 ↑1.046 [c]	70.7 ± 0.42 ↑101.9 [c]	14.5 ± 0.53 ↑97 [d]	24.7 ± 0.53 ↑85 [c]	8.8 ± 0.27 ↑1082 [e]
7	75.5 ± 0.17 ↑0.066 [b]	69.5 ± 0.23 ↑98.6 [b]	13.9 ± 0.39 ↑88 [c]	20.2 ± 0.85 ↑51 [b]	6.9 ± 0.34 ↑826 [d]

DPPH—2,2-diphenyl-1-picrylhydrazyl; ABTS-2,2′—Azinobis(3-ethylbenzothiazoline-6-sulphonic acid) diammonium salt; TAC-Total Antioxidant Capacity; HFRSA—Hydroxyl Free Radical Scavenging Activity; RPA—Reducing Power Activity. Means ± standard deviation, values followed by similar superscript within a column do not differ significantly ($p < 0.05$). Subscripts denote the percentage increase (↑) from unfermented sample for corresponding properties.

Scavenging activities were increased till the 4th day of fermentation and further increase in fermentation time reverse was observed. The values observed on 4th day were 85.4% and 82.7% with percentage increase of 12.6 and 136.5% for DPPH and $ABTS^+$ tests, respectively. Fermentation showed a positive effect on DPPH and $ABTS^+$ inhibitory effect on RB. Increase in antioxidant characteristics may be due to an increase in phenol and anthocyanin contents during fermentation [39]. Belefant-Miller et al. [40] suggested that the DPPH radical-scavenging activity of FRB was extremely correlated to the presence of antioxidant secondary metabolites. A significant increase in TPC was observed after the fermentation of mung bean cultivars [41]. The antioxidant activity of FRB was observed to be higher compared to URB when the HFRSA method was chosen for evaluating the antioxidant activity, and the values varied from 13.3¨ to 28.8%, respectively (Table 4). TAC is the total capacity of antioxidants for removing the free radicals in cells [42]. As shown in Table 4, the highest TAC was observed on the 4th day of fermentation of FRB sample as compared to the URB. The values for TAC ranged between 7.3 mg/g (URB) and 15.4 mg/g (FRB). TAC increased until the 4th of fermentation and subsequently declined on the 5th day. The release of flavonoids during the fermentation process enhances the anti-oxidative activity from plant-based foods, which may be a useful method of improving the supply of natural antioxidants [42]. The reducing power (RP) is a main indicator of the potential antioxidant activity of antioxidants. The antioxidant effect statistically increases as a function of the development of the RP, indicating that the antioxidant properties are related to the development of RP [28]. The highest RPA (16.5 mg/g) was observed on the 4th day of incubation of FRB extracts as compared to the URB (0.75 mg/g) (Table 4). A similar increase in RPA of rice fermented by *Phellinus linteus* was observed by Liang et al. [43]. During fermentation, phenolic content increased and these compounds can act as reducing agents and hydrogen donors [44]. Study of fermented products in different aspects proved to be helpful as it clearly demonstrates SSF effect whether positive or negative. Further, detailed analysis of fermented products helps to eradicate the negative aspects. As nutrients present in fermented products could be capable of combatting various medical problems and, hence, these products could be recommended to persons suffering from certain specific disorders.

4. Conclusions

The SSF of RB with *Aspergillus oryzae* significantly ($p < 0.05$) increased the TAC, TPC, CTC, RPA, DPPH, ABTS$^+$ and HFRSA. Except for CTC, the bioactive properties studied showed the maximum increase until 4th day of fermentation. Increase in SSF duration after a specific period resulted in the loss of important bioactive constituents. Four standards *viz.* gallic acid, ascorbic acid, catechin and vanillin were chosen to screen bioactive compounds using HPLC. SSF using *Aspergillus oryzae* thus can be effective method for the increment of antioxidants in rice bran. More research is required to optimize the size of inoculum, starter culture age and extraction parameters. The results of this study demonstrate that fermented rice bran would be an antioxidant rich and healthy food supplement as compared to non-fermented rice bran. After evaluating of the other nutritional components (shelf life and sensory analysis) under standardized conditions, the fermented rice bran could be used for formulation of different health benefiting food products. The SSF process could be recommended to modulate the bioactive profile and nutritional composition of industrial waste as well as eatable food materials. More research is also needed to standardize the fermentation process using other microorganisms so that their effect on nutritional quality may be evaluated.

Author Contributions: Conceptualization, S.P. and K.S.S.; methodology, S.P. and M.K. (Manoj Kumar); software, M.K. (Maninder Kaur); S.S.P.; validation, K.S.S., S.P. and S.S.P.; formal analysis, M.K. (Manoj Kumar) and A.K.S.; investigation, K.K. and V.K.; data curation, K.K. and V.K.; writing—original draft preparation, S.P. and S.S.P.; writing—review and editing, M.K. (Manoj Kumar) and S.G.; visualization, M.K. (Maninder Kaur) and A.K.S.; supervision, K.S.S. and S.P.; project administration, K.S.S.; funding acquisition, K.S.S. and S.G. All authors have read and agreed to the published version of the manuscript.

Funding: This research received no external funding.

Data Availability Statement: Not Applicable.

Conflicts of Interest: The authors declare no conflict of interest.

References

1. FAO (Food and Agricultural Organization of United Nations). 2017. Available online: http://faostat.fao.org/beta/en/#data/QC (accessed on 20 May 2019).
2. Pourali, O.; Asghari, F.S.; Yoshida, H. Production of phenolic compounds from rice bran biomass under subcritical water conditions. *Chem. Eng. J.* **2010**, *160*, 259–266. [CrossRef]
3. Parrado, J.; Miramontes, E.; Jover, M.; Gutierrez, J.F.; de Teran, L.C.; Bautista, J. Preparation of a rice bran enzymatic extract with potential use as functional food. *Food Chem.* **2006**, *98*, 742–748. [CrossRef]
4. Iqbal, S.; Bhanger, M.I.; Anwar, F. Antioxidant properties and components of some commercially available varieties of rice bran in Pakistan. *Food Chem.* **2005**, *93*, 265–272. [CrossRef]
5. Anal, A. (Ed.) *Food Processing By-Products and Their Utilization*; John Wiley & Sons: Hoboken, NJ, USA, 2018; Incorporated.
6. Tosuner, Z.V.; Taylan, G.G.; Özmıhçı, S. Effects of rice husk particle size on biohydrogen production under solid state fermentation. *Int. J. Hydrog. Energy.* **2019**, *44*, 18785–18791. [CrossRef]

7. Postemsky, P.D.; Bidegain, M.A.; Lluberas, G.; Lopretti, M.I.; Bonifacino, S.; Landache, M.I.; Omarini, A.B. Biorefining via solid-state fermentation of rice and sunflower by-products employing novel monosporic strains from Pleurotus sapidus. *Bioresour Technol.* **2019**, *289*, 121692. [CrossRef] [PubMed]
8. Sandhu, K.S.; Punia, S. Enhancement of bioactive compounds in barley cultivars by solid substrate fermentation. *Food Meas.* **2017**, *11*, 1355–1361. [CrossRef]
9. Salar, R.K.; Purewal, S.S.; Bhatti, M.S. Optimization of extraction conditions and enhancement of phenolic content and antioxidant activity of pearl millet fermented with Aspergillus awamori MTCC-548. *Resour. Effic. Technol.* **2016**, *2*, 148–157. [CrossRef]
10. Salar, R.K.; Purewal, S.S.; Sandhu, K.S. Fermented pearl millet (Pennisetum glaucum) with in vitro DNA damage protection activity, bioactive compounds and antioxidant potential. *Food Res. Int.* **2017**, *100*, 204–210. [CrossRef]
11. Sandhu, K.S.; Punia, S.; Kaur, M. Effect of duration of solid state fermentation by Aspergillus awamorinakazawa on antioxidant properties of wheat cultivars. *LWT Food Sci. Technol.* **2016**, *71*, 323–328. [CrossRef]
12. Abd Razak, D.L.; Abd Rashid, N.Y.; Jamaluddin, A.; Sharifudin, S.A.; Long, K. Enhancement of phenolic acid content and antioxidant activity of rice bran fermented with Rhizopus oligosporus and Monascus purpureus. *Biocatal. Agric. Biotechnol.* **2015**, *4*, 33–38. [CrossRef]
13. Schmidt, C.G.; Gonçalves, L.M.; Prietto, L.; Hackbart, H.S.; Furlong, E.B. Antioxidant activity and enzyme inhibition of phenolic acids from fermented rice bran with fungus Rizhopus oryzae. *Food Chem.* **2014**, *146*, 371–377. [CrossRef] [PubMed]
14. Bhanja Dey, T.; Chakraborty, S.; Jain, K.K.; Sharma, A.; Kuhad, R.C. Antioxidant phenolics and their microbial production by submerged and solid state fermentation process: A review. *Trends Food Sci. Technol.* **2016**, *53*, 60–74. [CrossRef]
15. Yasui, M.; Oda, K.; Masuo, S.; Hosoda, S.; Katayama, T.; Maruyama, J.; Takaya, N.; Takeshita, N. Invasive growth of Aspergillus oryzae in rice koji and increase of nuclear number. *Fungal Biol. Biotechnol.* **2020**, *7*, 8. [CrossRef] [PubMed]
16. Lee, D.E.; Lee, S.; Jang, E.S.; Shin, H.W.; Moon, B.S.; Lee, C.H. Metabolomic Profiles of Aspergillus oryzae and Bacillus amyloliquefaciens During Rice Koji Fermentation. *Molecules* **2016**, *21*, 773. [CrossRef] [PubMed]
17. Onuma, K.; Kanda, Y.; Ikeda, S.S.; Sakaki, R.; Nonomura, T.; Kobayashi, M.; Osaki, M.; Shikanai, M.; Kobayashi, H.; Okada, F. Fermented Brown Rice and Rice Bran with Aspergillus oryzae (FBRA) Prevents Inflammation-Related Carcinogenesis in Mice, through Inhibition of Inflammatory Cell Infiltration. *Nutrients* **2015**, *7*, 10237–10250. [CrossRef] [PubMed]
18. Trease, G.E.; Evans, E.C. *Pharmacognosy*, 13th ed.; Bailliere Tindall: London, UK, 1996; pp. 282–396.
19. Yu, L.; Haley, S.; Perret, J.; Harris, M. Comparison of wheat flours grown at different locations for their antioxidant properties. *Food Chem.* **2004**, *86*, 11–16. [CrossRef]
20. Julkunen-Tiitto, R. Phenolic constituents in the leaves of northern willows: Methods for the analysis of certain phenolics. *J. Agric. Food Chem.* **1985**, *33*, 213–217. [CrossRef]
21. Salar, R.K.; Purewal, S.S.; Sandhu, K.S. Bioactive profile, free-radical scavenging potential, DNA damage protection activity, and mycochemicals in Aspergillus awamori (MTCC 548) extracts: A novel report on filamentous fungi. *3 Biotech* **2017**, *7*, 164. [CrossRef]
22. Grobelna, A.; Kalisz, S.; Kieliszek, M. Effect of processing methods and storage time on the content of bioactive compounds in blue honeysuckle berry purees. *Agronomy* **2019**, *9*, 860. [CrossRef]
23. Yen, G.C.; Chen, H.Y. Antioxidant activity of various tea extracts in relation to their antimutagenicity. *J. Agric. Food Chem.* **1995**, *43*, 27–32. [CrossRef]
24. Re, R.; Pellegrini, N.; Proteggente, A.; Pannala, A.; Yang, M.; Rice-Evans, C. Antioxidant activity applying an improved ABTS radical cation decolorization assay. *Free Radic. Biol. Med.* **1999**, *26*, 1231–1237. [CrossRef]
25. Grobelna, A.; Kalisz, S.; Kieliszek, M. The effect of the addition of blue honeysuckle berry juice to apple juice on the selected quality characteristics, anthocyanin stability, and antioxidant properties. *Biomolecules* **2019**, *9*, 744. [CrossRef] [PubMed]
26. Smirnoff, N.; Cumbes, Q.J. Hydroxyl radical scavenging activity of compatible solutes. *Phytochemistry* **1989**, *28*, 1057–1060. [CrossRef]
27. Prieto, P.; Pineda, M.; Aguilar, M. Spectrophotometric quantitation of antioxidant capacity through the formation of a phosphomolybdenum complex: Specific application to the determination of vitamin E. *Anal. Biochem.* **1999**, *269*, 337–341. [CrossRef]
28. Oyaizu, M. Studies on products of browning reaction: Antioxidative activity of products of browning reaction. *Jpn. J. Nutr.* **1986**, *44*, 307–315. [CrossRef]
29. Shahidi, F. Functional foods: Their role in health promotion and disease prevention. *J. Food Sci.* **2004**, *69*, R146–R149. [CrossRef]
30. Bhat, T.K.; Singh, B.; Sharma, O.P. Microbial degradation of tannins–a current perspective. *Biodegradation* **1998**, *9*, 343–357. [CrossRef]
31. Hagerman, A.E.; Riedl, K.M.; Jones, G.A.; Sovik, K.N.; Ritchard, N.T.; Hartzfeld, P.W.; Riechel, T.L. High molecular weight plant polyphenolics (tannins) as biological antioxidants. *J. Agric. Food Chem.* **1998**, *46*, 1887–1892. [CrossRef]
32. Walter, M.; Marchesan, E. Phenolic compounds and antioxidant activity of rice. *Braz. Arch. Biol. Technol.* **2011**, *54*, 371–377. [CrossRef]
33. Sánchez-Moreno, C.; Larrauri, J.A.; Saura-Calixto, F. A procedure to measure the antiradical efficiency of polyphenols. *J. Sci. Food Agric.* **1998**, *76*, 270–276. [CrossRef]
34. Martins, S.; Mussatto, S.I.; Martínez-Avila, G.; Montañez-Saenz, J.; Aguilar, C.N.; Teixeira, J.A. Bioactive phenolic compounds: Production and extraction by solid-state fermentation. A review. *Biotechnol. Adv.* **2011**, *29*, 365–373. [CrossRef] [PubMed]

35. Schmidt, C.G.; Furlong, E.B. Effect of particle size and ammonium sulfate concentration on rice bran fermentation with the fungus Rhizopus oryzae. *Bioresour. Technol.* **2012**, *123*, 36–41. [CrossRef] [PubMed]
36. Singh, S.; Kaur, M.; Sogi, D.S.; Purewal, S.S. A comparative study of phytochemicals, antioxidant potential and in-vitro DNA damage protection activity of different oat (Avena sativa) cultivars from India. *Food Measure.* **2019**, *13*, 347–356. [CrossRef]
37. Kaur, P.; Dhull, S.B.; Sandhu, K.S.; Salar, R.K.; Purewal, S.S. Tulsi (Ocimumtenuiflorum) seeds: In vitro DNA damage protection, bioactive compounds and antioxidant potential. *Food Measure.* **2018**, *12*, 1530–1538. [CrossRef]
38. Pérez-Jiménez, J.; Arranz, S.; Tabernero, M.; Díaz-Rubio, M.E.; Serrano, J.; Goñi, I.; Saura-Calixto, F. Updated methodology to determine antioxidant capacity in plant foods, oils and beverages: Extraction, measurement and expression of results. *Food Res. Int.* **2008**, *41*, 274–285. [CrossRef]
39. Lee, I.H.; Hung, Y.H.; Chou, C.C. Solid-state fermentation with fungi to enhance the antioxidative activity, total phenolic and anthocyanin contents of black bean. *Int. J. Food Microbiol.* **2008**, *121*, 150–156. [CrossRef] [PubMed]
40. Belefant-Miller, H.; Grace, S.C. Variations in bran carotenoid levels within and between rice subgroups. *Plant Foods Hum. Nutr.* **2010**, *65*, 358–363. [CrossRef]
41. Randhir, R.; Shetty, K. Mung beans processed by solid-state bioconversion improves phenolic content and functionality relevant for diabetes and ulcer management. *Innov. Food Sci. Emerg. Technol.* **2007**, *8*, 197–204. [CrossRef]
42. Hur, S.J.; Lee, S.Y.; Kim, Y.C.; Choi, I.; Kim, G.B. Effect of fermentation on the antioxidant activity in plant-based foods. *Food Chem.* **2014**, *160*, 346–356. [CrossRef]
43. Liang, C.H.; Syu, J.L.; Mau, J.L. Antioxidant properties of solid-state fermented adlay and rice by Phellinus linteus. *Food Chem.* **2009**, *116*, 841–845. [CrossRef]
44. Ng, C.C.; Wang, C.Y.; Wang, Y.P.; Tzeng, W.S.; Shyu, Y.T. Lactic acid bacterial fermentation on the production of functional antioxidant herbal AnoectochilusformosanusHayata. *J. Biosci. Bioeng.* **2011**, *111*, 289–293. [CrossRef] [PubMed]

Article

Comparative Evaluation of the Nutritional, Antinutritional, Functional, and Bioactivity Attributes of Rice Bran Stabilized by Different Heat Treatments

Maria Irakli [1,*], Athina Lazaridou [2] and Costas G. Biliaderis [2]

1 Institute of Plant Breeding & Genetic Resources, Hellenic Agricultural Organization—Demeter, Thermi, 57001 Thessaloniki, Greece
2 Department of Food Science and Technology, School of Agriculture, Aristotle University of Thessaloniki, P.O. Box 235, 54124 Thessaloniki, Greece; athlazar@agro.auth.gr (A.L.); biliader@agro.auth.gr (C.G.B.)
* Correspondence: irakli@cerealinstitute.gr; Tel.: +30-231-047-1544

Abstract: The objective of this study was to evaluate the effects of different stabilization treatments—namely, dry-heating, infrared-radiation, and microwave-heating—on the nutritional, antinutritional, functional, and bioactivity attributes of rice bran (RB). Among the heating treatments, infrared-radiation exerted the strongest inactivation, resulting in 34.7% residual lipase activity. All the stabilization methods were found to be effective in the reduction of antinutrients, including phytates, oxalate, saponins, and trypsin inhibitors. No adverse effect of stabilization was noted on chemical composition and fatty acid profile of RB. Instead, stabilization by all heat treatments caused a significant decrease of vitamin E and total phenolics content in RB; the same trend was observed for the antioxidant activity as evaluated by the DPPH test. The antioxidant activity, as evaluated by ABTS and FRAP tests, and water absorption capacity were improved by the stabilization of RB, whereas the oil absorption capacity and emulsifying properties decreased. Microwave-heating enhanced the foaming properties, whereas infrared-radiation improved the water solubility index and swelling power of RB. Consequently, treatment of RB with infrared-radiation has a potential for industrialization to inactivate the lipase and improve some functional properties of this material for uses as a nutraceutical ingredient in food and cosmetic products.

Keywords: rice bran; stabilization; antioxidants; functional properties; bioactives; anti-nutritional components

1. Introduction

Rice bran (RB) is an agro-industrial by-product of rice kernel dry milling and consists mainly of the germ, pericarp, aleurone, and sub-aleurone layers [1]. It is a natural source of protein (14–16%), fat (12–23%), crude fiber (8–10%), other carbohydrates, vitamins, minerals, essential unsaturated fatty acids, and antioxidants, such as phenolics, tocopherols, tocotrienols, and γ-oryzanol with well-known beneficial effects in human health [2]. The predominant bioactive compound in the RB is γ-oryzanol due to its antioxidant [3], hypocholesterolemic [4], anti-inflammatory [5], anti-cancer [6], and anti-diabetic properties [7]. Furthermore, RB is gaining increased interest in the food, nutraceutical, and pharmaceutical industries, due to its high nutritional value, low cost, easy availability, high bioactivity potential, and the associated health benefits [2].

Despite the presence of various bioactives, the majority of RB is under-utilized as animal feed, fertilizer, or fuel in many developing countries. In Asian countries, it is used successfully for the production of RB oil with many health benefits. The main limitation in food industry applications is the requirement for quick stabilization of the RB, in order to reduce the antinutrients present and deactivate the action of lipase present in the outer layers of rice grain, which is primarily responsible for the hydrolysis of triglycerides into glycerol and free fatty acids following the milling process of the rice grains [8]. Immediately after milling, the free fatty acids level is increased, and the RB is unsuitable for human consumption, owing to rancid flavor and soapy taste. Moreover, lipoxygenase and peroxidase also contribute to the rancidity of RB although to a lesser extent [9].

Several stabilization treatments including dry and moist heating [10], extrusion cooking [11], microwave-heating [12,13], infrared-radiation [14,15], ohmic heating [16], parboiling or hydrothermal treatment [17], other physical methods of stabilization or refrigeration [18], enzymatic treatment [19], and chemical methods [20] have been employed immediately after milling to prevent the development of rancidity in RB. The selection of an optimized stabilization method is a crucial point for the food industry in order to ensure high component yield, low cost, and limited loss of bioactive constituents.

As a process, dry-heating is simple, convenient, and amenable to industrialization [21]. On the other hand, microwave-heating and infrared radiation are considered as alternative energy sources with high heat efficiency over shorter processing times without affecting RB quality. Microwave stabilization has been referred to as a quick heating method with high efficiency to inactivate lipase along with a better retention of bioactive compounds [12,22]. Similarly, infrared-radiation constitutes an alternative processing method to achieve efficient drying along with lipase inactivation in RB without affecting its quality [14,15,23]. In recent years, infrared-radiation heating has become an important technique in the food processing industry, because of its numerous advantages, such as the low capital cost and low energy cost, the simplicity of the required equipment, a lower drying time, the high quality of dried products, easy control of the process parameters, a uniform temperature distribution, and the clean operational environment, as well as space savings and easy accommodation with convective, conductive, and microwave heating [24]. As for the microwave heating, there are several major drawbacks that limit its application as the sole heating method for the drying process due to the high start-up costs, relatively complicated technology as compared to conventional convection drying despite the shorter drying time, improved product quality, and flexibility [25].

Although the efficiency of the microwave and infrared-radiation heating methods in stabilizing the RB, in relation to nutritional and bioactives profile, has been extensively studied by many researchers, to date, there are no reports dealing systematically with the effects of stabilization methods on functional properties from the technological point of view (physicochemical properties of RB) and their impact on anti-nutritional components. Relevant research works in this respect refer to protein isolates extracted from RB stabilized with different methods [26,27].

Therefore, the present study aimed to investigate the effects of dry-heating, microwave and infrared-radiation heating on the nutritional, antinutritional, and bioactive components of stabilized RB as well as on its functional properties and antioxidant capacities.

2. Materials and Methods

2.1. Sample Preparation

Fresh RB (10.93% moisture content) was collected from a local rice mill (Megas Alexandros, Sindos, Thssaloniki, Greece), and it was stabilized immediately using three heating processes as described below: (i) infrared-radiation: RB was transferred in a custom-made device consisting of a shaking aluminum tray, a radiator with two infrared lambs and a connected thermostat, as described by Irakli et al. [15] and it was heated at 140 °C for 15 min, (ii) dry-heating: RB was spread in thin layers to open pans and then was heated in an oven at 150 °C for 40 min and (iii) microwave-heating: RB was moistened up to 21% and then was placed in a glass plate (spread out in thin layer) and subjected to heating for 2 min at 650 W, corresponding to an approximately temperature of 160 °C. The stabilized RB samples were subsequently cooled at ambient temperature, packed in plastic containers, and stored at 4 °C for further use. An unstabilized RB sample was used as a control for comparative purposes, and it was stored under the same conditions as the stabilized RB samples.

2.2. Color Measurements

Color was measured using a colorimeter (HunterLab, model MiniScan XE Plus, Reston, VA, USA), following the CIE system defined by the L*, a*, and b* coordinates. The total color differences (ΔE) between the control and stabilized RB samples were calculated using the equation: $\Delta E = (\Delta L^{*2} + \Delta a^{*2} + \Delta b^{*2})^{1/2}$ [28].

2.3. Proximate Analysis

Moisture, protein, ash, and fat content were determined according to official methods [29]. Protein content was estimated by the Kjeldahl method, ash was determined by the dry ashing procedure, and fat content was determined using a Soxhlet apparatus, whereas total carbohydrates were calculated by difference.

2.4. Fatty Acid Composition

The fatty acid composition of the RB samples extracted with ether was determined using gas chromatography with flame ionization detection (Model Varian CP-3800, Middelburg, The Netherlands) based on the AOAC 996.06 method [30]. Fatty acid methyl esters were identified by comparison of their retention times with those of external standards (Supelco 37 Component FAME Mix) and the amount of individual fatty acids was expressed as percentages (%) of the total fatty acids determined.

2.5. Free Fatty Acids Content

Free fatty acids contents of the lipid fraction of RB samples were determined using a standard titration method [31], and the results were calculated as oleic acid equivalent, which was expressed as a percentage of total lipids.

2.6. Lipase Assay

The lipase activity in RB samples was determined according to Rose and Pike [32] with some modification. Briefly, 1 g of defatted RB was weighed into each of two test tubes: one blank and one sample. Then, 400 µL of olive oil and 200 µL of distilled water were added to both tubes and mixed. The lipids from the blank were immediately extracted with 5 mL of hexane (2 times), the supernatants after centrifugation at $1000 \times g$ for 5 min were pooled and evaporated using a rotary vacuum evaporator, and the residue was redissolved in 4 mL of isooctane. The other test tube (sample) was incubated for 4 h at 40 °C. After incubation, lipids were extracted as described for the blank, and both extracts were used for the lipase assay. An aliquot of 0.75 mL isooctane extract was mixed with 0.5 mL of 3% (v/v) pyridine in 5% (w/v) aqueous cupric acetate, the mixture was shaken for 1 min, centrifuged for 1 min, and the absorbance of supernatant was read at 715 nm and quantified against an external standard curve of oleic acid. The lipase activity was expressed as units per g RB (U/g), where 1 U was defined as the micromoles of fatty acid liberated per h, according to Equation (1):

$$\text{Lipase activity} = 1000\,(4+v)\,(As - AB)/\varepsilon t l s \qquad (1)$$

where v = volume of olive oil added (mL), As = absorbance of sample after incubation at 715 nm, AB = absorbance of blank at 715 nm, ε = molar absorptivity of oleic acid at 715 nm ($M^{-1} \cdot cm^{-1}$), t = incubation time (h), l = path length (1 cm for a standard cuvette), and s = sample weight (g).

2.7. Bioactive Components

Free phenolic compounds were extracted twice from 0.25 g sample with 5 mL of 60% aqueous ethanol in an ultrasound bath at room temperature for 10 min, followed by centrifugation at 10,000 rpm for 10 min at 4 °C. *Bound phenolic compounds* were recovered after the alkaline hydrolysis of the remained residue with 4N NaOH for 90 min under sonication, followed by acidification to pH 2.0 with concentrated HCl and finally extraction with ethyl acetate. The ethyl acetate fraction was vacuum-evaporated at 40 °C, and the residue was reconstituted in 4 mL of 70% aqueous methanol. Both free and bound fractions were stored at −25 °C until analysis.

Total phenolic content (TPC) of extracts was performed using the Folin–Ciocalteu's method according to Singleton et al. [33]. Briefly, extracts of 0.2 mL were mixed with 0.8 mL Folin–Ciocalteu reagent, 2 mL of sodium carbonate (7.5% w/v) solution, and distilled water until reaching a final volume of 10 mL. The absorbance at 725 nm was recorded after incubation for 60 min. The results were expressed as mg of gallic acid equivalents per g of sample on a dry basis (mg GAE/g dw).

Tocopherols and tocotrienols contents were determined as follows: 0.2g of sample was sonicated twice with 2.5 mL ethanol for 10 min, and the extract was collected after centrifugation at 1500× g for 10min. The combined supernatants were evaporated under the flow of nitrogen, the remaining residue was redissolved in 1 mL of a mixture acetonitrile/methanol (85:15, v/v), and finally, a 20 μL aliquot was injected into an HPLC system (Agilent Technologies, 1200 series, Urdorf, Switzerland) equipped with a YMC C_{30} column (250 × 4.6 mm id, 3 μm, MZ Analysentechnik, Mainz, Germany); the chromatographic conditions were as described by Irakli et al. [34].

γ-oryzanol content was determined in the same extract as with tocol analysis. The separation was carried out using a Target C_{18} (4.6 × 150 mm, 5 μm, MZ Analysentechnik, Mainz, Germany) column at 30 °C, with mobile phase consisted of acetonitrile/methanol/dichloromethane (40:45:15, $v/v/v$) under isocratic conditions at a flow rate of 1.5 mL/min. γ-oryzanol was detected at 330 nm. Quantitation was based on a linear calibration curve of the sum of the areas of four curves of γ-oryzanol standard, namely cycloartenyl ferulate, 24-methylene cycloartanyl ferulate, campesteryl ferulate and sitosteryl ferulate, as were analyzed by HPLC and confirmed by nuclear magnetic resonance (NMR) and MS by others researchers [35].

2.8. Antinutritional Composition

Oxalate content was determined using the titration method described by Oyeyinka et al. [36]. Firstly, 1 g of sample was extracted with 75 mL of 3M sulfuric acid under continuous mechanical stirring for 1 h, followed by filtration. Then, 25 mL of the filtrate was heated at 70 °C and was titrated steadily against 0.05 M potassium permanganate, until an extremely faint, pale pink end point color persistence was observed for 10 s. Oxalate content was calculated by the equation: *Oxalate (mg/g) = 1.33 x titer value*. The results were expressed on a dry basis.

Phytate content was evaluated using the method as described by Wheeler and Ferrel [37]. Firstly, 1 g of sample was macerated with 50 mL of 3% trichloroacetic acid (TCA) for 30 min with mechanical shaking. After centrifugation, a 10 mL aliquot of supernatant was mixed with 4 mL of $FeCl_3$ (2 mg ferric ion per mL in TCA), and the mixture was boiled for 45 min. After centrifugation, the precipitate was washed with 20–25 mL 3% TCA and was re-boiled for another 10 min. The washing was repeated with water, and the precipitate was dispersed in few mL water and 3 mL of 1.5N NaOH, and water was then added until the final volume of 30 mL. The suspension was boiled for 30 min, followed by filtration and washing with hot water, and the precipitate was dissolved with 40 mL of hot 3.2N HNO_3 into a 100-mL volumetric flask that was filled with the washing water; then, a 5 mL aliquot of the above extract was transferred in a 100-mL volumetric flask containing 70 mL water and 20 mL 1.5N KSCN was added, filled with water, and the absorbance was measured at 480 nm within 1 min. Iron content from $Fe(NO_3)_3$ was measured via a standard curve, and phytate phosphorus was calculated from the iron results assuming a 4:6 iron/phosphorus molecular ratio with the results being expressed on dry basis.

Total saponin content was determined using a spectrophotometric method, as described by Hierro et al. [38]. Firstly, 0.5 g of sample was extracted with 10 mL of absolute ethanol in an ultrasonic bath for 20 min at 70 °C. Then, the mixture was centrifuged at 4000 rpm for 10 min, and the above procedure was repeated. Combined supernatants were dried under vacuum, and the residue was reconstituted in 5 mL absolute ethanol. Aliquots of 250 µL were mixed with 250 µL of freshly prepared vanillin in ethanol (0.8%, w/v) and 2.5 mL of sulfuric acid in water (72%, v/v). Mixtures were vortexed, heated at 60 °C for 10 min, cooled in ice for 5 min, and the absorbance was recorded at 560 nm against the control sample containing ethanol. Results were expressed as mg diosgenin equivalent per g RB sample on dry basis.

Trypsin inhibitor activity was estimated by the method of Kakade et al. [39] as follows: 1 g of sample was extracted with 50 mL of 0.01 NaOH with continuous shaking for 1 h. The pH of the suspension was 9.5 to 9.8, and the suspension was diluted to the point that 1 mL produces trypsin inhibition of 40 to 60%. Then, portions of 0, 0.6, 1.0, 1.4, and 1.8 mL of the suspension were transferred into a double set of test tubes and adjusted to 2 mL with water. After the addition of 2 mL of trypsin (0.2 mg/mL), the tubes were boiled at 37 °C and 5 mL of BAPA solution (40 mg benzoyl-D L-arginine-p-nitroanilide hydrochloride dissolved in 1 mL of dimethyl sulfoxide and diluted to 100 mL with Tris-buffer, 0.05M, pH 8.2, containing 0.02M $CaCl_2$) was added, and the mixture was incubated for exactly 10 min. The reaction was terminated with 1 mL 30% acetic acid, and the absorbance was recorded at 410 nm against the reagent blank that was prepared by adding 1 mL of 30% acetic acid before the addition of BAPA. Trypsin inhibition activity is expressed in terms of trypsin units inhibited (TUI) taking into account that 1 TU is arbitrary defined as an increase of 0.01 absorbance units at 410 nm per 10 mL of extract on dry basis.

Tannins content was estimated as follows: Aliquots of about 400 µL of the phenolic extract prepared as described above were treated with polyvinylpolypyrrolidone (40 mg, 100 mg/mL) at 4 °C for 20 min. After centrifugation at 10,000 rpm (4 °C, 10 min), non-tannin phenolics (supernatant) were determined as above by the Folin–Ciocalteu method [40]. Total tannins were calculated by subtracting non-tannin phenolics from total phenolics.

2.9. Antioxidant Activity Assays

Three assays were employed to determine the antioxidant activities of the RB extracts: 2,2'-azinobis-(3-ethylbenzothiazoline-6-sulphonate) radical (ABTS·+) scavenging activity (ABTS assay) [41], 1,1-diphenyl-2-picrylhydrazyl radical scavenging activity (DPPH assay) [42] and ferric-reducing antioxidant power (FRAP) [43]. Trolox was used as the standard compound for calibration curves and the results were expressed in mg of trolox equivalents (TE) per g of RB on dry basis (mg TE/g dw).

2.10. Bulk Density

Bulk density (BD) of RB samples was determined by the method described by Kaushala et al. [44]: 25 g of flour was gently filled into 50 mL graduated cylinders that were previously tared. The bottom of each cylinder was gently tapped several times, and the volume was measured. BD was calculated as the weight of the sample per unit volume of sample (g/mL).

2.11. Functional Properties Determination

The water absorption capacity (WAC) of the RB samples was determined according to Abebe et al. [45]: RB samples were dispersed in distilled water and diluted to 10 g/mL. The dispersions were held at room temperature for 30 min with continuous stirring and followed by centrifugation for 30 min at $4000\times g$. Then, the supernatant was weighed, and the results were expressed as g of water retained per g of RB.

Oil absorption capacity (OAC) was determined according to Kaushala et al. [44]: A RB sample (0.5 g) was mixed with 6 mL of corn oil in pre-weighed tubes and centrifuged for 25 min at $3000\times g$. After stirring for 1 min with a thin brass wire, the tubes were allowed for 30 min and then were centrifuged for 25 min at $3000\times g$. The separated oil was removed with a pipette, and the tubes were inverted for 25 min to drain the oil prior to reweighing. The OAC was expressed as g of oil bound per g of the RB.

The water absorption index (WAI) and water solubility index (WSI) were measured according to Abebe et al. [45]. First, 2.5 g of RB sample was mixed with 30 mL of distilled water, vortexed, and cooked for 10 min in a 90 °C water bath. After centrifugation at 4000× g for 10 min, the supernatant was weighted and evaporated overnight at 110 °C to determine the soluble solids, and the sediment was weighed. The WAI was calculated by weighing the sediment and expressed as g of absorbed water per g of sample. WSI was calculated from the amount of soluble solids recovered from the supernatant divided by the sample weight × 100. The swelling power (SP) was calculated by dividing the weight of sediment per the difference between the initial sample weight and the amount of dry solids of the supernatant (g/g).

The foaming capacity (FC) and foaming stability (FS) were determined using the method described by Kaushala et al. [44]. The dispersions of the sample (1.5 g in 50 mL of distilled water) were homogenized, using an ultrasonic probe (Sonoplus, model HD 4100, Berlin, Germany) at high speed for 2–3 min. The suspension was immediately transferred into a graduated cylinder, and the homogenizer cup was rinsed with 10 mL of distilled water, which was then added to the graduated cylinder. The volume was recorded before and after whipping. FC was calculated by dividing the difference between the volume after and before the whipping per the volume before whipping × 100. Foam volume changes in the graduated cylinder were recorded at an interval of 60 min of storage in order to estimate the FS.

Emulsifying activity (EA) was determined by the method of Nazck et al. [46]. First, 3.5 g sample of RB was homogenized for 30 s with 50 mL of water using an ultrasonic probe (Sonoplus model HD 4100, Berlin, Germany) at high speed. After the addition of 25 mL of corn oil, the mixture was again homogenized for 90 s. The emulsion was centrifuged at 1100× g for 5 min. The EA was calculated by dividing the volume of the emulsified layer by the volume of emulsion before centrifugation × 100. The emulsion stability (ES) was determined by heating the above emulsion at 85 °C for 15 min, after which it was cooled and centrifuged. The ES was expressed as a percentage of the EA remaining after heating.

All results were expressed on dry basis to avoid the effect of different water content in the samples.

2.12. Statistical Analysis

Values were reported as the mean ± standard deviation of triplicate measurements. All parameters were subjected to one-way analysis of variance (ANOVA), and when ANOVA revealed significant differences between means, a Tukey's test at $p < 0.05$ was used to separate means by using the Minitab 17 (Minitab Inc., State College, PA, USA) software.

3. Results and Discussion

3.1. Free Fatty Acids and Lipase Activity

Heating treatments are the most widely used methods of RB stabilization. Stabilized RB offers multiple benefits including better quality characteristics and extended product shelf-life on storage, thus providing a wider range of RB fortified products for consumers. Novel technologies in food preservation have gained increased industrial interest as they are more environmentally and economically sustainable compared to conventional preservation methods. In this context, the effects of three different heating treatments, such as dry-heating, infrared-radiation, and microwave-heating, on the stabilization of RB were examined in the present study on a comparative basis. The applied conditions (temperature and time duration) for each heating treatment were selected according to preliminary results based on response surface methodology using relative lipase activity as a probe for the effectiveness of the treatment (data not shown). The free fatty acids (FFA) content and lipase activity of untreated and heat-treated RB samples were determined after 1 day of storage at room temperature. As shown in Figure 1, significant decreases ($p < 0.05$) on FFA content were observed in the heat-treated RB samples by all three methods of heat treatment employed. Among the heating treatments, dry-heating gave the lowest

FFA content (3.91%), whereas microwave-heating and infrared-radiation had similar FFA content with a mean value of 5.53%.

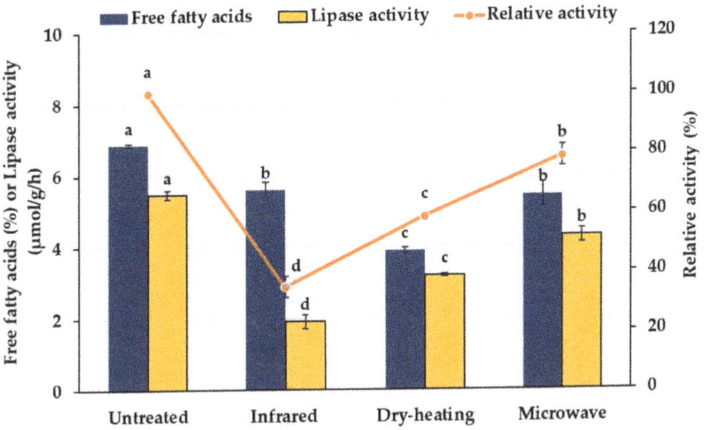

Figure 1. Effect of three different heating treatments (infrared-radiation, dry-heating, and microwave-heating) on free fatty acids content and relative lipase activity of RB samples; values followed by the same letter for the same estimated parameter are not significantly different ($p > 0.05$).

Lipase is the major cause of RB lipid deterioration i.e., hydrolytic oxidation. The changes in lipase activity as a result of stabilization treatments are also shown in Figure 1. Among the treatments, microwave-heating for 2 min at 650 Watt resulted in the most inefficient stabilization (78.6% relative enzymatic activity was remained). Dry-heating at 140 °C for 40 min showed better stabilization results (lowest content of liberated FA), with remaining lipase activity at ≈58.5%; moreover, infrared radiation at the same temperature for 15 min gave even lower residual lipase activity of 34.8%. Yu et al [47] compared 11 RB stabilization methods and found that among various heating treatments, autoclaving exerted the best inactivation effect, yielding ≈10.7% relative activity of lipase, whereas other heating methods resulted in about 30–35% residual lipase activities, except dry-heating (68.9%). However, although autoclaving was the most effective method to inactivate the lipase of RB, it is not easily applicable at an industrial scale. Furthermore, ultraviolet irradiation was found to be a promising alternative (convenient and energy-saving) stabilization method without influencing oil quality and nutrient content.

3.2. Nutritional Profile

The proximate analysis of untreated and heat-treated of RB samples, reflecting the major nutritional components, is presented in Table 1. As expected, the moisture content of all heat-treated RB samples is significantly lower ($p < 0.05$) than the untreated sample. Among the heat-treated samples, dry-heating significantly decreased the moisture content at the lowest level compared to microwave treatment, showing the highest level, whereas the infrared-radiation treatment gave an intermediate moisture content. The protein content was not affected significantly by the method of stabilization, ranging between 18.13 and 18.51%. However, stabilization by all heating methods increased significantly ($p < 0.05$) the fat and ash contents, when the data were expressed on a dry basis. The increased fat in heat-treated RB may be attributed to structure loosening of the treated samples, permitting a more efficient fat extractability by the organic solvent [21]. As for the carbohydrate content of the RB samples, there seems to be a very slight reduction ($p < 0.05$) in all heat-treated materials.

Table 1. Proximate composition (% db) of rice bran (RB) samples before and after heat treatments.

Samples	Moisture	Protein	Fat	Ash	Carbohydrates
Untreated	10.93 ± 0.07 [a]	18.39 ± 0.03 [ab]	17.81 ± 0.31 [b]	8.83 ± 0.03 [b]	55.01 ± 0.34 [a]
Infrared	5.48 ± 0.05 [c]	18.13 ± 0.10 [b]	18.76 ± 0.37 [a]	8.94 ± 0.04 [a]	54.42 ± 0.27 [b]
Dry-heating	2.76 ± 0.00 [d]	18.51 ± 0.10 [a]	18.53 ± 0.04 [a]	9.00 ± 0.02 [a]	53.97 ± 0.12 [b]
Microwave	8.30 ± 0.01 [b]	18.37 ± 0.10 [ab]	18.74 ± 0.18 [a]	8.96 ± 0.01 [a]	53.94 ± 0.02 [b]

Values are means ± standard deviation (n = 3), while different letters for values in each column indicate significant differences (p < 0.05); db: dry basis.

3.3. Fatty Acid Profile

The fatty acid composition of untreated and heat-stabilized RB oils (Table 2) has been determined to assess the effect of stabilization treatment. In general, RB oil is rich in polyunsaturated (41.70–42.60%), followed by monounsaturated (37.14–37.68%) and saturated fatty acids (19.50–20.03%). Palmitic (16.30–16.60%), oleic (36.30–36.80%), and linoleic acids (40.05–40.91%) were found to be the three dominant fatty acids, as previously reported in the literature [13–15]. The rations of polyunsaturated/saturated FAs (PUFA/SFA) and ω6/ω3 for untreated RB were 2.12 and 24.64, respectively, while the heat-treated RB samples were in the range of 2.08–2.19 and 24.35–26.66, respectively, without any significant change for the heated RB samples compared to the untreated RB. According to the nutritionally recommended values for the PUFA/SFA ratio (higher than 0.45) and ω6/ω3 ratio (a range between 4:1 and 10:1, due to differing opinions in the literature) [48], it can be concluded that the PUFA/SFA ratio of RB is beneficial to health, whereas the ω6/ω3 ratio does not comply with a desired good nutritional index [49].

Table 2. Fatty acid profile of RB samples before and after heat treatments (% of the total fatty acids).

Fatty Acids	Untreated	Infrared	Dry-Heating	Microwave
Myristic acid (C14:0)	0.36 ± 0.01 [c]	0.41 ± 0.01 [a]	0.38 ± 0.01 [b]	0.35 ± 0.01 [bc]
Palmitic acid (C16:0)	16.30 ± 0.30 [a]	16.30 ± 0.50 [a]	16.55 ± 0.15 [a]	16.60 ± 0.10 [a]
Stearic acid (C18:0)	1.30 ± 0.01 [a]	1.19 ± 0.03 [b]	1.31 ± 0.01 [a]	1.30 ± 0.01 [a]
Oleic acid (C18:1, cis-9)	36.75 ± 0.15 [a]	36.30 ± 0.70 [a]	36.70 ± 0.10 [a]	36.80 ± 0.10 [a]
Linoleic acid (C18:2 cis-9,12), n6	40.29 ± 0.22 [a]	40.91 ± 1.06 [a]	40.05 ± 0.25 [a]	40.25 ± 0.15 [a]
α-Linolenic acid (C18:3, cis-9,12,15), n3	1.64 ± 0.02 [a]	1.54 ± 0.12 [a]	1.64 ± 0.01 [a]	1.65 ± 0.02 [a]
Arachidic acid (C20:0)	0.64 ± 0.01 [a]	0.51 ± 0.01 [c]	0.58 ± 0.01 [b]	0.57 ± 0.01 [b]
cis-11-Eicosenoic acid (C20:1)	0.55 ± 0.01 [ab]	0.52 ± 0.01 [c]	0.55 ± 0.01 [a]	0.53 ± 0.01 [bc]
Behenic acid (C22:0)	0.27 ± 0.01 [b]	0.28 ± 0.01 [b]	0.28 ± 0.01 [a]	0.29 ± 0.01 [a]
Lignoceric acid (C24:0)	0.76 ± 0.01 [b]	0.71 ± 0.02 [c]	0.82 ± 0.02 [a]	0.75 ± 0.01 [b]
Saturated fatty acids (SFA)	19.74 ± 0.28 [a]	19.50 ± 0.55 [a]	20.03 ± 0.18 [a]	20.02 ± 0.09 [a]
Monounsaturated fatty acids	37.64 ± 0.14 [a]	37.14 ± 0.70 [a]	37.63 ± 0.08 [a]	37.68 ± 0.10 [a]
Polyunsaturated fatty acids (PUFA)	41.93 ± 0.20 [a]	42.60 ± 0.98 [a]	41.70 ± 0.25 [a]	41.92 ± 0.17 [a]
PUFA/SFA	2.12 ± 0.04 [a]	2.19 ± 0.11 [a]	2.08 ± 0.03 [a]	2.09 ± 0.02 [a]
ω6	40.36 ± 0.16 [a]	40.92 ± 1.04 [a]	40.21 ± 0.17 [a]	40.30 ± 0.13 [a]
ω3	1.64 ± 0.02 [a]	1.54 ± 0.12 [a]	1.64 ± 0.01 [a]	1.65 ± 0.02 [a]
ω6/ω3	24.64 ± 0.40 [a]	26.66 ± 2.66 [a]	24.59 ± 0.07 [a]	24.35 ± 0.15 [a]

Minor fatty acids were lauric (C12:0), pentadecanoic (C15:0), heptadecanoic (C17:0), cis-10 heptadecenoic (C17:1), linolelaidic (C18:2, trans-9,12), erucic (C22:1, cis-13), tricosanoic (C23:0). Means with same superscript letter within the same line are not significantly different (p > 0.05).

The effect of heat treatments on the fatty acid composition of RB was found to be statistically insignificant (p > 0.05), as already observed by Yilmaz et al. [14] and Irakli et al. [15]. However, for infrared-radiation treatment, a slight increase in polyunsaturated and a decrease in monounsaturated and saturated fatty acids were noted compared to the untreated sample. Similarly, Ramezanzadeh et al. [13] found no significant differences for the fatty acids between stabilized RB by microwave-heating and the raw RB.

3.4. Color Parameters

Color is an important quality determinant, influencing consumer preference for certain food products. The effects of different heating treatments on the color changes of RB are shown in Figure 2. The color values (L* and b*) were affected significantly ($p < 0.05$) by all three heating methods; however, no differences were noted among the heating treatments ($p > 0.05$). The RB stabilized with microwave-heating exhibited the brightest brown color, as evidenced from the highest ΔE* values among all stabilized RB samples. The color parameters of the stabilized RB samples were comparable to those of RB stabilized by other methods in previous studies; Rodchuajeen et al. [50] reported color values in the range of 62–64 for L*, 5.5–5.9 for a*, 18.2–19.1 for b*, and 6.1–8.2 for ΔE* for stabilized RB samples by different moving-bed drying methods.

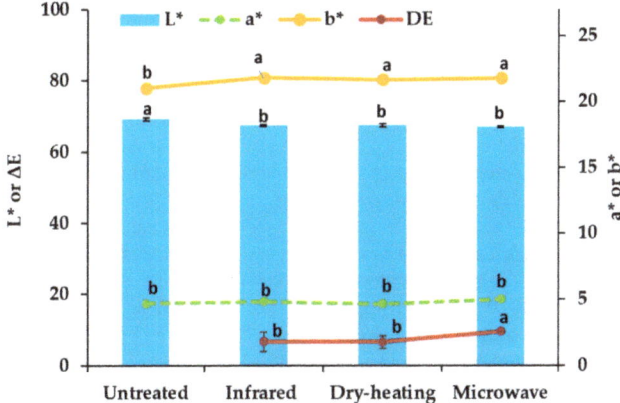

Figure 2. Effect of three different heating treatments (infrared-radiation, dry-heating, and microwave-heating) on color values of RB samples; values on bars or lines with specified by different letter are significantly different from each other ($p < 0.05$).

3.5. Antinutritional Components

Although the increased consumption of cereal bran in the diet is encouraged by nutritionists to increase overall fiber intake, no consideration of the antinutritional components has ever been, although antinutrients are known to impair food digestion and absorption. The antinutritional composition of the RB samples before and after heat treatments is presented in Table 3. Stabilization treatments significantly reduced the antinutritional components. Overall, the highest phytate, oxalate, saponins, and trypsin inhibition levels were found in untreated RB, whereas heating treatments reduced all these antinutritional components. Phytate has been considered as an antinutrient because of its ability to interact with minerals, proteins, and starch, resulting in insoluble complexes that modify the functionality, digestion, and absorption of these food components [51]. The phytate content of untreated and treated RB samples ranged between 20.04 and 27.08 mg/g dw. Similar values are reported by Kaur et al. [52], although a large variation in phytate content is often found in the literature due to genotypic and environmental effects or different extraction rates adopted upon milling of rice grains. In the present study, the greatest reduction in phytate content was noted for the microwave stabilized RB (≈26.0%) followed by the infrared-radiation treated (≈22.2%) and the dry-heating treated RB (≈19.9%); however, no differences were observed among the heated-treated RB samples ($p > 0.05$). Previous work by Sharma et al. [21] showed a significant reduction in phytate content during the extrusion processing of RB. Similarly, Khan et al. [26] reported that several stabilization treatments significantly reduce the phytate content of RB. The phytic acid reduction by heating treatments may be partially due to the heat labile nature of phytic acid and the formation of insoluble complexes between phytate and other components [53].

Table 3. Antinutritional composition of RB samples before and after heat treatments.

Samples	Phytate (mg/g dw)	Oxalate (mg/g dw)	Saponins (mg/g dw)	Tannins (mg/g dw)	Trypsin Inhibition (TUI/g dw)
Untreated	27.08 ± 0.73 [a]	6.52 ± 0.15 [a]	89.70 ± 0.28 [a]	3.39 ± 0.16 [a]	14.93 ± 0.79 [a]
Infrared	21.06 ± 0.46 [b]	4.96 ± 0.07 [c]	78.98 ± 1.59 [b]	3.27 ± 0.02 [a]	10.84 ± 0.37 [b]
Dry-heating	21.70 ± 0.11 [b]	5.80 ± 0.04 [b]	62.35 ± 2.57 [c]	3.37 ± 0.06 [a]	10.08 ± 0.10 [b]
Microwave	20.04 ± 1.00 [b]	4.25 ± 0.07 [d]	65.87 ± 0.82 [c]	3.44 ± 0.14 [a]	10.31 ± 0.38 [b]

Values are mean ± standard deviation, while same letter along a column indicates no significant differences at $p > 0.05$; dw, dry-weighted.

Oxalate is also considered as an antinutritional component, as it can form non-absorbable salts with sodium, calcium, and ammonium ions, rendering these minerals unavailable [54]. High intakes of soluble oxalate may cause calcium oxalate crystallization and the formation of kidney stones (nephrolithiasis) in the urinary tract [55]. The effect of heating treatments on the oxalate content of RB is indicated in Table 3. Significant variation was noted in oxalate content among the RB samples. The oxalate content of untreated RB was 6.52 mg/g dw, which was reduced to 4.96, 4.25, and 5.80 mg/g dw with the infrared-radiation, microwave-heating, and dry-heating, respectively. Among the different heating treatments, microwave-heating was found most effective in reducing the oxalate content in RB by 34.9%. Kaur et al. [52] also reported a lower level of oxalate content subjected to extrusion cooking, compared to its untreated counterpart.

Trypsin inhibitors, being low molecular weight proteins, were significantly inactivated ($p < 0.05$) by all heat treatments employed, as illustrated in Table 3. The trypsin inhibition was minimum in the dry-heated RB, followed by microwave-heated and infrared-radiation-treated RB samples; i.e., dry-heating, microwave-heating, and infrared-radiation was effective in reducing the trypsin inhibitor activity by ≈32.5, 31.0, and 27.4%, respectively. It has been reported that heat treatments such as microwave treatment, cooking, and autoclaving inactivate the trypsin inhibitors as a result of denaturation of these heat-labile proteins [56]. It has been suggested that reactions involving deamidation (splitting of covalent bonds) and the destruction of disulfide bonds might be responsible for the thermal inactivation process [52].

Saponins have the ability to bind proteins, enhancing protein stability against heat denaturation and decreasing the susceptibility of proteins to proteases. They may also cause gastrointestinal lesions, entering into the blood stream and hemolyzing the red blood cells [57]. The data displayed in Table 3 show the effect of stabilization on the saponin content of RB. All heating treatments resulted in a significant reduction of the saponin content, with the dry and microwave-heating presenting the maximum percentage reduction in saponin content at ≈30.5 and 26.6%, respectively, whereas the infrared radiation was less effective (reduction ~12.0%). The loss of saponins during microwave heating might be attributed to the thermo-labile nature of these compounds.

Although tannins form tannin–protein complexes leading to the inactivation of digestive enzymes and protein insolubility, they are considered effective in lowering blood glucose levels by delaying intestinal glucose absorption, thus delaying the onset of insulin-dependent diabetes mellitus [58]. In the present study, dry-heating, microwave processing, and infrared-radiation did not seem to reduce the total tannin content, in contrast to all other antinutritional components studied. Similarly, Sahni and Sharma et al. [59] also found that thermal processing treatments on alfalfa seeds showed less reduction in tannin content than other treatments. Moreover, Osman [60] found that the tannins content increased in roasted or cooked lablab beans due to the inhibition of polyphenol oxidase after heat treatment. However, Deng et al. [61] reported a significant reduction of tannins in buckwheat grains after cooking and microwave treatment.

3.6. Bioactive Components

The principal bioactive compounds that contribute to the promising health-related benefits of RB comprise of phenolic compounds, tocols, and sterol derivatives (particularly γ-oryzanol). The phenolic compounds in RB exhibit antioxidant activity and may reduce free radical-mediated cellular damage. The TPC of each of the stabilized RB samples was determined using the Folin–Ciocalteu method in both free and bound forms. Our results demonstrate that total TPC decreased ($p < 0.05$) as a result of the thermal stabilization of RB (Table 4), with significant differences ($p < 0.05$) among heat treatment applied. Untreated RB had the highest total TPC, followed by microwave-heated, infrared-radiated, and dry-heated RB samples. The free and bound TPC decreased by 16.5% and 10.3% after dry-heating, by 12.3% and 11.1% after infrared-radiation, and by 11.4% and 8.1% after microwave-heating treatment, respectively. Similar observations have been made by Rodchuajeen et al. [50], showing that heat processing can reduce the polyphenolic components of RB. In contrast, Saji et al. [62] found that stabilized RB by dry-heating had higher TPC than the untreated material. This discrepancy may be due to the fact that different factors, such as extraction parameters, varietal differences, bran fraction, and environmental conditions during the growing season may have a compositional impact and modulate the effects of the stabilization treatments.

Table 4. Bioactive components of RB samples before and after heat treatments.

Samples	Fractions	Untreated	Infrared	Dry-Heating	Microwave
TPC (mg GAE/g dw)	Free	5.99 ± 0.05 [a]	5.15 ± 0.06 [c]	5.00 ± 0.01 [d]	5.31 ± 0.01 [b]
	Bound	1.80 ± 0.02 [a]	1.60 ± 0.06 [b]	1.62 ± 0.03 [b]	1.65 ± 0.04 [b]
	Total	7.79 ± 0.03 [a]	6.75 ± 0.01 [c]	6.61 ± 0.02 [d]	6.96 ± 0.02 [b]
Tocols (mg 100/g dw)	T3s	13.45 ± 0.20 [a]	11.42 ± 0.31 [b]	10.96 ± 0.23 [b]	11.59 ± 0.41 [b]
	Ts	3.22 ± 0.03 [a]	2.54 ± 0.01 [b]	2.58 ± 0.01 [b]	2.57 ± 0.26 [b]
	Total	16.67 ± 0.16 [a]	13.96 ± 0.31 [b]	13.54 ± 0.24 [b]	14.16 ± 0.67 [b]
γ-Oryzanol (mg/g dw)		2.44 ± 0.04 [a]	2.37 ± 0.03 [ab]	2.40 ± 0.01 [ah]	2.32 ± 0.13 [ab]
DPPH (mg TE/g dw)	Free	7.75 ± 0.06 [a]	6.79 ± 0.14 [b]	6.69 ± 0.05 [b]	6.85 ± 0.09 [b]
	Bound	2.72 ± 0.19 [a]	2.17 ± 0.02 [b]	2.39 ± 0.15 [b]	2.79 ± 0.08 [a]
	Total	10.46 ± 0.25 [a]	8.96 ± 0.12 [c]	9.08 ± 0.20 [c]	9.64 ± 0.01 [b]
ABTS (mg TE/g dw)	Free	14.73 ± 0.11 [a]	12.92 ± 0.18 [c]	12.69 ± 0.30 [c]	13.61 ± 0.11 [b]
	Bound	5.35 ± 0.33 [c]	5.05 ± 0.49 [c]	12.65 ± 0.41 [b]	13.03 ± 0.29 [a]
	Total	20.73 ± 0.48 [c]	18.27 ± 0.34 [d]	25.70 ± 0.72 [b]	27.82 ± 0.43 [a]
FRAP (mg TE/g dw)	Free	6.61 ± 0.04 [b]	7.13 ± 0.17 [a]	6.58 ± 0.06 [b]	5.01 ± 0.01 [c]
	Bound	2.46 ± 0.08 [c]	2.55 ± 0.03 [c]	3.69 ± 0.03 [b]	3.98 ± 0.16 [a]
	Total	9.08 ± 0.12 [c]	9.68 ± 0.20 [b]	10.27 ± 0.10 [a]	8.99 ± 0.14 [c]

Values are mean ± standard deviation, while same letter along a line indicates no significant differences ($p > 0.05$). TPC: total phenolic content.

The major lipophilic fractions of RB are tocopherols, tocotrienols (known as tocols), and γ-oryzanol that are characterized as the strongest antioxidants in RB [9]. According to the results presented in Table 4, the contents of tocotrienols and tocopherols in heat-treated RB were significantly lower than the untreated material ($p < 0.05$), which was presumably due to the degradation of the heat-sensitive antioxidants, with a loss of 15.8 and 20.5% for tocopherols and tocotrienols, respectively. On the contrary, γ-oryzanol was not reduced significantly ($p < 0.05$) by all heat treatments, as it has been reported to be a relatively thermostable antioxidant [63]. Untreated RB had the highest γ-oryzanol content, whereas the microwave-heated RB the lowest concentration among all the heat-treated RB samples. Similar results have been reported in the study of Lakkakula et al. [64] in which ohmic heat processing was adopted for RB stabilization.

The effects of heat treatments on the antioxidant activity of RB were evaluated by three different methods, and the results are given in Table 4. Generally, greater antioxidant activity was observed in the free phenolic extract, as evaluated by the three tests (DPPH, ABTS, and FRAP). The DPPH free radical method is being used extensively to evaluate reducing substances. It has been noted that the DPPH radical scavenging activity of free RB phenolics was on average 2.6 times higher than those of the extracts of bound phenolics. A significant decrease in DPPH radical scavenging activity of the free phenolic fractions as well as total phenolics appeared upon heat treatment; nevertheless, the DPPH values of the bound fractions were not altered significantly, except in the case of the infrared-radiation and the dry-heating treatments. Such reductions go in parallel with the reduction in TPC and could be attributed to the formation of an irreversible covalent bond between liberated phenolic compounds and proteins [65].

A similar trend was observed for ABTS radical scavenging activity in the case of extracts of free phenolics, as the ABTS scavenging capacity decreased for all heat-stabilized RB samples (Table 4). However, for the bound phenolic extracts, the ABTS scavenging capacity increased after heat-stabilization, particularly for samples subjected to dry and microwave heat treatments ($\approx$2.4 folds). This could be due to the differences in the type of radicals and phenolic compounds in both extracts. The presence of other phenolic compounds and non-phenolic antioxidants in the free extracts could also contribute to the antioxidant potential of these materials. Despite the large differences observed between free and bound phenolic extracts, which is in contrast to the DPPH assay, higher antioxidant activity was also noted in a study by Saji et al. [62] in which microwave and dry-heating methods were adopted for the RB stabilization. Overall, among all stabilization treatments, dry-heating and microwave-heating led to 23.9 and 34.2% increase in total ABTS value, respectively, while the infrared-radiation did not bring a significant change.

For the bound phenolic extracts, the antioxidant capacity of all stabilized RB samples, using the FRAP test, was increased compared to the untreated RB, and this was more pronounced to microwave-treated RB sample (Table 4). Some differences in the antioxidant activity values were also noted among the free phenolic extracts; the sample treated with infrared radiation exhibited the highest value for antioxidant capacity. The enhanced antioxidant activity of heat-treated RB samples may be due to the generation and accumulation of Maillard-derived melanoidins at high temperatures [66].

3.7. Functional Properties

The bulk density (BD) is a key factor for the storage, transportation, and processing of dry powders in food product formulations. In our study, the BD of RB samples varied from 0.33 to 0.36 g/mL. The stabilization by all heat treatments slightly decreased the BD of RB. The lower values for BD ($p < 0.05$) were observed for the infrared-treated and dry-heated RB (Figure 3a), which was presumably due to some structural collapse of the material, leading to decreased porosity. A slightly higher BD in the microwave-heated RB reflects a more compact structure for these materials.

The functional properties of RB are important for its technological interest and physiological effects. The hydration properties of the RB dietary fiber determine their optimal usage levels in composite food matrices, since a desirable texture should be always retained [67]. The water absorption capacity (WAC) of RB samples ranged from 3.42 to 4.29 g/g dw (Figure 3a), which is within the range of 1.49–4.72 g/g, which is considered critical in viscous food dispersions [68]. The results indicate that the WAC of RB was affected significantly ($p < 0.05$) by the application of different methods of stabilization. Specifically, the WAC of stabilized RB by dry-heating and microwave-heating improved by 8.4 and 14.8% ($p < 0.05$), respectively, making it a more suitable ingredient in foods where a greater water absorption is targeted (e.g., baked goods). However, the infrared-radiation treatment decreased significantly the WAC of RB compared to the untreated sample. In general, the relatively high WAC of the RB materials might be due to the presence of the hydroxyl group bearing polysaccharide components and the polar amino acids at the bran particle–water interface [27]. Similarly, Rafe et al. [68] found that the stabilization of RB by extrusion improved up to 20% the WAC; however, Khan et al. [27] indicated that the WAC of RB stabilized by dry-heating and microwave-heating was not significant ($p > 0.05$) as a result of the applied thermal treatment.

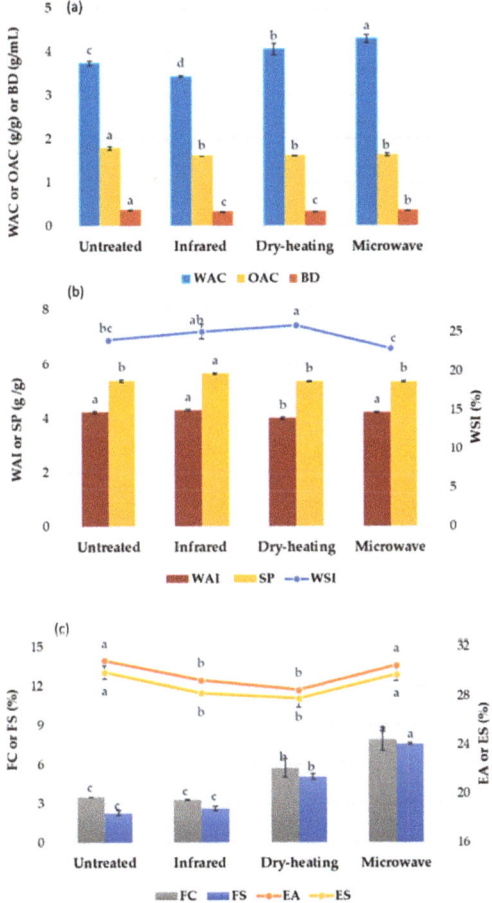

Figure 3. Effect of three different heating treatments (infrared-radiation, dry-heating, and microwave-heating) on WAC, OAC and BD (**a**), WAI, WSI and SP (**b**) and FC, FS, EA and ES (**c**) of RB samples; values followed by the same letter for the same physicochemical parameter are not significantly different ($p > 0.05$). WAC: Water Absorption Capacity, OAC: Oil Absorption Capacity, BD: Bulk Density, WAI: Water Absorption Index, SP: Swelling Power, WSI: Water Solubility Index, FC: Foaming Capacity, FS: Foaming Stability, EA: Emulsifying Activity, ES: Emulsion Stability.

Oil absorption capacity (OAC) is another important functional property, which is related to mouthfeel perception of the final product. A high OAC is essential in food systems such as processed meats (sausages), cake batters, mayonnaise, and salad dressings. According to the results of Figure 3a, the heat-treated RB samples showed slightly less strength for binding oil as compared to their counterpart. This might imply decreased surface hydrophobicity following the heat treatment. A lower hydrophobicity of the stabilized RB particles would not enhance the interactions between the fiber matrix of the RB and the oil, resulting in decreased oil absorption capacity [69].

The water absorption index (WAI) is the amount of water absorbed by starch or other particulate materials after swelling in an excess of water. In the case of starch, it can be related to starch gelatinization, which is greatly affected by the temperature and moisture content of the heated raw material [70]. The WAI was not affected significantly ($p > 0.05$) with the infrared-radiation and microwave-heating of RB stabilization (Figure 3b). Among all heat treatments applied, the dry-heating brought about the highest influence on WAI, leading to a 5.4% decrease in WAI of stabilized RB. The water solubility index (WSI) measures the amount of solubles (mostly low molecular weight material and water-soluble polysaccharides) released from a starch-containing material after thermal treatment. It is generally used as an indicator of material phase change (e.g., starch gelatinization) and/or some degradation of molecular components present in the heated particles [69]. The increase of WSI with heat treatment was more pronounced in the dry-heated RB, which was followed by the infrared-radiated RB, whereas the microwave-heated RB showed similar WSI with the untreated sample, as shown in Figure 3b. This may be due to the higher moisture content of RB treated by microwaves compared to the other samples [70]. Similar results were also reported for RB stabilized by extrusion cooking [71] and by microwave-heating [72]. A similar trend to WAI was also observed for swelling power (SP), indicating rather insignificant effects ($p > 0.05$) for the SP in all thermally stabilized RB, compared with the untreated RB; however, infrared radiation treatment increased significantly ($p < 0.05$) the SP of RB.

The foaming capacity (FC) and foaming stability (FS) of untreated and heat-stabilized RB samples are presented in Figure 3c. Generally, the RB has a low FC that may be associated with the presence of amphiphilic lipids in the RB, since these are more easily adsorbed at the interface than proteins, influencing the strength and elasticity of the film and, therefore, its ability to incorporate air [73]. It is noticed that the heat stabilization of RB improved the FC compared to the untreated RB, particularly in the case of dry-heating and microwave treated samples; i.e., FC values were improved from 3.51% (untreated) to 5.70 and 7.92% after treatment by dry-heating and microwave-heating, respectively. Proteins also play a significant role in forming a stable air bubble because of their amphiphilic nature; i.e., the protein structure gets rapidly unfolded to form a cohesive macromolecular layer at the air–water interface [72]. Heat treatment could partially unfold the protein chains, making them easier to absorb at the air–water interface, and as a result, it leads to an increase in FC. Similarly, Rafe et al. [68] reported that extruded RB had a higher FC than unstabilized RB, and Zhu et al. [74] also found that the RB protein following high-pressure treatment enhanced its FC. Instead, Khan et al. [27] found that extended heat application impairs the foaming properties of protein products, whereas partial enzymatic hydrolysis improves the interfacial properties of these biomolecules. The decrease in the FC of the RB protein by infrared treatment might be due to its thermal denaturation, making the diffusion and adsorption of the protein components at the air–water interface more difficult [75]. FS refers to the ability to maintain the air bubble against breaking or collapsing. Similar to FC, FS increased with heat treatment following dry-heating and microwave-heating of the RB with an increase of 121 and 232%, respectively, as compared with the untreated RB. However, there were no significant differences ($p > 0.05$) in the FS between untreated and stabilized RB by infrared-radiation. An improvement in FS might be related to enhanced protein–protein interactions (aggregation) upon heat treatment, bringing about a thick proteinaceous film around the air bubbles [76].

The emulsifying activity (EA) is mainly dependent on the diffusion of surface active constituents at the oil–water interfaces. The effect of heat stabilization of RB on EA and emulsion stability (ES) of RB is shown in Figure 3c. Dry-heating and infrared-radiation treatments of RB decreased the EA of RB ($p < 0.05$). Instead, microwave-heating treatment did not lead to significant changes in the EA of RB. The emulsions of all RB samples were very stable in the present study; the ES showed a similar trend among all samples to that of EA. In general, heating treatments decreased the surface activity and emulsion properties, which was probably due to protein denaturation in high temperatures [77].

Similar observations have been also reported by Capellini et al. [73] for rice bran defatted with alcoholic solvents.

4. Conclusions

The present study has demonstrated that infrared-radiation heating among other heat treatments such as dry-heating and microwave-heating could be an efficient method for inactivating lipase activity and prolonging the shelf-life of RB. It took a shorter processing time to effectively stabilize this dietary fiber source without affecting the nutritional profile, γ-oryzanol content, and ABTS radical scavenging activity while reducing the antinutrients and improving the ferric reducing antioxidant power. It also enhanced some functional properties (WSI and SP) of the heat-treated RB that may improve its use as a functional ingredient in food and cosmetics formulations. Instead, microwave stabilization was an effective treatment for improving the foaming and emulsifying capacities as well as water absorption of the RB. In addition, microwave treatment resulted in the largest reductions of antinutrients present in RB among all heat stabilization methods tested. Overall, among all studied methods, infrared-radiation heating appears a promising processing alternative to conventional dry-heating to stabilize RB, provided that appropriate optimization studies will be undertaken to examine the impact of the various operation parameters and conditions on the physicochemical, bioactivity, and functional properties of this material at a commercial scale level.

Author Contributions: Conceptualization, M.I. and A.L.; methodology, M.I.; investigation, M.I.; resources, M I. and A.L.; data curation, M.I.; writing—original draft preparation, M.I.; writing—review and editing, A.L. and C.G.B.; supervision, C.G.B.; funding acquisition, A.L. All authors have read and agreed to the published version of the manuscript.

Funding: This research was funded by EYDE-ETAK through the Operational Program Competitiveness, Entrepreneurship and Innovation, which is co-financed by the European Union and Greek national funds, under the call RESEARCH-CREATE-INNOVATE, grant number T1EDK-01669 and the APC was funded by the same grant (T1EDK-01669)

Conflicts of Interest: The authors declare no conflict of interest.

References

1. Sharif, M.K.; Butt, M.S.; Anjum, F.M.; Khan, S.H. Rice bran: A novel functional ingredient. *Crit. Rev. Food Sci. Nutr.* **2014**, *54*, 807–816. [CrossRef]
2. Sohail, M.; Rakha, A.; Butt, M.S.; Iqbal, M.J.; Rashid, S. Rice bran nutraceutics: A comprehensive review. *Crit. Rev. Food Sci. Nutr.* **2017**, *57*, 3771–3780. [CrossRef]
3. Xu, Z.; Hua, N.; Godber, J.S. Antioxidant activity of tocopherols, tocotrienols, and γ-oryzanol components from rice bran against cholesterol oxidation accelerated by 2,20-azobis(2-methylpropionamidine) dihydrochloride. *J. Agric. Food Chem.* **2001**, *49*, 2077–2081. [CrossRef] [PubMed]
4. Revilla, E.; Maria, C.S.; Miramontes, E.; Bautista, J.; Martínez, A.M.; Olga Cremades, O.; Cert, R.; Parrad, J. Nutraceutical composition, antioxidant activity and hypocholesterolemic effect of a water-soluble enzymatic extract from rice bran. *Food Res. Intern.* **2009**, *42*, 387–393. [CrossRef]
5. Debnath, T.; Park, S.R.; Kim, D.H.; Jo, J.E.; Lim, B.O. Anti-Oxidant and anti-inflammatory activities of inonotus obliquus and germinated brown rice extracts. *Molecules* **2013**, *18*, 9293–9304. [CrossRef] [PubMed]
6. Yasukawa, K.; Akihisa, T.; Kimura, Y.; Tamura, T.; Takido, M. Inhibitory effect of cycloartenyl ferulate, a component of rice bran, on tumor promotion in two stage carcinogenesis in mouse skin. *Biol. Pharm. Bull.* **1998**, *21*, 1072–1076. [CrossRef] [PubMed]
7. Ghatak, S.B.; Panchal, S.S. Anti-Diabetic activity of oryzanol and its relationship with the anti-oxidant property. *Int. J. Diabetes Devel. Countr.* **2012**, *32*, 185–192. [CrossRef]
8. Saunders, R.M. Rice bran: Composition and potential food uses. *Food Rev. Int.* **1985**, *1*, 465–495. [CrossRef]
9. Orthoefer, F.T. Rice bran oil: Healthy lipid source. *Food Technol.* **1996**, *50*, 62–64.
10. Thanonkaew, A.; Wongyai, S.; Mcclements, D.J.; Decker, E.A. Effect of stabilization of rice bran by domestic heating on mechanical extraction yield, quality, and antioxidant properties of cold-pressed rice bran oil (*Oryza saltiva* L.). *LWT Food Sci. Technol.* **2012**, *48*, 231–236. [CrossRef]
11. Kim, C.J.; Byun, S.M.; Cheigh, H.S.; Kwon, T.W. Optimization of extrusion rice bran stabilization process. *J. Food Sci.* **1987**, *52*, 1355–1357. [CrossRef]

12. Patil, S.S.; Kar, A.; Mohapatra, D. Stabilization of rice bran using microwave: Process optimization and storage studies. *Food Bioprod. Proces.* **2016**, *99*, 204–211. [CrossRef]
13. Ramezanzadeh, F.M.; Rao, R.M.; Prinyawiwatkul, W.; Marshall, W.E.; Windhauser, M. Effects of microwave heat, packaging, and storage temperature on fatty acid and proximate compositions in rice bran. *J. Agric. Food Chem.* **2000**, *48*, 464–467. [CrossRef] [PubMed]
14. Yilmaz, N.; Tuncel, B.N.; Kocabiyik, H. Infrared stabilization of rice bran and its effects on gamma-oryzanol content, tocopherols and fatty acid composition. *J. Sci. Food Agric.* **2014**, *94*, 1568–1576. [CrossRef]
15. Irakli, M.; Kleisiaris, F.; Mygdalia, A.; Katsantonis, D. Stabilization of rice bran and its effect on bioactive compounds content, antioxidant activity and storage stability during infrared radiation heating. *J. Cereal Sci.* **2018**, *80*, 135–142. [CrossRef]
16. Dhingra, D.; Chopra, S.; Rai, D.R. Stabilization of raw rice bran using ohmic heating. *Agric. Res.* **2012**, *1*, 392–398. [CrossRef]
17. Pradeep, P.M.; Jayadeep, A.; Guha, M.; Singh, V. Hydrothermal and biotechnological treatments on nutraceutical content and antioxidant activity of rice bran. *J. Cereal Sci.* **2014**, *60*, 187–192. [CrossRef]
18. Amarasinghe, B.M.W.P.K.; Kumarasiri, M.P.M.; Gangodavilage, N.C. Effect of method of stabilization on aqueous extraction of rice bran oil. *Food Bioprod. Proces.* **2009**, *87*, 108–114. [CrossRef]
19. Vallabha, V.S.; Indira, T.N.; Lakshmi, A.J.; Radha, C.; Tiku, P.K. Enzymatic process of rice bran: A stabilized functional food with nutraceuticals and nutrients. *J. Food Sci. Techn.* **2015**, *52*, 8252–8259. [CrossRef]
20. Gopinger, E.; Ziegler, V.; Catalan, A.A.D.S.; Krabbe, E.L.; Elias, M.C.; Xavier, E.G. Whole rice bran stabilization using a short chain organic acid mixture. *J. Stored Prod. Res.* **2015**, *61*, 108–113. [CrossRef]
21. Sharma, H.R.; Chauhan, G.S.; Agrawal, K. Physico-Chemical characteristics of rice bran processed by dry heating and extrusion cooking. *Intern. J. Food Prop.* **2004**, *7*, 603–614. [CrossRef]
22. Zigoneanu, I.G.; Williams, L.; Xu, Z.; Sabliov, C.M. Determination of antioxidant components in rice bran oil extracted by microwave-assisted method. *Bioresour. Technol.* **2008**, *99*, 4910–4918. [CrossRef] [PubMed]
23. Yan, W.; Liu, Q.; Wang, Y.; Tao, T.; Liu, B.; Liu, J.; Ding, C. Inhibition of lipid and aroma deterioration in rice bran by infrared heating. *Food Bioproc. Technol.* **2020**, *13*, 1677–1687. [CrossRef]
24. Riadh, M.H.; Ahmad, S.A.B.; Marhaban, M.H.; Soh, A.C. Infrared heating in food drying: An overview. *Dry. Technol.* **2015**, *33*, 322–335. [CrossRef]
25. Zhang, M.; Tang, J.; Mujumdar, A.S.; Wang, S. Trends in microwave related drying of fruits and vegetables. *Trends Food Sci. Technol.* **2006**, *17*, 524–534. [CrossRef]
26. Khan, S.H.; Butt, M.S.; Anjum, F.M.; Jamil, A. Antinutritional appraisal and protein extraction from differently stabilized rice bran. *Pakistan J. Nutr.* **2009**, *8*, 1281–1286. [CrossRef]
27. Khan, S.H.; Butt, M.S.; Sharif, M.K.; Sameen, A.; Mumtaz, S.; Sultan, M.T. Functional properties of protein isolates extracted from stabilized rice bran by microwave, dry heat, and parboiling. *J. Agric. Food Chem.* **2011**, *59*, 2416–2420. [CrossRef]
28. Yueh, J.Y.; Phillips, R.D.; Resurreccion, A.V.A.; Hung, Y.C. Physicochemical and sensory characteristic changes in fortified peanut spreads after 3 months of storage at different temperatures. *J. Agric. Food Chem.* **2002**, *50*, 2377–2384. [CrossRef]
29. AOAC. *International Official Methods of Analysis*, 20th ed.; Association of Official Agricultural Chemists: Rockville, MD, USA, 2016.
30. AOAC Method 996-06. Fats (total, saturated and unsaturated) in foods. In *AOAC Official Methods of Analysis*; Association of Official Analytical Chemists Inc.: Gaithersburg, MD, USA, 2002.
31. AACC. Method 02-01 Fat acidity—General method. In *Approved Methods of Analysis*; AACC International: St. Paul, MN, USA, 2000.
32. Rose, D.J.; Pike, O.A. A simple method to measure lipase activity in wheat and wheat bran as an estimation of storage quality. *J. Am. Oil Chem. Soc.* **2006**, *83*, 415–419. [CrossRef]
33. Singleton, V.L.; Orthofer, R.; Lamuela-Raventos, R.M. Analysis of total phenols and other oxidation substrates and antioxidants by means of Folin-Ciocalteu reagents. *Methods Enzymol.* **1999**, *299*, 152–178.
34. Irakli, M.; Chatzopoulou, P.; Kadoglidou, K.; Tsivelika, N. Optimization and development of a high-performance liquid chromatography method for the simultaneous determination of vitamin E and carotenoids in tomato fruits. *J. Sep. Sci.* **2016**, *39*, 3348–3356. [CrossRef] [PubMed]
35. Fang, N.; Yu, S.; Badger, T.M. Characterization of triterpene alcohol and sterol ferulates in rice bran using LC-MS/MS. *J. Agric. Food Chem.* **2003**, *51*, 3260–3267. [CrossRef] [PubMed]
36. Oyeyinka, B.O.; Afolayan, A.J. Comparative evaluation of the nutritive, mineral, and antinutritive composition of musa sinensis l. (banana) and musa paradisiaca l. (plantain) fruit compartments. *Plants* **2019**, *8*, 598. [CrossRef] [PubMed]
37. Wheeler, E.L.; Ferrel, R.E. A method for phytic acid determination in wheat and wheat products. *Cereal Chem.* **1971**, *48*, 312–320.
38. Hierro, J.N.; Herrera, T.; García-Risco, M.R.; Fornari, T.; Reglero, G.; Martina, D. Ultrasound-Assisted extraction and bioaccessibility of saponins from edible seeds: Quinoa, lentil, fenugreek, soybean and lupin. *Food Res. Intern.* **2018**, *109*, 440–447. [CrossRef] [PubMed]
39. Kakade, M.L.; Rackis, J.J.; McGhee, J.E.; Puski, G. Determination of trypsin inhibitor activity of soy products: A collaborative analysis of an improved procedure. *Cereal Chem.* **1974**, *51*, 376–381.
40. Makkar, H.P.S.; Bluemmel, M.; Borowy, N.K.; Becker, K. Gravimetric determination of tannins and their correlation with chemical and protein precipitation method. *J. Sci. Food Agric.* **1993**, *61*, 161–165. [CrossRef]
41. Re, R.; Pellegrini, N.; Proteggente, A.; Pannala, A.; Yang, M.; Rice-Evans, C.A. Antioxidant activity applying an improved ABTS radical cation decolorization assay. *Free Rad. Biol. Med.* **1999**, *26*, 1231–1237. [CrossRef]

42. Yen, G.C.; Chen, H.Y. Antioxidant activity of various tea extracts in relation to their antimutagenicity. *J. Agric. Food Chem.* **1995**, *43*, 27–32. [CrossRef]
43. Benzie, F.; Strain, J. Ferric reducing/antioxidant power assay: Direct measure of total antioxidant activity of biological fluids and modified version for simultaneous measurement of total antioxidant power and ascorbic acid concentration. *Methods Enzymol.* **1999**, *299*, 15–23.
44. Kaushala, P.; Kumara, V.; Sharma, H.K. Comparative study of physicochemical, functional, antinutritional and pasting properties of taro (*Colocasia esculenta*), rice (*Oryza sativa*) flour, pigeonpea (*Cajanus cajan*) flour and their blends. *LWT Food Sci. Techn.* **2012**, *48*, 59–68. [CrossRef]
45. Abebe, W.; Collar, C.; Ronda, F. Impact of variety type and particle size distribution on starch enzymatic hydrolysis and functional properties of tef flours. *Carbohydr. Polym.* **2015**, *115*, 260–268. [CrossRef] [PubMed]
46. Nazck, M.; Diosady, L.L.; Rubin, L.J. Functional properties of canola meals produced by two-phase solvent extraction systems. *J. Food Sci.* **1985**, *50*, 1685–1692.
47. Yu, C.W.; Hu, Q.R.; Wang, H.W.; Deng, Z.Y. Comparison of 11 rice bran stabilization methods by analyzing lipase activities. *J. Food Process. Preserv.* **2020**, *44*, 14370–14384. [CrossRef]
48. Oliveira, M.S.; Feddern, V.; Kupski, L.; Cipolatti, E.P.; Badiale-Furlong, E.; Souza-Soares, L.A. Changes in lipid, fatty acids and phospholipids composition of whole rice bran after solid-state fungal fermentation. *Bioresource Technol.* **2011**, *102*, 8335–8338. [CrossRef] [PubMed]
49. Martin, C.A.; Almeida, V.V.; Ruiz, M.R.; Visentainer, J.E.L.; Matshushita, M.; Souza, N.E.; Visentainer, J.V. Omega-3 and omega-6 polyunsaturated fatty acids: Importance and occurrence in foods. *Rev. Nutr.* **2006**, *19*, 761–770. [CrossRef]
50. Rodchuajeen, K.; Niamnuy, C.; Charunuch, C.; Soponronnarit, S.; Devahastin, S. Stabilization of rice bran via different moving-bed drying methods. *Drying Technol.* **2016**, *34*, 1854–1867. [CrossRef]
51. Oatway, L.; Vasanthan, T.; Helm, J.H. Phytic acid. *Food Rev. Int.* **2001**, *17*, 419–431. [CrossRef]
52. Kaur, S.; Sharma, S.S.; Singh, B.; Dar, B.N. Effect of extrusion variables (temperature, moisture) on the antinutrient components of cereal brans. *J. Food Sci. Technol.* **2015**, *52*, 1670–1676. [CrossRef]
53. Kakati, P.; Deka, S.; Kotoki, D.; Saikia, S. Effect of traditional methods of processing on the nutrient contents and some antinutritional factors in newly developed cultivars of green gram [*Vigna radiata* (L.) Wilezek] and black gram [*Vigna mungo* (L.) Hepper] of Assam, India. *Intern. Food Res. J.* **2010**, *17*, 377–384.
54. Savage, G.P.; Vanhanen, L.; Mason, S.M.; Rose, A.B. Effect of cooking on the soluble and insoluble oxalate content of some New Zealand foods. *J. Food Comp. Anal.* **2000**, *13*, 201–206. [CrossRef]
55. Morozumi, M.; Ogawa, Y. Impact of dietary calcium and oxalate ratio on urinary stone formation in rats. *Mol. Urol.* **2000**, *4*, 313–320. [PubMed]
56. Embaby, H.E. Effect of soaking, dehulling, and cooking methods on certain antinutrients and in vitro protein digestibility of bitter and sweet lupin seeds. *Food Sci. Biotechnol.* **2010**, *19*, 1055–1062. [CrossRef]
57. Bissinger, R.; Modicano, P.; Alzoubi, K.; Honisch, S.; Faggio, C.; Abed, M.; Lang, F. Effect of saponin on erythrocytes. *Int. J. Hematol.* **2014**, *100*, 51–59. [CrossRef] [PubMed]
58. Serrano, J.; Puupponen-Pimiä, R.; Dauer, A.; Aura, A.M.; Saura-Calixto, F. Tannins: Current knowledge of food sources, intake, bioavailability and biological effects. *Mol. Nutr. Food Res.* **2009**, *53*, S310–S329. [CrossRef]
59. Sahni, P.; Sharma, S. Influence of processing treatments on cooking quality, functional properties, antinutrients, bioactive potential and mineral profile of alfalfa. *LWT Food Sci. Technol.* **2020**, *132*, 109890. [CrossRef]
60. Osman, M.A. Effect of different processing methods on nutrient composition, antinutritional factors, and in vitro protein digestibility of dolichos lablab beans (LabLab purpuresus (L) sweet). *Pakistan J. Nutr.* **2007**, *6*, 299–303. [CrossRef]
61. Deng, Y.; Padilla-Zakour, O.; Zhao, Y.; Tao, S. Influences of high hydrostatic pressure, microwave heating, and boiling on chemical compositions, antinutritional factors, fatty acids, in vitro protein digestibility, and microstructure of buckwheat. *Food Bioprocess. Technol.* **2015**, *8*, 2235–2245. [CrossRef]
62. Saji, N.; Schwarz, L.J.; Santhakumar, A.B.; Blanchard, C.L. Stabilization treatment of rice bran alters phenolic content and antioxidant activity. *Cereal Chem.* **2020**, *97*, 281–292. [CrossRef]
63. Laokuldilok, T.; Rattanathanan, Y. Protease treatment for the stabilization of rice bran: Effects on lipase activity, antioxidants, and lipid stability. *Cereal Chem.* **2014**, *91*, 560–565. [CrossRef]
64. Lakkakula, N.R.; Lima, M.; Walker, T. Rice bran stabilization and rice bran oil extraction using ohmic heating. *Bioresour. Technol.* **2004**, *92*, 157–161. [CrossRef] [PubMed]
65. Yeo, J.; Shahidi, F. Effect of hydrothermal processing on changes of insoluble-bound phenolics of lentils. *J. Funct. Foods* **2017**, *38*, 716–722. [CrossRef]
66. Miranda, M.; Maureira, H.; Rodriguez, K.; Vega-Galvez, A. Influence of temperature on the drying kinetics, physicochemical properties, and antioxidant capacity of Aloe Vera (Aloe Barbadensis Miller) gel. *J. Food Eng.* **2009**, *91*, 297–304. [CrossRef]
67. Daou, C.; Zhang, H. Study on functional properties of physically modified dietary fibres derived from defatted rice bran. *J. Agric. Sci.* **2012**, *4*, 85–97. [CrossRef]
68. Rafe, A.; Sadeghian, A.; Hoseini-Yazdi, S.Z. Physicochemical, functional, and nutritional characteristics of stabilized rice bran form tarom cultivar. *Food Sci. Nutri.* **2017**, *5*, 407–414. [CrossRef]
69. Britten, L.L.M. Foaming properties of proteins as affected by concentration. *J. Food Sci.* **1992**, *57*, 1219–1241. [CrossRef]

70. Tang, S.; Hettiarachchy, N.S.; Horax, R.; Eswaranandam, S. Physicochemical properties and functionality of rice bran protein hydrolyzate prepared from heat-stabilized defatted rice bran with the aid of enzymes. *J. Food Sci.* **2003**, *68*, 152–157. [CrossRef]
71. Sharma, R.; Srivastava, T.; Saxena, D.C. Valorization of deoiled rice bran by development and process optimization of extrudates. *Engineer. Agric. Envi. Food* **2019**, *12*, 173–180. [CrossRef]
72. Villanueva, M.; Lamo, B.D.; Harasyma, J.; Ronda, R. Microwave radiation and protein addition modulate hydration, pasting and gel rheological characteristics of rice and potato starches. *Carboh. Polym.* **2018**, *201*, 374–381. [CrossRef]
73. Capellinia, M.C.; Novais, J.S.; Monteiro, R.F.; Veiga, B.Q.; Osiro, D.; Rodrigues, C.E.C. Thermal, structural and functional properties of rice bran defatted with alcoholic solvents. *J. Cereal Sci.* **2020**, *95*, 103067. [CrossRef]
74. Zhu, S.M.; Lin, L.; Ramaswamy, H.S.; Yu, Y.; Zhang, Q.T. Enhancement of functional properties of rice bran proteins by high pressure treatment and their correlation with surface hydrophobicity. *Food Bioprocess. Technol.* **2017**, *10*, 317–327. [CrossRef]
75. LV, S.H.; He, L.Y.; Sun, L.H. Effect of different stabilisation treatments on preparation and functional properties of rice bran proteins. *Czech J. Food Sci.* **2018**, *36*, 57–65. [CrossRef]
76. Andersen, O.S.; Fuchs, M. Preparation and some functional properties of rice bran protein concentrate at different degree of hydrolysis using bromelain and alkaline extraction. *Prepar. Biochem. Biotechnol.* **2009**, *39*, 183–193.
77. Zhang, Y.; Wang, B.; Zhang, W.; Xu, W.; Hu, Z. Effects and mechanism of dilute acid soaking with ultrasound pretreatment on rice bran protein extraction. *J. Cereal Sci.* **2019**, *87*, 318–324. [CrossRef]

Article

Consumers' Perspectives on Eggs from Insect-Fed Hens: A UK Focus Group Study

Sabrina Spartano [1] and Simona Grasso [2,*]

[1] Department of Applied Economics and Marketing, School of Agriculture, Policy and Development, University of Reading, Reading RG6 6EU, UK; sabrina.spartano@hotmail.it

[2] School of Agriculture, Policy and Development, Institute of Food, Nutrition and Health, University of Reading, Reading RG6 6EU, UK

* Correspondence: simona.grasso@ucdconnect.ie; Tel.: +44-118-378-6576

Abstract: In recent years, there has been growing interest in insects as an alternative to soybean meal as laying hen feed due to nutrition, sustainability, and animal welfare benefits. Although some studies have investigated consumer acceptance and intentions towards insect-fed foodstuffs, no studies are available on eggs from insect-fed hens. This qualitative study aimed to explore consumers' attitudes and perceptions towards eggs from insect-fed hens and factors influencing intentions to consume and purchase the product. Three focus group discussions were employed with a total of 19 individuals from the UK. Results showed that the environmental, animal welfare, and food waste benefits of feeding hens with insects positively influenced attitudes. Results also indicated price and disgust towards insects as feed were the main barriers, while enhanced welfare standards (e.g., free-range labelling) and information on benefits were main drivers. Therefore, the study suggests that educating and informing consumers about the benefits of feeding hens with insects may increase intentions to consume and purchase eggs from insect-fed hens. Given this emerging area of research, this study contributes to the limited literature on insect-fed foodstuffs and paves the way for further research on the topic.

Keywords: animal welfare; circular economy; consumer acceptance; consumer attitudes; food waste; insects as feed; Nvivo; poultry; qualitative study; sustainability

Citation: Spartano, S.; Grasso, S. Consumers' Perspectives on Eggs from Insect-Fed Hens: A UK Focus Group Study. *Foods* **2021**, *10*, 420. https://doi.org/10.3390/foods10020420

Academic Editor: María Del Mar Campo Arribas

Received: 31 January 2021
Accepted: 9 February 2021
Published: 14 February 2021

Publisher's Note: MDPI stays neutral with regard to jurisdictional claims in published maps and institutional affiliations.

Copyright: © 2021 by the authors. Licensee MDPI, Basel, Switzerland. This article is an open access article distributed under the terms and conditions of the Creative Commons Attribution (CC BY) license (https://creativecommons.org/licenses/by/4.0/).

1. Introduction

In the last decade, edible insects have received growing interest as a sustainable alternative source of protein, as both food and animal feed, because of their environmental, nutritional, and animal welfare benefits [1]. In the context of poultry farming and egg production, the use of insects reduces the environmental burden associated with producing traditional feed such as soya, utilises food that would otherwise go to waste, and increases welfare by encouraging natural behaviour without affecting egg quality or taste [2].

Eating insects is a natural behaviour for chickens. When raised in a natural or semi-natural environment (e.g., free-range), chickens spend part of their time foraging and eating insects [2,3].

The European Union has allowed the use of live insects as feed for poultry since 2017 [4]. Although some insect-fed animal foodstuffs have entered the European market (e.g., Oerei eggs in the Netherlands), these products are still considered niche [5]. In the UK, there is a great deal of interest from scientists and companies to produce insects for animal feed. The UK thus represents a potential market for insect-fed eggs and other animal products, although there will likely be a need for regulatory clarification in a post-EU environment [6].

In recent years, consumers have shown increasing acceptance for insect-fed foodstuffs, in particular for insect-fed fish and chicken [7–15]. However, given the novelty of insects as feed, it is perhaps unsurprising that research in this area is limited. The available literature embodies mainly studies on insect-fed fish [16–20], studies on insects as feed in general [8,14], and only a few studies that considered insect-fed chickens as a part of broader research [7,9,10,12–14].

Bazoche and Poret [20] found neophobia and disgust as some of the main barriers to consumer acceptance of insect-fed animals (aquaculture). This French study showed that while less neophobic consumers were more likely to accept insects as feed, part of the sample appeared disgusted by the idea of eating insect-fed fish. This result is in line with the study of Szendrő, Nagy, and Tóth conducted in Hungary [14]. Despite disgust and neophobia negatively influencing acceptance and intentions, studies agreed that this effect appeared to be less strong for insects as feed than as food [8,13] and could be overcome by informing consumers about the benefits of eating insect-fed foodstuffs [13–15,20].

Based on the study of Szendrő, Nagy, and Tóth [14], knowledge of the enhanced animal welfare associated with insect-fed animals may increase its acceptance by Hungarian consumers. Furthermore, as stated by Roma, Palmisano, and De Boni [13] and Bazoche and Poret [20], information on environmental benefits may increase positive intentions among Italian and French consumers. Likewise, Naranjo-Guevara et al. [15] mentioned the importance of providing information not only about environmental benefits but also about the enhanced nutritional content of insects as food and feed.

Among determinants of consumer acceptance, the study of Popoff, MacLeod, and Leschen [17] showed that among UK consumers, nutritional content and taste of insect-fed salmon were more important than price. Conversely, Mancuso, Baldi, and Gasco [16] suggested safety requirements and price as the main factors affecting acceptance of insect-fed fish among Italian consumers. Price also appeared to be an important factor in other studies. According to Bazoche and Poret [20], French consumers were not willing to pay a premium price for insect-fed trout. Similar results were shown by Popoff, MacLeod, and Leschen [17], confirming that although UK consumers considered quality important, they were not willing to pay more for insect-fed salmon. This result was further confirmed by the study of Ankamah-Yeboah, Jacobsen, and Olsen [18], showing that Danish consumers had greater intentions to purchase insect-fed fish at a lower price. In contrast with the aforementioned studies, Ferrer Llagostera et al. [19] showed that Spanish consumers were more likely to purchase insect-fed fish at a premium price.

Although the available research provides insight into consumer acceptance of insects as feed and insect-fed foodstuffs, to the best of our knowledge, there is no available research on consumers' attitudes and acceptance of eggs from insect-fed hens.

This study is the first one to explore through a qualitative approach consumers' attitudes and perceptions towards eggs from insect-fed hens and factors influencing intentions.

2. Materials and Methods

2.1. Focus Groups

Due to the uniqueness of the product under study, the novel topic, and the limited knowledge that consumers have, a qualitative approach based on focus groups was used for the study. This methodology is particularly appropriate in marketing and consumer studies where nothing or little is known about a topic. Focus group discussions provide a more natural environment where respondents share ideas as in real-life situations, making it possible to examine consumers' perceptions, thoughts, feelings, and beliefs [21,22].

Numerous consumer studies have adopted qualitative research as a data collection method to provide insights into consumers' motivations, acceptance of, and intentions towards insects as food [23–31]. However, to the best of our knowledge, no focus group studies are available on consumer acceptance of insect-fed foodstuffs. Therefore, the aim of the focus groups for this study was to generate insights into the diversity of consumers' attitudes, opinions, and perceptions towards eggs from insect-fed hens.

2.2. Participants

Three focus groups were conducted in the UK in June 2020. Due to recent COVID-19 developments, face-to-face focus groups were replaced with online focus groups undertaken throughout the online platform Zoom, adapting the methodology to the new study design.

A sample of 19 participants (9 male and 10 female) aged from 18 to 56 was recruited and then divided into three groups based on age and gender (see Table 1). Each group participated in a discussion session lasting around 125 min with a 10-min break. Focus group interviews were audio and video recorded.

Table 1. Socio-demographics of focus group participants.

Characteristics		Frequency
Gender	Male	9
	Female	10
Age		
	18–24	6
	25–34	8
	Over 34	5
Education level	High School	2
	Bachelor's degree	9
	Master's degree	7
	Doctorate	1
Income	Less than £20,000	6
	£20,000 to £39,999	6
	£40,000 to £59,999	4
	£60,000 and higher	3
Diet	Omnivore	16
	Vegetarian	3

In the sample, 18.7% identified themselves as vegetarians. We did not distinguish between more specific dietary patterns such as flexitarian, pescatarian, etc; 18.48% of UK consumers indicate having a diet that is neither "meat eater" nor "vegan" [32], suggesting that vegetarians in the sample can be considered a sufficiently representative sub-subset of the population.

Participants were recruited utilizing a non-stochastic sampling and then a snowball sampling technique. The recruitment was carried out throughout a short questionnaire posted on social media, and only participants with eligible criteria were selected. Based on inclusion and exclusion criteria, participants were required to be 18 or over, currently living in the UK for at least 6 months, at least partially responsible for food purchase in their household, egg eaters, and egg buyers.

Before the starting date, ethical clearance was obtained by the University of Reading (UK).

2.3. Focus Group Structure

A semi-structured protocol with open ended and follow up questions was used to encourage spontaneous answers from participants. The discussion was organized into five main sections and different sub-sections (Table 2).

Table 2. Focus group structure.

Sections	Questions (sub-sections)
1: Introduction	Introduction of moderator, assistant moderator, and participants. Elucidation of the discussion topic and ground rules. Engagement (ice-braker) question
2: Egg consumption patterns	Exploratory questions (possible use of probing and follow up questions): Egg attributes preferred, and related motivation
3: Attitudes	Provision of information: Card Exploratory questions: Attitudes and perceptions towards eggs from insect-fed hens and factors related
4: Packaging	Provision of information: Packaging Exploratory questions: Reactions and opinions towards the packaging
5: Willingness to buy	Exploratory questions: Willingness to buy Exit questions: Asking for questions and thank participants

Section 1: Participants were welcomed on the online platform and introduced to the research project and participation rules.

Section 2: Participants were asked to discuss their egg consumption patterns and attributes they look at when they buy eggs. This stage aimed to identify consumers' egg preferences in terms of product features and explore the reasons behind their choices.

Section 3: Participants were provided with an information card about the benefits of feeding hens with insects (Figure 1). The provision on the information card allowed participants to generate and explore knowledge and opinions about a new product that was not yet in the UK market. After the provision of the information, participants were asked to express their perceptions towards the product and related benefits to identify their level of acceptance.

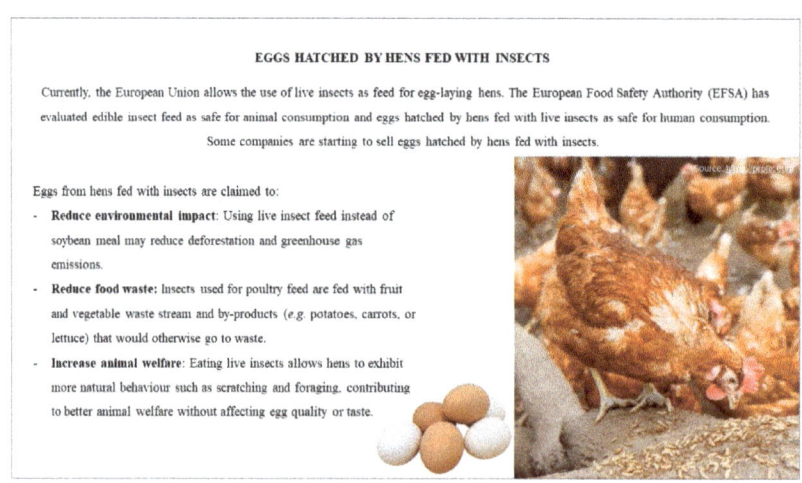

Figure 1. Information card provided to participants during Section 3 of the focus group.

Section 4: Participants were provided with a translated packaging design of insect-fed eggs that is currently available on the Dutch market. After the provision of the packaging, participants were asked to share opinions about the packaging. The main reason was to understand how packaging information and design may affect consumer perceptions and intentions.

Section 5: Participants were asked to express their intention to consume and purchase the product. This question was meant to explore how perceptions and attitudes towards the product, as well as interest in the benefits, may affect intentions.

At the very end of the discussion, participants were thanked for their participation and their useful contribution.

2.4. Data Analysis

Data analysis was carried out on the software Nvivo v. 12 [33]. Audio and video recordings of the focus groups were transcribed and coded using a thematic analysis that is particularly effective to identify and describe themes within a dataset and find patterns among the sample [34].

The qualitative approach used for the analysis incorporated both inductive and deductive methodologies [35]. For this purpose, at first, data were organized in common themes and sub-themes by coding. Themes were generated based on the research questions and recorded under five different nodes. Inductive codes were generated by looking at data and by identifying possible recurring topics. Participants' quotations belonging to the same themes were codified and recorded under the same node [36]. Each node was successively ramified in sub-nodes, allowing a better distribution and classification of themes and sub-themes. The validity of themes and codes was double-checked by a second researcher. The literature review was used deductively to answer the research questions.

Finally, project maps were used to visualize themes and code relationships. Direct quotes from participants and the names of participants on the project maps are presented with a number (n.1–n.19), gender identification (M or F), and age group (e.g., 18–24 years old).

3. Results

Following the focus group sub-sections, results are serially presented as such: (1) egg attributes preferred and related motivations; (2) attitudes and perceptions towards eggs from insect-fed hens; (3) factors influencing attitudes towards eggs from insect-fed hens; (4) reactions and opinions towards the packaging; (5) willingness to buy.

3.1. Egg Attributes Preferred and Related Motivations

For most participants, the free-range label (11 cases) was the first attribute they looked at when buying eggs. However, for some of them, organic was always their first choice (6 cases). As a motivation driving them to free-range and/or organic eggs, participants mainly mentioned better animal welfare, better taste, better egg quality, higher nutritional content, and high-quality feed.

The second most important attribute participants looked at when buying eggs was the price (8 cases). While some participants (2 cases) declared they always chose the cheapest eggs, for other participants (5 cases) budget constraints usually drove them to choose a cheaper choice than free-range and organic eggs.

Other attributes frequently mentioned were size (7 cases), local (6 cases), eco-friendly eggs (6 cases), specialized eggs (4 cases), not broken (4 cases), and British Lion mark (food safety standard) (3 cases) (Figure 2).

Figure 2. Packaging provided to participants during Section 4 of the focus groups (source: https://oerei.nl/ (accessed on 12 February 2021)).

3.2. Attitudes and Perceptions towards Eggs from Insect-Fed Hens

The majority of participants showed positive attitudes towards eggs hatched by hens fed with insects. In particular, the benefits provided by the products in terms of higher animal welfare, reduction of environmental impact, contribution to reducing food waste, and the perception of a natural method played an important role in increasing intentions to consume and purchase. However, some of the participants revealed a lack of trust in the product (4 cases). They seemed suspicious about the benefits provided and producers' aims (Table 3).

Table 3. Attitudes and perceptions towards eggs from insect-fed hens.

Main Factors	Main Quotes	Participant
Benefits	"For me, the fact that you can fed insects with food-by-product is a massive benefit and if it was marketed right, in the same way free-range is marketed, I think lots and lots of people would go for that" "I would be encouraged mainly for the environmental benefit and animal welfare aspect. It's quite an important selling point for me!" "I like the fact you can feed eggs with insects because the way in which you feed hens naturally may indirectly help the environment too"	(P13, F, 25–34 y) (P17, F, over 34 y) (P12, F, 25–34 y)
Lack of trust	"I'm slightly suspicious they are fed with food waste. It sounds good but again, is it actually like this? I do just not trust so much when it comes to marketing." "My concern is, is a company doing this for improving the system or for their bottom line? It is a cost-cutting or is actually related to the welfare of the hen!?" "You are never going to develop a hen feeding system in which there is no impact."	(P8, F, 25–34 y) (P11, M, 25–34 y) (P17, F, over 34 y)

3.3. Factors Influencing Attitudes towards Eggs from Insect-Fed Hens

Other factors that might encourage/discourage participants to consume and purchase the product were mainly price (19 cases), quality and quantity of information provided on the packaging (7 cases), availability (5 cases), nutritional content (5 cases), and taste (4 cases). The feeling of disgust and the rejection of insects seemed to slightly moderate consumers' intentions. However, only one participant revealed that the negative feeling for insects reduced their level of acceptance (Table 4).

Table 4. Factors influencing attitudes towards eggs from insect-fed hens.

Main Factors	Main Quotes	Participant
Price	"Price should be reasonable. If the product costs too much, more than free-range, I wouldn't give it a go."	P1, M, 18–24 y
Information	"I have to find out more about the product. I have to be sure these benefits exist for sure."	P8, F, 25–34 y
Availability	"For me is the availability on the market the most important point. I can say I would try them, but I don't think I would go out of my way to buy them."	P13, F, 25–34 y
Nutritional content	"I will be very happy to try these eggs for the fact that insects have good protein level. So, I would not be surprised if research said that eggs fed with insects are better in quality and nutritional content."	P17, F, over 34 y
Taste	"Quality would be at least the same as the others in order for me to buy them. It can be the best product in the world but If it doesn't taste good, I wouldn't support them."	P4, M, 18–24 y P8, F, 25–34 y
Disgust	"Taste and nutritional content. These attributes would be important only if you cover the animal welfare and the other issues we discussed."	P3, F, 18–24 y

3.4. Reactions and Opinions towards the Packaging

All participants showed positive reactions towards the packaging design. Participants mentioned that the box was very "engaging" and immediately brought them to the concepts of "sustainability", "welfare", and "natural". Even though these participants showed an intention to try, the word "insects" written on the packaging would discourage some participants to purchase the product (4 cases). The word "natural", instead, would encourage some participants to try the product (3 cases).

All participants mentioned the importance of the free-range label on the packaging (19). Everyone agreed that to effectively improve animal welfare, eggs hatched by hens fed with insects should be produced according to high animal welfare standards, and thus, these eggs should be marketed with a free-range label on the packaging.

Again, participants mentioned the importance of clearer information on the packaging (6 cases) (Table 5).

Table 5. Reactions and opinions towards the packaging.

Main Factors	Main Quotes	Participant
Disgust	"I like the message on the back of the packaging, but on the front, the word 'insect' is written three times and that kind of puts me off."	(P4, M, 18–24 y)
Natural aspect	"Taste is important. So for me just saying 'natural taste' for me is fine, I like it, it's a good selling point. I'm convinced."	(P18, F, over 34 y)
Animal welfare standards	"There is nothing bad about increasing animal welfare. I still want a free-range egg. I wouldn't buy eggs from insect-fed caged hens"	(P5, F, 18–24 y)
Information	"I would like to see more evidence about the benefits. Why is this product sustainable, etc.?"	(P1, M, 18–24 y)

3.5. Willingness to Buy for Insect-Fed Eggs

Most participants revealed they were willing to buy the product (18 cases). However, they required knowing the price before making any choice.

Only one participant showed less willingness to buy (1 case) due to the perceived negative feeling towards insects (Table 6).

Table 6. Willingness to buy insect-fed eggs.

Main Factors	Main Quotes	Participant
Price	"I would probably give this product a go, but I don't know if I would buy it again. It really depends on the price and how many are in the pack." "It depends on the price. How much do these eggs cost?"	(P1, M, 18–24 y) (P8, F, 25–34 y)
Disgust	"I might buy them. The only problem for me is that, there are insects in it!"	(P3, F, 18–24 y)

4. Discussion

Focus group results showed that among attributes participants look at when buying eggs, the free-range and the organic label were the most important ones. Among motivations driving to free-range and/or organic eggs, participants mainly mentioned the perception of higher animal welfare, better taste, better egg quality, higher nutritional content, and high-quality feed. This result is in accordance with studies showing that consumption of free-range and organic eggs, especially in the UK, is mainly related to animal welfare concerns. Moreover, consumers who purchase cage-free eggs also associate the enhanced animal welfare standards with higher quality, safety, and better taste of eggs [37,38].

Although, as stated by Bennett et al. [37] and Pettersson et al. [38], consumers with animal welfare concerns care little about the price of free-range and organic eggs, price in this study was still an important determinant of purchase decision-making, acting as a mediator between production method and affordability. Moreover, based on Fearne and Lavelle [39], UK egg consumers can be segmented into two major clusters according to price and production method. Our study confirmed that although for some participants the production method was more important than price, for others, a low price was the only attribute to consider.

Participants overall showed positive attitudes towards eggs from insect-fed hens [7,9,12,13,15]. In particular, with the provision of information, participants demonstrated that environmental, food waste, and animal welfare benefits positively influenced intentions to consume and purchase the product [5,10–12]. In line with the attributes that consumers looked at when purchasing eggs, participants revealed that besides the benefits mentioned, factors such as production method, taste, quality, and nutritional value were important for the evaluation of the product and its potential consumption [4,14,16]. Although many participants believed that by eating insects, hens show higher welfare, they also expected that these eggs should be at least produced and marketed as free-range. Following Bennett et al. [37] and Rondoni, Asioli, and Millan [40], trust in the certification institution was an important factor when evaluating the product.

Price, like certification, was an important determinant for purchasing. As already found in many studies on insect-fed fish, participants were willing to purchase the product [16–20]. However, assessing the price before purchasing appeared to be essential.

Some of the participants revealed disgust and rejection towards insects as feed, demonstrating that this factor may act as a barrier to consumer's intentions [9,13,14,20]. However, only for very few participants did disgust strongly reduce the intention to purchase the product.

The lack of trust in the benefits of feeding hens with insects appeared to be a potential barrier to acceptance. Consumers had limited knowledge on this product; therefore, they required more detailed information to evaluate it. According to Grunert [41], eco- and animal-friendly food products have attributes of sustainability and animal welfare that can be evaluated by consumers neither before nor after the purchase. Consequently, consumers need to be provided with reliable information to increase their confidence and trust in the product. In line with other studies, the provision of comprehensive information about the benefits of feeding hens with insects may enhance awareness and, in turn, increase consumers' intentions to consume and purchase the product [13–15,20].

Based on our findings, several implications for businesses and stakeholders may be highlighted for the introduction of these eggs to the UK market.

Considering animal welfare concerns, the large market share of free-range eggs, and the importance of labelling accredited by reliable institutions among UK consumers, stakeholders should consider producing and selling eggs from insect-fed hens with enhanced animal welfare standards (e.g., free-range and organic labelling). Insects as feed should be therefore considered more as an additional benefit associated with cage-free eggs (e.g., free-range and organic). In contrast, insect-fed cage hens would be expected to produce too few additional benefits to be accepted and consumed.

The lack of trust in the product and the limited knowledge about insects as feed among consumers suggest that the provision of information about the benefits of feeding hens with insects may increase trust in the product and enhance intentions. However, the negative effect of the word "insects" on the packaging suggests the need to position the product on the market specifying the use of insects as feed but without drawing excessive attention to them on the packaging. In contrast, the positive effect of the word "natural" and the positive attitudes towards benefits highlight the need for companies to position these eggs as a natural, eco- and animal-friendly food product.

The study presents some limitations that should be addressed in future research or a follow-up study. First, due to the restricted sample size, these findings are considered to be exploratory research. Further research is required among a larger sample in order to better define the determinants of consumer acceptance and marketing strategies. Secondly, considering that participants were not provided with information about quality, nutritional content, and taste, further research is recommended in order to understand how the product's intrinsic attributes may affect consumers' perceptions. Moreover, price is an important determinant of purchasing. Given the increase in price from the benefits provided, consumers are willing to purchase the product. However, whether consumers are willing to pay a premium price for these eggs needs further investigation.

5. Conclusions

This is the first qualitative study exploring consumers' attitudes and perception towards eggs from insect-fed hens in the UK. It provides preliminary evidence regarding factors affecting acceptance and consumption intentions towards eggs from insect-fed hens.

The study found that UK consumers have positive attitudes towards eggs from insect-fed hens. Consumer acceptance appears to be driven by mainly the environmental, animal welfare, and food waste benefits associated with feeding hens with insects. However, other egg attributes such as price, production method, taste, quality, and nutritional value affect intentions. Price was an important factor when considering a potential purchase. Therefore, whether consumers are willing to pay a premium price for these eggs still needs to be clarified. Although disgust may negatively influence consumer acceptance and intentions towards insects as feed, the lack of awareness about the product and the limited knowledge about its benefits appear to be stronger barriers. With this in mind, our study suggests that educating and informing consumers about the benefits of feeding hens with insects may increase intentions to consume and purchase eggs from insect-fed hens.

The study also suggests the importance of trust in the egg certification on the packaging. Based on the results, we can conclude that replacing soybean meal with insects in hen feed has little effect on consumers' preferences by itself. In order to encourage consumer consumption, eggs from insect-fed hens should be produced and marketed under enhanced animal welfare standards (e.g., at least free-range labelling).

Despite its exploratory nature, this study contributes to the limited literature on insect-fed foodstuffs. Given the emerging area of research, this study may contribute and pave the way for further qualitative and quantitative studies on the topic.

Author Contributions: Conceptualization, S.S. and S.G.; methodology, S.S.; software, S.S.; validation, S.S.; formal analysis, S.S.; investigation, S.S.; resources, S.S. and S.G.; data curation, S.S.; writing—original draft preparation, S.S. and S.G.; writing—review and editing, S.S. and S.G.; visualization, S.S.; supervision, S.S. and S.G.; project administration, S.S. and S.G.; funding acquisition, S.G. All authors have read and agreed to the published version of the manuscript.

Funding: This research received no external funding. The APC was funded by the University of Reading Open Access Fund.

Institutional Review Board Statement: The study was conducted according to the guidelines of the Declaration of Helsinki, and approved by the Institutional Review Board (or Ethics Committee) of the University of Reading (protocol code 1280C, 14/04/20).

Informed Consent Statement: Informed consent was obtained from all subjects involved in the study.

Acknowledgments: The authors would like to acknowledge the technical support (packaging and materials used for experiments) provided by Protix. Also, we would like to express our thanks to Agnese Rondoni for her assistance in double-checking the validity of themes and codes in Nvivo.

Conflicts of Interest: The authors declare no conflict of interest.

References

1. Verneau, F.; La Barbera, F.; Kolle, S.; Amato, M.; Del Giudice, T.; Grunert, K. The effect of communication and implicit associations on consuming insects: An experiment in Denmark and Italy. *Appetite* **2016**, *106*, 30–36. [CrossRef] [PubMed]
2. Star, L.; Arsiwalla, T.; Molist, F.; Leushuis, R.; Dalim, M.; Paul, A. Gradual Provision of Live Black Soldier Fly (Hermetia illucens) Larvae to Older Laying Hens: Effect on Production Performance, Egg Quality, Feather Condition and Behavior. *Animals* **2020**, *10*, 216. [CrossRef] [PubMed]
3. Gasco, L.; Biasato, I.; Dabbou, S.; Schiavone, A. Quality and consumer acceptance of products from insect-fed animals. In *Edible Insects Food Sect*, 1st ed.; Sogari, G., Mora, C., Menozzi, D., Eds.; Springer: Berlin, Germany, 2019.
4. European Union. Commission Regulation (EU) 2017/1017 of 15 June 2017 amending Regulation (EU) No 68/2013 on the Cat-alogue of feed materials. *Off. J. Eur. Union.* **2017**, *L 159*, 48–119.
5. Derrien, C.; Boccuni, A. Current status of the insects producing industry in Europe. In *Edible Insects in Sustainable Food Systems*, 1st ed.; Halloran, A., Flore, R., Vantomme, P., Roos, N., Eds.; Springer: Berlin, Germany, 2018; pp. 471–479.
6. Insect Biomass Task, Group F. The Insect Biomass Industry for Animal Feed—The Case for UK-based and Global Business. 2019. Available online: https://www.fera.co.uk/media/wysiwyg/Final_Insect_Biomass_TF_Paper_Mar19.pdf (accessed on 9 February 2021).
7. Verbeke, W.; Spranghers, T.; De Clercq, P.; De Smet, S.; Sas, B.; Eeckhout, M. Insects in animal feed: Acceptance and its determinants among farmers, agriculture sector stakeholders and citizens. *Anim. Feed. Sci. Technol.* **2015**, *204*, 72–87. [CrossRef]
8. Laureati, M.; Proserpio, C.; Jucker, C.; Savoldelli, S. New sustainable protein sources: Consumers' willingness to adopt insects as feed and food. *Ital. J. Food Sci.* **2016**, *28*, 652–668.
9. Kostecka, J.; Konieczna, K.; Cunha, L. Evaluation of insect-based food acceptance by representatives of Polish consumers in the context of natural resources processing retardation. *J. Ecol. Eng.* **2017**, *18*, 166–174. [CrossRef]
10. Onwezen, M.; Puttelaar, J.V.D.; Verain, M.; Veldkamp, T. Consumer acceptance of insects as food and feed: The relevance of affective factors. *Food Qual. Prefer.* **2019**, *77*, 51–63. [CrossRef]
11. Sogari, G.; Amato, M.; Biasato, I.; Chiesa, S.; Gasco, L. The Potential Role of Insects as Feed: A Multi-Perspective Review. *Animals* **2019**, *9*, 119. [CrossRef]
12. De Faria Domingues, C.H.; Rossi Borges, J.A.; Ruviaro, C.F.; Freire Guidolin, D.G.; Mauad Carrijo, J.R. Understanding the factors influencing consumer willingness to accept the use of insects to feed poultry, cattle, pigs and fish in Brazil. *PLoS ONE* **2020**, *15*, e0224059. [CrossRef]
13. Roma, R.; Palmisano, G.O.; De Boni, A. Insects as Novel Food: A Consumer Attitude Analysis through the Dominance-Based Rough Set Approach. *Foods* **2020**, *9*, 387. [CrossRef] [PubMed]
14. Szendrő, K.; Nagy, M.Z.; Tóth, K. Consumer Acceptance of Meat from Animals Reared on Insect Meal as Feed. *Animals* **2020**, *10*, 1312. [CrossRef]
15. Naranjo-Guevara, N.; Fanter, M.; Conconi, A.M.; Floto-Stammen, S. Consumer acceptance among Dutch and German students of insects in feed and food. *Food Sci. Nutr.* **2021**, *9*, 414–428. [CrossRef] [PubMed]
16. Mancuso, T.; Baldi, L.; Gasco, L. An empirical study on consumer acceptance of farmed fish fed on insect meals: The Italian case. *Aquac. Int.* **2016**, *24*, 1489–1507. [CrossRef]
17. Popoff, M.; MacLeod, M.; Leschen, W. Attitudes towards the use of insect-derived materials in Scottish salmon feeds. *J. Insects Food Feed.* **2017**, *3*, 131–138. [CrossRef]
18. Ankamah-Yeboah, I.; Jacobsen, J.B.; Olsen, S.B. Innovating out of the fishmeal trap: The role of insect-based fish feed in consumers' preferences for fish attributes. *Br. Food J.* **2018**, *120*, 2395–2410. [CrossRef]

19. Ferrer Llagostera, P.; Kallas, Z.; Reig, L.; Amores de Gea, D. The use of insect meal as a sustainable feeding alternative in aquaculture: Current situation, Spanish consumers' perceptions and willingness to pay. *J. Clean Prod.* **2019**, *229*, 10–21. [CrossRef]
20. Bazoche, P.; Poret, S. Acceptability of insects in animal feed: A survey of French consumers. *J. Consum. Behav.* **2020**, *2020*. [CrossRef]
21. Krueger, R.A.; Casey, M.A. *Focus Groups: A Practical Guide for Applied Research*, 3rd ed.; Sage: Thousand Oaks, CA, USA, 2000.
22. Rabiee, F. Focus-group interview and data analysis. *Proc. Nutr. Soc.* **2004**, *63*, 655–660. [CrossRef]
23. Sogari, G. Entomophagy and Italian consumers: An exploratory analysis. *Prog. Nutr.* **2015**, *17*, 311–316.
24. Tan, H.S.G.; Fischer, A.R.; Tinchan, P.; Stieger, M.; Steenbekkers, L.; Van Trijp, H.C. Insects as food: Exploring cultural exposure and individual experience as determinants of acceptance. *Food Qual. Prefer.* **2015**, *42*, 78–89. [CrossRef]
25. Balzan, S.; Fasolato, L.; Maniero, S.; Novelli, E. Edible insects and young adults in a north-east Italian city an exploratory study. *Br. Food J.* **2016**, *118*, 318–326. [CrossRef]
26. Pambo, K.; Mbeche, R.; Okello, J.; Kinyuru, J.; Mose, G. Consumers' salient beliefs regarding foods from edible insects in Kenya: A qualitative study using concepts from the theory of planned behaviour. *Afr. J. Food, Agric. Nutr. Dev.* **2016**, *16*, 11366–11385. [CrossRef]
27. Marberg, A.; Van Kranenburg, H.; Korzilius, H. The big bug: The legitimation of the edible insect sector in the Netherlands. *Food Policy* **2017**, *71*, 111–123. [CrossRef]
28. Sogari, G.; Menozzi, D.; Mora, C. Exploring young foodies' knowledge and attitude regarding entomophagy: A qualitative study in Italy. *Int. J. Gastron. Food Sci.* **2017**, *7*, 16–19. [CrossRef]
29. Clarkson, C.; Mirosa, M.; Birch, J. Consumer acceptance of insects and ideal product attributes. *Br. Food J.* **2018**, *120*, 2898–2911. [CrossRef]
30. Myers, G.; Pettigrew, S. A qualitative exploration of the factors underlying seniors' receptiveness to entomophagy. *Food Res. Int.* **2018**, *103*, 163–169. [CrossRef] [PubMed]
31. Sogari, G.; Bogueva, D.; Marinova, D. Australian Consumers' Response to Insects as Food. *Agriculture* **2019**, *9*, 108. [CrossRef]
32. Statista. Available online: https://www.statista-com.eu1.proxy.openathens.net/statistics/1066772/main-dietary-habits-in-the-united-kingdom/ (accessed on 6 February 2021).
33. *NVivo Qualitative Data Analysis Software*; Version 12 2018; QSR, QSR International Pty Ltd Australia: Chadstone, Australia, 2018.
34. Nowell, L.S.; Norris, J.M.; White, D.E.; Moules, N.J. Thematic Analysis: Striving to Meet the Trustworthiness Criteria. *Int. J. Qual. Methods* **2017**, *16*, 1–13. [CrossRef]
35. Gale, N.K.; Heath, G.; Cameron, E.; Rashid, S.; Redwood, S. Using the framework method for the analysis of qualitative data in multi-disciplinary health research. *BMC Med Res. Methodol.* **2013**, *13*, 117. [CrossRef]
36. Fereday, J.; Muir-Cochrane, E. Demonstrating rigor using thematic analysis: A hybrid approach of inductive and deductive coding and theme development. *Int. J. Qual. Methods* **2006**, *5*, 80–92. [CrossRef]
37. Bennett, R.; Jones, P.; Nicol, C.; Tranter, R.; Weeks, C.A. Consumer attitudes to injurious pecking in free-range egg production. *Anim. Welf.* **2016**, *25*, 91–100. [CrossRef]
38. Pettersson, I.C.; Weeks, C.A.; Wilson, L.R.M.; Nicol, C.J. Consumer perceptions of free-range laying hen welfare. *Br. Food J.* **2016**, *118*, 1999–2013. [CrossRef]
39. Fearne, A.; Lavelle, D. Segmenting the UK egg market: Results of a survey of consumer attitudes and perceptions. *Br. Food J.* **1996**, *98*, 7–12. [CrossRef]
40. Rondoni, A.; Asioli, D.; Millan, E. Consumer behaviour, perceptions, and preferences towards eggs: A review of the literature and discussion of industry implications. *Trends Food Sci. Technol.* **2020**, *106*, 391–401. [CrossRef]
41. Grunert, K.G. Sustainability in the Food Sector A Consumer Behaviour Perspective. *Int. J. Food Syst. Dyn.* **2011**, *2*, 207–218.

MDPI
St. Alban-Anlage 66
4052 Basel
Switzerland
Tel. +41 61 683 77 34
Fax +41 61 302 89 18
www.mdpi.com

Foods Editorial Office
E-mail: foods@mdpi.com
www.mdpi.com/journal/foods